THE HISTORY OF
COAST ARTILLERY IN THE
BRITISH ARMY

General Eliott on the King's Bastion, Gibraltar. September 13th, 1782.
General Eliott (The Governor) stands in the foreground, with Koehler (hat off) talking to him.
Drawn by G. F. Koehler. Lieutenant, Royal Artillery.

THE HISTORY
OF
COAST ARTILLERY
IN THE
BRITISH ARMY

COL. K. W. MAURICE-JONES
D.S.O., *late* R.A.

with a foreword by
GENERAL SIR CAMERON NICHOLSON
G.C.B., K.B.E., D.S.O., M.C.
Master-Gunner, St. James's Park

The Naval & Military Press Ltd
published in association with
FIREPOWER
The Royal Artillery Museum
Woolwich

Published by
The Naval & Military Press Ltd
Unit 10 Ridgewood Industrial Park,
Uckfield, East Sussex,
TN22 5QE England
Tel: +44 (0) 1825 749494
Fax: +44 (0) 1825 765701
www.naval-military-press.com

in association with

FIREPOWER
The Royal Artillery Museum, Woolwich
www.firepower.org.uk

In reprinting in facsimile from the original, any imperfections are inevitably reproduced and the quality may fall short of modern type and cartographic standards.

FOREWORD

BY

GENERAL SIR CAMERON NICHOLSON, G.C.B., K.B.E., D.S.O., M.C.
Master Gunner St. James's Park; Col. Comdt. Royal Horse Artillery.

On the 17th February, 1956, the Army Council announced their intention of discontinuing Coast Artillery in the British Army. This was an historic decision and one which could not fail to have far-reaching repercussions throughout the Commonwealth and Empire, the Army as a whole and in particular the Royal Regiment of Artillery.

The decision was historic because it was a complete break with the organization and traditions of the past in that the defence of our ports and bases by Coast Artillery had been for centuries an integral part of our imperial strategy. But in the light of scientific and technical advances it was clear that there was no longer any justification for maintaining Coast Artillery and as a result the decision was made and the disbandment of Coast Artillery was ordered and completed within the year. Territorial Regiments changed to other duties, and the Regular element in the defences was posted elsewhere and the guns and equipment disposed of or withdrawn to be put to other uses.

During the period of disbandment the thoughts of many coast gunners turned introspectively to the passing for ever of this historic branch of the Regiment and to considering whether the opportunity should not be taken now to write a comprehensive history of Coast Artillery. It was true that individual actions and episodes had been recorded but a consecutive history had never been attempted. It was thereupon decided that a short history should be written forthwith, and the result is this excellent book by Colonel K. W. Maurice-Jones which in under 300 pages highlights the vicissitudes through which Coast Artillery passed from its earliest days until its recent disbandment.

I commend this short history to military and civilian readers alike, not only because it is a well written and graphic story of the tribulations through which our sea defences have passed over the centuries but because it also throws into relief a number of minor episodes of historical interest which occurred during the many wars in which we were involved during the past 400 years. Not least it is a well deserved tribute to those officers and men who have steadfastly done their duty in all parts of the world, and maintained watch and ward over our imperial communications.

Master Gunner.

PREFACE

WHEN I was given the task of compiling this history, I was asked to produce a book of not more than 300 pages, taking no longer than six months over the writing of it. This book therefore does not pretend to be an exhaustive history of Coast Artillery in the British Army, neither time nor space would allow me to collect the material for such a work. The sources for a detailed history of Coast Artillery lie hidden deep in the mass of Board of Ordnance and War Department papers at the Public Record Office, and it would have taken many years to go through them all and to pick out those which refer to Coast Artillery, its story and its development. Nevertheless I have done my best in the time allowed to go to the original sources and have spent many hours—most interesting and fascinating ones I might say—at the Public Record Office, delving into the records of the Board of Ordnance and War Department.

I was surprised, when I began this work, to find that I was apparently the first ever to set about writing the history of Coast Artillery in the British Army. Most of the historians of the Royal Artillery have concentrated on the story of the mobile guns, horse, field, mountain, heavy, and siege, and, except for the great siege of Gibraltar, have paid very little attention to the branch which manned what was only too frequently called "the Fixed Defences". Only Callwell and Headlam, in their History of the Royal Artillery from 1860 to 1914, have given Coast Artillery its fair share and place. I therefore had to start right from the beginning, and plough a field which, as far as I could ascertain, had never been ploughed before.

A word about the sources which I did employ, and firstly those for the guns and garrisons which form such an important and essential part of the history of Coast Artillery. The details of the guns up to 1815 were obtained from the Board of Ordnance ledgers kept at the Public Record Office; during the nineteenth century from the annual "Distribution of the Army" kept in the War Office Library; and for the period 1939-45 from the returns in the War Office files. The guns at the outbreak of war in August 1914 were compiled from a variety of sources, not perhaps so reliable as the others. The official War Office returns of that period were unfortunately lost or destroyed when the department of the Master General of the Ordnance was transferred from the War Office to the Ministry of Supply in 1938. As for the garrisons, those before 1771 were compiled from the State Papers Domestic at the Public Record Office, from the Journals of the House of Commons at the War Office Library, and from "Military Pamphlets and Papers" held in the Royal Artillery Institution. For the period 1771 to 1859 they were gathered from the Muster Rolls of the Invalid Companies, Royal Artillery, at the Public Record Office, and from Lieut.-Colonel M. E. S. Laws' most accurate and wonderful volume "Battery Records of the Royal Artillery 1716-1859", a book which no historian of the Royal Artillery can possibly do without. After 1859 the garrisons were obtained from the official Army Lists.

There is a fiercesome bibliography printed in the book which gives all the sources I have used. Much of the story up to 1815 was gleaned from the correspondence and papers of the Board of Ordnance and War Department kept at the Public Record Office. Also for this period I was again greatly assisted by Lieut.-Colonel Laws' book. Many of the stories of Coast Artillery actions have been taken from articles published in the Journal of the Royal Artillery for which I am much indebted to the authors. All the material for the Second World War was collected direct from the War Office files. It was unfortunate that I had to write of this war before publication of most of the official histories. I trust I have not fallen into error nor omitted anything important because of this.

The question of maps to accompany this volume has caused me some worry. The original arrangement was to provide ten maps and six illustrations, but the period to be covered was too long and the area so large that it was soon evident that ten maps would not be sufficient. After some discussion it was decided to have seventeen maps and one illustration (a frontispiece only), six of the maps being small scale showing large areas, and eleven of them large scale showing comparatively small areas where operations described in the text have occurred. As this decision was taken only at the last moment, it is possible that some of the places mentioned in the text have not been entered on the appropriate map. If this has happened I must apologise for the omissions. I must also express my heartfelt gratitude to Sergeant-Major J. C. Camies, Royal Artillery Clerks' School, who so kindly and readily volunteered to design and draw the maps. Without his help I do not know what we should have done.

I owe my thanks to a number of people for their kind help and assistance in producing this work. Firstly to the Coast Gunners themselves who so willingly subscribed the money without which this book could never have been written and published, and to their Brigadiers who chose me for the task and organised the whole business. Next to the Secretary (Brigadier J. H. Frowen) and Staff of the Royal Artillery Institution who, despite their efforts to destroy my poor writings by fire and water, have borne with me for six months and during that period given me every possible support and co-operation. Also to Mr. D. W. King and Mr. C. A. Potts of the War Office Library who never failed to produce the answer no matter how difficult the conundrum. To Lieut.-Colonel M. E. S.

Laws for his most helpful advice and guidance concerning my researches at the Public Record Office, and to the staff of that admirable institution who so quickly and willingly always produced the documents I was seeking. To Brigadier F. W. Rice of the Royal Artillery Directorate at the War Office who was my mentor, counsellor, and ally when I was carrying out my researches at that formidable establishment, and to Major J. H. N. Weston R.A. of the Seaward Defence Wing of the School of Anti-Aircraft Artillery who so kindly aided me with the technical side of my work. To Mr. N. Lush and Mr. C. B. Bramley of the Directorate of Fortifications and Works at the War Office who volunteered so cheerfully and with such good nature to produce the answer to a problem of which I had almost given up hope of finding the solution. And finally to my old friend and fellow-countryman, Mr. G. J. Roberts of the Central Registry at the War Office, who never gave up helping me, who pursued files for me to the very depths of the archives, and who always produced a cup of tea for me when I was feeling depressed and exhausted. My thanks are also due to the proprietors of the *Gunner* Magazine for allowing me to reproduce the article by "Cascable".

In conclusion let me tell you my own story of Coast Artillery. William Hickey, the Calcutta lawyer, tells us in his Memoirs how during the War of American Independence, while he was at Madras, a French Squadron under the famous Admiral de Suffren appeared off the port one evening and leisurely bombarded Fort St. George, the guns of the fort failing to reply with one single round. Next morning the worthy citizens and merchants of Madras went to the Governor in a body and complained bitterly of this failure on the part of the Coast Artillery manning the guns of the fort. The Governor sadly agreed with the complaints, but said the guns could not answer the fire of the enemy warships as there was no ammunition available, the magazine being locked and the master gunner in the bazaar with the key in his pocket. During the second World War, close on 200 years later, I was sent to Madras to command the Coast Artillery at that important port. Shortly after my arrival, I went to inspect the 6 inch battery, the principal battery of the coast-defences. Having finished my round of the guns etc., I wished to look at the magazine. I was unable to do so, the magazine was locked, and the master-gunner in the bazaar with the key in his pocket.

Woolwich,
May, 1957. K.W.M-J.

LIST OF MAPS

MAP I.—British Isles & Eire ... ... *Cover pocket*
MAP II.—The Mediterranean ... ... *Cover pocket*
MAP III.—North American Seaboard & West Indies *Cover pocket*
MAP IV.—The Atlantic Ocean ... ... *Cover pocket*
MAP V.—The Indian Ocean ... ... *Cover pocket*
MAP VI.—The South China Sea & East Indies ... *Cover pocket*

	Page
MAP VII.—St. Lucia	57
MAP VIII.—Minorca	64
MAP IX.—Gibraltar	67
MAP X.—Toulon	84
MAP XI.—Haiti (San Domingo)	109
MAP XII.—Guadeloupe	111
MAP XIII.—Dominica	116
MAP XIV.—Crete	244
MAP XV.—Hong-Kong	260
MAP XVI.—Malaya	265
MAP XVII.—Singapore	271

CONTENTS

	Page
FOREWORD	v
PREFACE	vii
BIBLIOGRAPHY	xii
LIST OF ABBREVIATIONS	xv
INTRODUCTION	1
CHAPTER I.—1540 to 1603	5
CHAPTER II.—1603 to 1667	10
CHAPTER III.—1667 to 1716	16
CHAPTER IV.—1716 to 1748	24
CHAPTER V.—1748 to 1763	32
CHAPTER VI.—1763 to 1774	41
CHAPTER VII.—1774 to 1783	48
CHAPTER VIII.—1774 to 1783 (concluded)	63
CHAPTER IX.—1784 to 1793	75
CHAPTER X.—1793 to 1815	82
CHAPTER XI.—1793 to 1815 (continued)	91
CHAPTER XII.—1793 to 1815 (continued)	108
CHAPTER XIII.—1793 to 1815 (concluded)	120
CHAPTER XIV.—1815 to 1856	133
CHAPTER XV.—1856 to 1914	147
CHAPTER XVI.—1856 to 1914 (continued)	159
CHAPTER XVII.—1856 to 1914 (concluded)	167
CHAPTER XVIII.—1914 to 1918	181
CHAPTER XIX.—1914 to 1918 (concluded)	192
CHAPTER XX.—1918 to 1939	202
CHAPTER XXI.—1919 to 1939 (concluded)	210
CHAPTER XXII.—1939 to 1945	219
CHAPTER XXIII.—1939 to 1945 (continued)	243
CHAPTER XXIV.—1939 to 1945 (continued)	255
CHAPTER XXV.—1939 to 1945 (concluded	264
CHAPTER XXVI. Conclusion	275

BIBLIOGRAPHY

(1) *Original Sources.*

State Papers (Domestic)
Board of Ordnance Correspondence and Papers
War Office Correspondence and Papers
Royal Artillery Pay and Muster Rolls
— at the Public Record Office.

Journals of the House of Commons at the War Office Library.
Files and Returns at the War Office.

(2) *Official Publications.*

Organisation and Fighting of the Fixed Armament of a Coast Fortress or Defended Port (1911).
Internal Defence Arrangements against the Spanish Armada (1798).
Lists of the Several Regiments and Corps of Fencible Cavalry and Infantry: of the Militia: of the Corps and Troops of Gentlemen and Yeomanry: and of the Corps and Companies of Volunteer Infantry. (1797-1805).
Militia Forces of the Colonies and Protectorates. (1902).
Report of Royal Commission appointed to consider the Defences of the United Kingdom. (1860).
Garrison Artillery Training. (1911).
Handbooks of the 9.2 inch, 6 inch, 4.7 inch Q.F., and 12 pdr. Q.F. Guns (1912).
Regulations for the Territorial Force. (1912).
Garrison Artillery Training. (1914).
Operations in Hong Kong from 8th to 25th December 1941.
        (Supplement to *London Gazette*, 27th June 1948.)
Operations of Malaya Command from 8th December 1941 to 15th February, 1942.
        (Supplement to *London Gazette*, 20th February, 1948.)
Annual Distribution of the Army.
Monthly Army Lists.

(3) *Service Publications.*

Military Pamphlets and Papers. Royal Artillery Institution.
Minutes of Proceedings of the Royal Artillery Institution.
Journal of the Royal Artillery.
Journal of the Society for Army Historical Research.
Army Review.

Brassey's Naval Annual.
"Gunner" Magazine.
Regimental History of 415 (Thames and Medway) Coast Regiment R.A., T.A.
History of Orkney Artillery Volunteers.
History of Lancashire and Cheshire R.G.A. (T.F.)
Royal Artillery Commemoration Book 1939-1945.

(4) *Reference Books.*

Battery Records of the Royal Artillery 1716-1859—Laws. (1952).
Biographical Dictionary of Jersey—Balleine. (1948).
Short Guide to Edinburgh—Muirhead. (1953).
Encyclopædia Britannica, Eleventh Edition. (1910).

(5) *Official Histories.*

Naval Operations 1914-1918—Corbett. (1921).
Defence of the United Kingdom 1939-1945—Collier. (1957).
The War at Sea 1939-1945—Roskill. (1954).

(6) *Other Books, etc.*

Notes on the Early History of the Royal Regiment of Artillery—Cleaveland. (1840).
Notes on the Board of Ordnance—Hogg. (1933).
Naval Strategy—Mahan. (1911).
The Service of Coast Artillery—Hines and Ward. (1910).
Castles and Fortresses—Sellman. (1954).
The Castles of Great Britain—Toy. (1953).
Castles—O'Neil. (1953).
History of the Royal Navy—Clowes. (1897).
History of the British Army—Fortescue. (1899-1930).
The West Kent Militia—Barhote. (1909).
History of Landguard Fort—Leslie. (1898).
History of the Great Civil War—Gardiner. (1894).
History of Portsmouth—Anonymous. (*circa* 1830).
Gibraltar and Its Sieges—Mann. (1870).
History of Gibraltar—Lapez de Ayala. (1845).
Le Siege de Gibraltar (1704-1705)—Revue d'Artillerie. (1930).
Journal of the Siege of Gibraltar 1727—Anonymous. (1727).
Artillery through the Ages—Manucy. (1949).
The Story of the Gun—Wilson. (1944).
A History of Firearms—Carman. (1955).
History of the Royal Regiment of Artillery—Duncan. (1872).
War of the Austrian Succession—Rolt. (1750).
England in the Seven Years War—Corbett. (1907).
Histoire des Descentes—La Grave. (1799).

History of the Royal Irish Regiment of Artillery.—Crooks. (1914).
Origins of Artillery—Hime. (1915).
The History of the Volunteer Forces to 1860—Sebag-Montefiore. (1908).
The Sailor whom England Feared—MacDermot Crawford. (1913).
History of Gibraltar—Sayer. (1862).
History of the Royal Naval Reserve—Bowen. (1926).
The Last Invasion of Britain—Stuart Jones. (1950).
Transactions of the Historical Society of West Wales. (1929).
The Martello Towers of England—Mead. (1948).
Constitutional History of Jersey—Le Quesne. (1856).
Royal Guernsey Militia—Groves. (1895).
Eastern Approaches—Fitzroy Maclean. (1949).
History of the Royal Regiment of Artillery 1815-1853—Hime. (1908).
History of the Royal Artillery, Crimean Period—Jocelyn. (1911).
Naval Gunnery—Howard Douglas. (1855).
Modern Artillery—Owen. (1873).
Fortification—Howard Douglas. (1859).
History of the Royal Artillery 1860-1914—Callwell and Headlam. (1931).
Military Forces of the Crown—Darnel. (1901).
Evolution of Naval Armament—Robertson. (1921).
The Grand Fleet 1914-1916—Jellicoe. (1919).
Germany's High Sea Fleet in the World War—Scheer. (1920).
Our own Times—King-Hall. (1938).
Drama in Malta—Weldon. (1948).
Coast Artillery in Sicily and Italy 1943-1945—Tomlin. (1945).
The Second World War 1939-1945—Fuller. (1948).

ABBREVIATIONS

A.A.	Anti-Aircraft.
A.P.	Armour Piercing.
A.T.C.	Anti-Torpedo Craft.
B.C.	Battery Commander.
B.L.	Breech Loading.
B.R.A.	Brigadier, Royal Artillery.
C.A.	Coast Artillery.
C.A.T.C.	Coast Artillery Training Centre.
C.C.C.A.	Commander Corps Coast Artillery.
C.D.	Coast Defence(s).
CD/CHL.	Early Warning Radar.
C.R.A.	Commander Royal Artillery.
C.R.H.	Calibre Radius Head.
D.A.G.R.A.	Deputy Adjutant General, Royal Artillery.
D.C.M.	Distinguished Conduct Medal.
D.E.L.	Defence Electric Light.
D.P.F.	Depression Position Finder.
D.R.F.	Depression Range Finder.
D.S.O.	Distinguished Service Order.
F.C.	Fire Commander.
F.O.O.	Forward Overvation Officer.
G.H.Q.	General Head Quarters.
G.O.C.	General Officer Commanding.
H.E.	High Explosive.
H.K.S.R.A.	Hongkong-Singapore Royal Artillery.
H.M.S.	His or Her Majesty's Ship.
H.Q.	Headquarters.
I.G.	Instructor of Gunnery.
M.D.	Cordite. Modified Design.
M.Ex.B.	Motor Explosive Boat.
M.L.	Muzzle Loading.
M.M.	Military Medal.
M.T.B.	Motor Torpedo Boat.
N.C.O.	Non-Commissioned Officer.
O.C.T.U.	Officer Cadet Training Unit.
O.P.	Observation Post.
P.F.	Position Finder.
Pdr.	Pounder.
P.W.S.S.	Port-War Signal Station.
Q.F.	Quick Firing.
R.A.	Royal Artillery.
R.A.F.	Royal Air Force.
R.B.L.	Rifled Breech Loading.

R.E.	Royal Engineers.
R.G.A.	Royal Garrison Artillery.
R.H. & R.F.A.	Royal Horse and Royal Field Artillery.
R.I.A.	Royal Irish Artillery.
R.M.A.	Royal Malta Artillery.
R.M.L.	Rifled Muzzle Loading.
R.N.	Royal Navy.
S.P.	State Papers (Domestic).
S.R.	Special Reserve.
T.A.	Territorial Army.
T.B.D.	Torpedo Boat Destroyer.
T.F.	Territorial Force.
W.O.	War Office (Public Record Office).

INTRODUCTION

IT was not until 1911 that the War Office laid down clearly and definitely the object, role and functions of Coast Artillery. In that year was published a manual entitled *"The Organisation and Fighting of the Fixed Armament of a Coast Fortress or Defended Port"*. This important text book stated that, while the Navy was carrying out its customary business of seeking out the enemy's fleets and ships wherever they were to be found, there might be occasions when there would be a temporary loss of command in certain waters left uncovered by our main fleets and squadrons. Fixed defences, manned by Coast Artillery, were therefore necessary for the protection of Naval Bases, to secure harbours whose positions were of strategic value, and to protect commercial ports so that trade might be disturbed as little as possible.

From their very designations, Coast or Fixed Defences have always been considered a strictly defensive arm, but this is not fundamentally true. A coast fortress or defended port executed its role mainly through the fleet or warships for whom it provided a safe base or protected harbour. For four centuries, the batteries and guns, served by Coast Artillery, enabled the British Navy to carry out its proper offensive duties by guaranteeing to it supplies, repairs, and shelter whenever it required them. It is therefore clear that coast defenses were really implements of offence or, as Admiral Mahan so concisely put it "The essential function of a coast fortress is offensive because it conduces to defence only by facilitaitng offence". (*Naval Strategy*, Chapter XV.) There has, of course, always been a school of thought, chiefly in the Navy, which has considered coast defence to be unnecessary. The Blue Water School, as it was known, asserted that coast defences only served for defence, that the Navy was able to defend more efficiently than any coast defences, and therefore money lavished on such defences was wasted and should be put to much better use by being spent on the Navy itself. This, of course, was obviously wrong. It was the coast defences that made it possible for the Navy to enact its offensive role by sustaining and securing that service in time of war. Strategically therefore, coast defences were maintained for offence.

Unfortunately, the views of the Blue Water School, especially as they came mainly from the Navy itself, have ever since the sixteenth century, carried great weight with the British Government. There never has been sufficient money available to cover the whole defence requirements of the British Isles, and the Government of the day has only too often been ready to listen to a doctrine which would enable it to cut down expenditure on the Army and spend the money instead on the Navy or even save it altogether. Moreover the Army itself has never been very enthusiastic about the coast defences for which it was responsible so that the combined economies of both Government and Army have only too frequently down the centuries resulted in neglected defences, out-of-date guns, and unfit and senile garrisons.

The object of coast defences was always to prevent hostile men-of-war from engaging or attacking our own ships while at at anchor in port, or from bombarding the wharves, go-downs, dockyards and such like installations which usually surround important harbours. Furthermore, the coast batteries were expected to deter enemy warships from supporting a landing operation directed at the port or harbour itself and to destroy boats carrying the attackers. In times of threatened invasion, it has even been attempted to defend with coast-defence batteries almost the whole of the coast line of Great Britian at points suitable for landing. The object of these batteries was to engage and sink the transports and small craft bearing the invading army. However, it is obvious that coast defence on this scale could only be provided at times of great national danger, and normally it was confined to important naval bases, ports and harbours.

The strength of the defences, the size and number of the guns defending any one place, depended upon the scale of attack to be expected against it and on its importance and value as a naval base, harbour or commercial port. As the chief danger to be feared was always attack by hostile warships, the type and calibre of the guns employed in coast defences were strictly tied to those used in Naval ships of the period, but, as the Navy had first call on the country's purse, and as it was accepted as an axiom right from the earliest days of coast defence that a gun on a steady base had great superiority over one on a pitching and rolling platform, the guns in coast defence batteries were normally somewhat smaller and of older pattern than those carried by their opponents.

The history of any portion of our fighting services should consist mainly of the records of the campaigns, battles, and fights in which it has taken part, of the dangers, trials and hardships which the men composing it have undergone, and of their valour and bravery in face of the enemy. However, throughout almost the whole of the period during which coast artillery existed as an integral part of the British forces—from about 1540 to 1956— Great Britain has held command of the seas, and the coast batteries guarding our ports and harbours have very rarely been called upon to open fire on hostile men-of-war. It has only been during those times when we lost control of certain areas of the ocean that coast artillery has really been called upon to fight face to face with the enemy. The loss of the command of the North Sea to the Dutch in 1667 brought about the attacks on the defences of Sheernees and Harwich; of the Western Mediterranean to the French in 1756, the first assault on and capitulation of Minorca; of the Mediterranean and Mid-Atlantic to the French, Spanish and Dutch from 1779 to 1782, the famous siege of Gibraltar and the second and final surrender of Minorca; and of the Western Pacific to the Japanese in 1941, the fall of the great coast fortresses of Hong Kong and Singapore. There have also, of course, been innumerable minor attacks and raids on ports and harbours protected by coast defences, and countless actions between warships and coast batteries, but for most of those long days and nights when Great Britain was at war during the four centuries between 1540 and 1956, the coast artillery has spent its time manning the batteries and ramparts, looking grimly out to sea for those enemy warships which never came. In addition, there have been those periods when Great Britain was threatened with invasion, 1588, 1756, 1779, 1797-98, 1801, 1803-05, and 1940-41. At these times of crisis vigorous efforts were made to defend all the likely coves and beaches where the enemy might make a landing, and vast numbers of temporary batteries and defence works were rapidly constructed to house the guns which were to protect and cover these possible lines of invasion. However, Britain's command of the narrow seas—and in the last case, command of the air over them as well—prevented invasion from materialising, the enemy ships and landing-craft never closed our shores, Philip of Spain, Louis of France, Napoleon and Hitler were never able to land their Armies in our country, and the coast gunners still continued to gaze out over the waters of the Channel for those hostile flotillas which never even set out from their ports of embarkation.

So the history of Coast Artillery must largely be one of lonely garrisons vegetating for years in distant ports and batteries, of changes in organisation, methods, armament and fortification, and very rarely of the roar and smoke of battle and sudden death. It must also be remembered that, until the industrial revolution and the gradual introduction of steam and armoured warships mounting rifled guns, methods, armament and fortifications altered very little and very slowly. The target was always the wooden man-of-war, carrying smooth bore, muzzle loading guns firing round-shot, and depending upon the wind for its motive power. The coast gunner who manned a culverin on the rampart of Pendennis Castle, Falmouth, in 1588 would have found but slight change in method if called upon to fire a 24 pdr. from the same spot in 1805.

Indeed if he had looked around at the fortifications themselves, he would have had no difficulty in recognising them. It was not until the latter half of the nineteenth century when steam-powered armoured warships mounting rifled guns firing tubular and pointed shell filled with high explosive made their appearance, that coast artillery really began to develop. To deal with these fast potent opponents, fresh methods had to be brought into use, novel instruments designed, new and more powerful guns mounted, and much stronger fortifications constructed, with the result that Coast Artillery became the leaders in accurate and effective shooting and the pioneers in scientific gunnery. Coast Artillery, which had marked time for so long, suddenly sprung to life and became a vigorous and virile branch of the Royal Artillery for what was to prove to be the last century of its existence.

CHAPTER 1

1540 to 1603

IT will be recalled that the manual of "*The Organisation and Fighting of the Fixed Armament of a Coast Fortress or Defended Port*, 1911" stated that fixed defences, manned by coast artillery, were necessary to provide for the protection of naval bases, to secure harbours whose positions were of strategic value, and to protect commercial ports so that trade might be disturbed as little as possible. And it was exactly for these reasons that the government of Henry VIII found it necessary to originate coast artillery with its forts, guns, and gunners towards the middle of the sixteenth century. The discovery of the New World was turning the English into a nation of sea-farers, and they were pushing out westwards and southwards in their small ships to discover new lands, to open up fresh opportunities for trade and commerce and to seize a share in the treasure and loot which were pouring forth from the Americas and Indies in such astonishing quantities. But the English were not the only nation to be out and about on the high seas on the make. Spanish, Portuguese and French were there also, and were more than prepared to struggle desperately to hold what they had already and to grab more if opportunity offered.

Moreover the foreign policy of Henry VIII made it necessary for England to be strong in the narrow seas. Henry had quarrelled irrevocably with the Pope and fallen out at different times with France and Spain, and either of these powers could threaten our overseas wool trade with Flanders which was the very life-blood of England's existence at that time. Henry therefore built and maintained great ships of war to protect his trade-routes so that the narrow seas were full of English ships—the King's ships going about their lawful occasions, ships of merchant adventurers, armed to the teeth, setting out for the high-seas and newly discovered lands, fierce privateers preying on whatever French and Spanish ships could be found, and capacious traders sailing to and fro' Flanders, the Hanseatic Towns and the Continental ports. And all these ships needed safe bases and secure harbours in which to refit, take on supplies, load and unload and find shelter.

By the third decade of the sixteenth century all ships of war and most others were mounting guns, and the mediaeval castles which had stood on guard so long over the narrow creeks and havens were not of sufficient strength to stand up to the battering of round-shot, nor in many cases were the waterways over which they watched deep enough to take the new sea-going ships. So Henry VIII set about building a series of forts (often called castles) which would carry guns, would be strong enough to withstand the shock of round-shot, and so could protect the more important ports and harbours around the coasts of his realm, from Hull to Milford Haven. These were begun about 1540 and work continued on them throughout Henry's reign. The most imposing of them were constructed at Tilbury and Gravesend—to guard the entrance of the *Thames*—Deal, Walmer, Dover, Sandgate, Rye (*Camber Castle*) Portsmouth (*Southsea Castle*), Sandown, Cowes, Hurst, Yarmouth (I.O.W.), Calshot, Portland, Dartmouth, Falmouth (*Pendennis and St. Mawes Castles*), and Guernsey (*Castle Cornet*). In some cases, such as Dover, Dartmouth and Guernsey, mediaeval castles, already standing, were adapted to suit the demands of the new conditions.

These forts formed the hard cores around which our coast defences gradually developed through the next four centuries. They were of course, being the first of their kind, experiments. The basis of their design was that the whole fort should be built in one compact block with all-round defence. They were constructed of stone with thick walls and rounded bastions and parapets, and were low and squat. The guns were mounted on emplacements in tiers, those on the top platforms firing through embrasures and those on the lower tiers through gunloops with wide openings, sloping rapidly down at the sills and splayed outwards at the sides. These forts of course, varied very much as to size, Deal, about one of the largest, measuring approximately 50 yards by 70 at the base. All the forts were surrounded by ditches and moats.

However, stone was not a good material to meet round-shot; it splintered, and its fragments caused casualties. Casemates were not suitable either for the firing of big guns; their enclosed spaces filled with fumes and smoke and almost suffocated the gunners while the gunloops were easy marks for close-range fire. As soon as these defects were appreciated, the Tudor engineers set to work to correct them. They began to construct ramparts or bulwarks of earth, closely packed and held together by turf, around the outside of the central keep. Earth formed a better protection than stone for it absorbed round-shot and could be rapidly repaired. On the rampart top the big guns were mounted under cover of the bulwark and, firing over the parapet, thus had a much wider field of fire. During the reigns of Edward VI and Mary some of the original forts were altered and had bulwarks added to them, but it was the war with Spain and the threat of invasion during the reign of Elizabeth I which set the engineers once again building new forts or reconstructing old ones. Forts were erected or new fortifications raised at Berwick-on-Tweed, Tynemouth, Chatham (*Upnor Castle*), Sandwich, Plymouth and many other harbours, and the basis of their designs was a central massif surrounded by earth or rubble ramparts and bulwarks with angled bastions at the corners.

The guns mounted on the Tudor forts were of many types, patterns, calibres and weights and were made of both brass and iron. The main types were:-

Type	Calibre in inches.	Approx. weight of shot in lbs.	Approx. extreme range in yards.
Cannon	8	60	2,000
Serpentine	7.5	53	2,000
Demi-Cannon	6.5	30	1,700
Culverin	5.5	17	2,500
Basilisk	5	15	2,000
Demi-Culverin	4.5	9	2,500
Saker	3.5	5	1,700
Minion	3.25	4	1,600
Falcon	2.33	3	1,500
Falconet	2	1	1,800

These guns were smooth-bore, muzzle-loaders which fired solid iron round-shot, cannister, grape, chain, or bar shot, propelled by a charge of gunpowder. They had no sights, and were laid for line by looking along the top of the barrel and for elevation by using the "Gunner's quadrant", wedges being driven under the breach according to the elevation required. It is difficult to discover how many guns these various forts carried, but it is known that in 1547 the Portsmouth Defences mounted 6 Cannons, 2 Serpentines, 4 Demi-Cannons, 4 Culverins, 16 Sakers, 24 Falcons, and 2 Bombards while *Yarmouth Castle* (I.O.W.) held 12 pieces of brass and iron ordnance of different calibres, one piece being unserviceable.

Now these coast-defence forts had to be manned continuously, their guns efficiently maintained and always ready for action so that permanent garrisons of gunners and infantry had to be found for them. Thus somewhere about 1540 there came into existence the very first and earliest establishment of regular artillery in England. The personnel were paid direct by the Exchequer but were under the Master-General and Board of Ordnance for appointments, efficiency, discipline and administration. This was a wise arrangement as the Master-General and Board were also responsible for the forts themselves, their fortifications, guns, ammunition, arms, stores etc. and it was one of the duties of the Master Gunner of England—unfortunately only too rarely carried out—to inspect the forts and the detachments of gunners stationed in them. These detachments consisted of a Master-Gunner or Chief-Gunner in charge—who was responsible to the Commander, usually known as the Captain of the Fort—and a squad of gunners (gunners, gunner's mates, quarter-gunners and matrosses.) At the time of the *Armada* (1588), the permanent artillery garrisons of some of the more important coast forts were as follows:-

Fort	Master Gunner or Chief Gunner	Gunners
Calshot	1	7
Camber (Rye)	1	16
Deal	1	16
Dover	1	47
Gravesend	1	4
Hurst	1	10
Plymouth	1	8
Portland	1	6
Portsmouth	1	16
Falmouth	1	7
Sandgate	1	7
Sandown	1	7
Tilbury	1	8
Tynemouth	1	5
Upnor (Chatham)	1	6
Walmer	1	16

It is obvious from these numbers that there were not sufficient gunners to find the detachments for the guns. In fact the permanent party was only a "District Establishment", responsible for the maintenance of the guns, ammunition, and stores, and possibly for providing the layers in action. The detachments were found from the infantry garrison, or from militia and trained-band soldiers brought into the fort specially for the purpose. They were usually quite untrained and could not have been very efficient.

It was war with Spain and the threat of Spanish invasion that furnished coast artillery with its first real alarm and call to action. By 1586 it had become clear to the Queen's Government that Spain was preparing a great effort so that her fleet could obtain command of the narrow seas and open the way for her army to invade England. The Queen's advisers considered that the enemy would have to seize a good, sheltered port immediately after the initial landing and were of the opinion that the most threatened harbours and havens were, from west to east, Falmouth, Plymouth, Dartmouth, Southampton, Isle of Wight, Portsmouth, The Downs, Sheppey and the Thames Estuary, and Harwich. Orders were therefore sent out to Lord Lieutenants of Counties and Captains of Forts to strengthen their existing defences and construct new ones to cover all exposed points and likely landing places. Bulwarks were to be constructed, pits and trenches dug, parapets raised, and stockades planted, especially at places where the depth of water would admit hostile vessels. The batteries on St. Nicholas (later Drake's) Island were not considered sufficient to safeguard *Plymouth Sound* and plans were made to build a new fort on the *Hoe* (It was not begun until after the defeat of the *Armada*), blockhouses and bulwarks were ordered to be erected along the south coast and at special points of danger such as around Dover and the *Cinque* Ports, at Clay, Milton and Higham on the coast of Kent, on Sheppey Island, and at Harwich. Unfortunately, although it was fairly easy to find the labour and materials for these works, the provision of guns for them was quite another matter, and one lieutenant complained bitterly that he had only bowmen and billmen to man the bulwark he had been detailed to hold. Finally Captains of Forts were told that every harbour and haven must be guarded by adequate fortifications and defended to the utmost but, if the harbour or haven did fall into enemy hands, the fort and fortifications were to be abandoned, the guns withdrawn inland, and the fight continued.

On July 20th, the great *Armada* was sighted approaching the Channel, the beacons were aflame, and the coast defences were alerted from Falmouth to Tynemouth. The Spanish plan was to gain control of the Channel, the Straits of Dover, and the southern narrows of the North Sea by defeating the English Fleet, and then to cover the crossing to England of the Spanish Army under Alexander Farnese, Duke of Parma, which was gathered, ready to embark, at the ports of Flanders. The expected chief opponent of the coast artillery was the Spanish Galleon, a four-masted ship, rather long in the beam, somewhat straight and flat, varying between 1,000 and 500 tons burthen, and mounting between 50 and 25 guns according to its size. There were 20 of these galleons with the *Armada* which totalled 126 ships. The Spanish Army, which was to be landed in England at the mouth of the *Thames* and march directly on London, consisted of about 5,000 horse and 20,500 foot and was to cross from Ghent, Bruges, Nieuport, Gravelines and Dunkirk in 70 landing craft and 200 flat-bottomed boats besides many other small ships. Fortunately the *Armada* and not the English fleet was defeated, the Spanish Army was never able to embark, and the Coast Artillery, in spite of many false alarms and excursions, were not called upon to enter into combat with the Spanish Galleons nor to defend their ports, harbours and havens.

CHAPTER II

1603 to 1667

THE Tudors passed away and were succeeded by the Stuarts who, quarrelling with their Parliaments, were always short of money so that defence was neglected, and coast defence, only too often the Cinderella of the forces, almost forgotten altogether. England of course, was frequently at war with either France or Spain, but Europe was distracted by the Thirty Years War, and the defeat of the *Armada* had drawn the teeth of Spain at sea. Therefore, it was possible to economise on defence, and, as the result of a commission set up by the King (James I) in 1623, it was decided to demolish most of the temporary bulwarks and defence works which had been erected during the war against Spain and to cease to make arrangements for manning them. On the other hand, the Commission did recommend that certain new permanent forts should be constructed at places where there had been only temporary works previously, amongst which was Harwich (*Landguard Fort*) later to become famous in the Second Dutch War.

From 1623 to the outbreak of the Civil War (1642) the coast defences and their garrisons were allowed to rot. The Stuart Kings were attempting to govern the country without Parliament, the forces were administered by Commissioners of the Treasury, and these found it extremely difficult to raise money with the result that the coast artillery and defences got nothing at all. The members of the permanent detachments in the forts were never discharged nor even paid so that they grew old, unfit, and disabled, were forced to take part-time—and even whole-time—employment outside their forts to keep bodies and souls together, ran into debt, had no decent clothes to wear and were often in the greatest misery. The ordnance, equipment and fortifications suffered equally, the guns being unmounted because their wooden carriages rotted, the equipment becoming unserviceable for lack of repair, and the ramparts and curtains falling into decay for insufficient money to pay for their renovation.

Such was the state of the coast artillery and defences when the great Civil War broke out in 1642. The Navy sided with Parliament, and each coast fort declared its allegiance according to the politics of its commander or of the port over which it stood guard. Neither side indulged in amphibious operations, and the coast defences were rarely called upon to take action. However, some of the forts had to defend their land fronts, being involved in the defence of their parent ports. Portsmouth was loyal to the King but, after a very short siege by Waller's Army and Warwick's warships, capitulated on 7th September 1642, *Southsea Castle* apparently being taken on the night of the 3rd while the Captain of the fort "had more drink in his head than was befitting such time and service". Plymouth on the other hand declared for Parliament and successfully survived a three year blockade by the Royalists. *Pendennis Castle*, Falmouth, was one of the last strongholds in the west-country to hold out for the King, and Charles, Prince of Wales, took refuge there when fleeing from Ireton's troopers in February 1646. The forts at Walmer, Deal and Sandgate revolted against Parliament in 1648 when a portion of the fleet in the Downs deserted under Admiral Batten and sailed for Holland to join the Royalists, but were very soon retaken.

With the King executed and Parliament purged and submissive, Cromwell set to work to govern England. He was determined to make his country a first-class, strong military power, and the coast artillery and defences were not forgotten, with the result that the Commonwealth Government took considerable interest in its coast forts, which were reported to be much decayed and ruined after the Civil War, and set about doing what it could to repair them. At the time of the First Dutch War (1652-1654) the artillery garrisons and armaments of some of the leading coast defences were:-

Defences	Garrison		Armament										
	Master Gunner or Chief Gunner	Gunners	Demi-Cannon	Culverins	12 pdrs.	Demi-Culverins	8 pdrs.	6 pdrs.	Sakers	Minions	3 pdrs.	Falcons	Falconets
Berwick ...	1	26		4		4			10	4	2	2	
Cowes ...	1	6		2	4				1				
Deal	1	8		3		11			7	5			
Dover ...	1	21		4		28			18				
Hull	1	19	2	9	3	23			32		2		
Plymouth ...	1	20	10	49	6	55	23	6	25	28	7		5
Portsmouth ...	1	23	21	45	6	78			31	4	8		
Scarborough	1	8		2		2		2	13	1			
Tilbury ...	1	16	48	4		13			6	3			
Tynemouth ...	1	20		1				3	2		2		
Upnor (Chatham)	1	8	15	19		9			13		2		

This list of armaments shows that a change was taking place in the naming and grading of guns. The old guns were still called by their picturesque names of *Cannon, Culverin, Saker,* etc. but the new ones were graded by the weight of ball they fired. This reflects the influence of Gustavus Adolphus who was the first to standardise his artillery by weight of shot. Otherwise guns had altered very little. They were still made of cast-iron and mounted on wooden carriages which, on account of long exposure to the elements, were only too liable to rot. On coast forts, the guns and carriages were usually emplaced in embrasures on sloping wooden platforms to take up the recoil with a skid under the rear axletree to assist. Gun-drill was being introduced, and the gun was loaded in 13 movements, *viz*:-

1. Put back your piece.
2. Order your piece to load.
3. Search your piece.
4. Sponge your piece.
5. Fill your ladle.
6. Put in your powder.
7. Empty your ladle.
8. Put up your powder.
9. Thrust home your rod.
10. Regard your shot.
11. Put home your shot gently.
12. Thrust home your rod with three strokes.
13. Gauge your piece.

Gunnery had not progressed, and it was always hoped to engage enemy ships at close or point-blank range.

Nor had there been any alterations in administration. The forts and their permanent artillery garrisons were still under the Master-General and Board of Ordnance for all purposes except pay which was issued by the Exchequer Office. During the reign of Charles I, the post of captain commanding a fort had been abolished, and a governor was appointed instead. This governor would usually be a great lord or person of political importance, very often both, and in time of peace would rarely take up his residence in the fort. The Lieutenant-Governor—who was always a military officer—was the actual executive commander on the spot and did all the work.

The art of fortification had not stood still. *Landguard Fort*, Harwich, which was built in 1628, was a typical example of the bastioned-trace *i.e.* flanks facing each other and connected by curtains. Built round a hollow square, in which were the administrative buildings, stores, and barracks, the earth bulwarks or ramparts were 200 feet long, 18 feet high, and 32 feet thick with an angled bastion at each corner 36 feet square. The parapet around the whole of the ramparts was 6 feet high and 6 feet thick, and the total exterior circuit of the fort nearly 1,000 yards. On completion this fort was armed with:-

Demi-Cannons	2
Culverins ...	28
Basilisks ...	1
Demi-Culverins	17
Sakers ...	10

In fact the bastioned-trace, once fully developed, remained the dominant form of fortification right through the 17th and 18th centuries. Earth was still the chief material for the construction of ramparts etc., but defences and forts of stone, or earth faced with stone, such as the *Royal Citadel* at Plymouth which was erected during the reign of Charles II, were still sometimes built.

In 1660 the King came into his own again, and once more a Stuart sat on the throne with the usual shortage of money for defence and scant consideration for the Navy and Army. The Second Dutch War, during which England was at war with both Holland and France, broke out in 1665 and lasted for two years. The Navy gained victories over the Dutch in the battles of Lowestoft and North Foreland and fought out a hard draw in "The Four Days Fight". By the spring of 1767, Charles II considered he had defeated the Dutch and expected them to sue for peace, so that, instead of commissioning the main fleet for the summer, he ordered most of the ships to be dismantled and laid up to save money. But the Dutch had other ideas, and, having obtained secret information concerning the shoals and channels at the mouth of the *Thames*, decided to deal a sudden blow at England. A fleet of 100 sail under the command of Admiral de Ruijter, which had been collected in the *Texel*, sailed for England on June 4th and anchored off the mouth of the *Thames* on the evening of the 7th. A squadron of 27 ships of the line, with auxiliaries and fire-ships, was then detached under Rear-Admiral van Ghent, with orders to move up-stream, enter the *Medway*, and do as much damage as possible to shipping and property. However, in order to pass successfully from the *Thames* into the *Medway*, it would be necessary first to silence the coast defences of Sheerness, so the Dutch Admiral decided to capture these defences to ensure they did not interfere with his ships when entering the *Medway*.

The coast-fort (*Garrison Point*) at Sheerness had only been begun earlier in the year (1667), it was indeed the threat of Dutch attack that had decided Charles II—who, with his brother James, Duke of York, had personally chosen the site—that a fort should be built there. By June, the work had not been completed, but the fort was hurriedly put in a state of defence, 16 guns mounted, and a garrison consisting of a detachment of permanent gunners, a party of seamen (from *H.M.S. Monmouth*) to assist with service of the guns, and a company of Lord Douglas' Regiment (1st Foot, Royal Scots) were quartered in it. When intelligence of the approach of the Dutch Fleet was received, a company of the West Kent Trained Bands was added to the garrison, making a total of about 250 men, all under the command of Sir Edward Spragge, the Governor.

At about 5 o'clock in the evening of June 10th, the enemy squadron appeared off the entrance to the *Medway*, and the Dutch Admiral detailed three ships of the line, two of 46 and one of 40 guns, to engage the fort, while 800 soldiers and marines, under Colonel Delman, an English renegade Roundhead, were landed in small boats. This operation having been successfully carried out, Delman advanced on the fort itself.

The garrison at first held out manfully, but, after about an hour's resistance and only when 9 guns had been put out of action by the fire of the enemy warships, was finally forced to evacuate the place and retire, under the direction of Spragge, up the *Medway* to Gillingham. The Dutch remained in possession of Sheerness fort until 21st June when, after utterly demolishing the defence works, they re-embarked and sailed away.

Meanwhile the enemy squadron had moved up the *Medway* where for three days the Dutch took complete charge of the river below Chatham and Rochester, burning the warships at anchor, capturing the largest, the *Royal Charles*, and destroying the dockyard and naval shore establishments. The enemy ships would have moved farther up the *Medway* had they not been prevented from doing so by the fire of the guns of *Upnor Castle*. At last the Dutch had done enough, and by the 22nd June de Ruijter had again concentrated his fleet at the mouth of the *Thames*. After a reconnaisance up the river had been foiled by the guns of Gravesend and Tilbury forts, de Ruijter determined to capture and destroy *Landguard Fort* at the entrance to Harwich Harbour.

This attack was not altogether unexpected. Suspicion had been aroused sometime previously by the appearance of two small Dutch vessels taking soundings below the fort, and precautions had therefore been taken. The Suffolk Trained Bands had been called out and, horse and foot some 2,000 strong, taken position at Walton under the command of Lord Suffolk, four companies of the Duke of York's Maritime Regiment under Colonel Legge had been stationed at Harwich, and eight bulky colliers moored at the entrance of the harbour, ready to be sunk at the first threat of an attack. *Landguard Fort* at this time mounted 59 guns, 18 culverins, 23 demi-culverins, 9 sakers, 4 minions, and 5—3 pdrs. and was garrisoned by a permanent artillery detachment of a Chief Gunner and 6 Gunners, and two companies of the Duke of York's Maritime Regiment, all under the command of Captain Nathaniel Darell of the same Regiment who had just taken over the Governorship of the fort from Lord Suffolk.

At about 11 a.m. on July 2nd, a fleet of 47 Dutch warships appeared off Harwich, 8 of which moved in close and opened a heavy fire upon the fort. Under cover of this bombardment, a force of some 1,200 men under the same Colonel Delman was landed on the Felixstowe beach, about a mile north of the fort. A beach-head was then formed, and a party of about 400 sent along the foreshore to attack the fort. This party, armed with muskets, cutlasses, and grenades and carrying scaling ladders ran along the beach, and, planting their ladders firmly in the ditch, attempted to storm the ramparts. "They came briskly up with their cutlasses drawn upon their arms and their muskets and came up close to the fort, whose reception to them, when discovered was as brisk. This assault, with a continual playing of small shot, lasted about half-an-hour, and they were repulsed. About an hour after, they tried again, but were presently discouraged, and in disorder ran away, leaving some of their ladders, their hand grenades, and a case of very handsome pistols". (Extract from letter from Silas Taylor, Keeper of the King's Store at Harwich, dated 3rd July 1667.) The casualties to the fort's garrison were small, being only about 8 men including their gallant Governor, Captain Darell, who was wounded in the shoulder.

Meanwhile, Lord Suffolk had come up with his trained-bands and set about attacking the beach-head but could make no impression on it, the ground being unsuitable for the use of his horse and his infantry being held off by the fire of two small guns which the Dutch had managed to land. About 9 p.m. the discomforted party returned from the fort, and the whole beach-head was successfully evacuated between 11 o'clock and 2. Unfortunaetly, Colonel Legge only arrived with his four companies of the Maritime Regiment from Harwich after the evacuation had been carried out: had he come earlier it is doubtful whether Colonel Delman could have re-embarked his force at all. By 6 o'clock next morning the Dutch Fleet was under sail and soon disappeared to the northwards. So ended the famous attack on *Landguard Fort.*

CHAPTER III

1667 to 1716.

THE comparatively brief period between the restoration of Charles II in 1660 and the death of Queen Anne in 1714 covered no fewer than four great wars. The second Dutch War (which has already been dealt with), the Third Dutch War (1672-73), the War of the Grand Alliance (1689-97) and the War of Spanish Succession (1702-13). The Navy, by the great victory over the French of La Hogue (Barfleur) in May 1692, settled the question of command of the sea until the Peace of Utrecht in 1713. Therefore the coast defences of Britain were as usual neglected and allowed to fall into decay. We hear that "Berwick is getting more defenceless every year and will take £31,000 to be spent at once to prevent the place from being insulted." "For six years Hull has been going to ruin: the earthworks have been abused by the garrison who have suffered all sorts of cattle to tread down the fencings." "This place (Portsmouth) should be made as strong as Toulon. We are suffering from the pernicious and mistaken notion of England's safety being wholly in wooden walls. This and other parts are utterly neglected and gone to ruin. This harbour is like a gate without locks, bolts, and bars, the fortifications being now hardly fit to resist battering rams and arrows." "Most of the buildings in this fortress (*Pendennis Castle*, Falmouth) are entirely gone to ruin." "This place (*Landguard Fort*) is in the most miserable condition of any fort in Europe. Everyone who sees it and considers its importance wonders that no great care is taken to secure it." (These extracts are taken from the reports of engineers to the Board of Ordnance sent to inspect the various fortifications during this period.) A Commission was therefore set up by Queen Anne in 1704 to put into execution "the Act for fortifying the harbours of Portsmouth, Chatham, and Harwich," with the result that during the next few years over £100,000 was spent on Portsmouth, £32,000 on Chatham, and the old fort at Landguard was demolished altogether and a new one built in 1715 in its place.

Moreover, the Board now had in addition overseas commitments. In the course of the War of the Spanish Succession, Gibraltar, Minorca, Nova Scotia and Newfoundland were acquired by conquest, and detachments of permanent gunners had to be found for them. The detachment for Gibraltar was sent out immediately after that stronghold had been captured in 1704 and consisted of 55 gunners with 2 fireworkers and 6 bombardiers for handling the mortars, while those for Minorca (Port Mahon) Nova Scotia (Annapolis) and Newfoundland (Placentia) were despatched in 1709, 1710, and 1713 respectively. The detachments for these defended ports were as follows:-

Rank	Port Mahon	Annapolis	Placentia
Captain of Gunners		1	
Lieutenant		1	
Master Gunner	1		1
Sergeants		2	
Corporals		2	
Gunners	20	11	20
Mattrosses	60	40	
Fireworkers ⎱ for mortars	3	1	
Bombardiers ⎰	9	3	

The distribution of ranks is curious, especially when it is borne in mind that 55 gunners were consigned to Gibraltar with neither captain, lieutenant, nor master-gunner to take charge of them, and it is to be noted that a commissioned officer appears for the first time among the ranks of the "permanent artillerymen" allotted to forts and garrisons.

The detachment at Gibraltar had scarcely arrived before its members were pitched into the first siege which that fortress suffered after its capture and occupation by the British. This had taken place on 24th July (1707) when a British fleet under Admiral Sir George Rooke, assisted by some 1,800 British and Dutch marines under Prince George of Hesse-Darmstadt, had seized the place after a short resistance. The French and Spanish were not slow to realize the importance of the stronghold which they had so carelessly allowed to fall into the hands of the British, and by the beginning of October a Franco-Spanish force some 12,000 strong had been concentrated under the Marquis of Villadarias opposite the isthumus, assisted by a French fleet of 22 sail to enforce the blockade from the sea side. Besides the gunner detachment, the garrison of the Rock at this time consisted of 3 battalions of marines (about 2,000) and a considerable party of seamen to help serve the guns, all under the command of Prince George of Hesse-Darmstadt. The armament of the fortress at this first siege would seem to have been about 150 guns—varying from 42 to 6 pdrs.—and 6 mortars. Most of the guns had been landed from warships and were still on their ship-carriages, but a large number of them must have been those taken from the Spaniards when the fortress was captured in the previous July.

By the 9th October the Rock was truly besieged. The French and Spanish made their major effort on the landside as the French fleet was unable to keep permanent station in Gibraltar Bay due to fear of the British Mediterranean fleet under Sir John Leake which was refitting at Lisbon. To hold off the land attack Prince George had a series of redoubts constructed to cover the Landport and armed them with 20 of his guns. Once the enemy had pushed forward their trenches within striking distance of the fortifications, it was around these redoubts that the main fighting took place. The enemy made several strong attacks on them but on each occasion failed to take them. By the end of November the garrison was rapidly becoming exhausted, but Leake arrived with the British fleet, chased away the French ships, and was able to disembark about 2,000 reinforcements and a quantity of stores and food.

Thus refreshed the garrison was able to withstand a new series of assaults made on the landward defences. One of the generals of Louis XIV, Marshal Tesso, had now arrived to take over direction of the siege, and the enemy redoubled his efforts. During January (1705) his guns opened a heavy bombardment on the redoubts, especially on the Round Tower which formed the bastion at their northern end. A breach having been made in the Tower, on January 27th the enemy launched a great assault with some 1,300 men, captured the Tower, and advanced towards the inner defences of the fortress. However, a counter-attack, promptly delivered, drove back the intruders, retook the Round Tower, and re-established the line of redoubts. After this the siege languished, and on March 10th Leake arrived with the British fleet and, defeating the French blockading squadron, raised the siege. The Rock had successfully survived the first attempt to recapture it.

The artillery detachments at Port Mahon, Annapolis, and Placentia seem to have been left at peace for some years after their first arrival, but foreign service was very unpopular with the British soldier during the eighteenth century, especially on islands. Once shipped off to some far-away isolated sea-girt colony, the Board of Ordnance was apt to forget its gunners, pay being slow and late in arriving, clothing unreplaced, and money for rations, stores, and maintenance unforthcoming. But above all it was death from sickness that the soldier most feared. In those days of little knowledge of hygiene, tropical diseases, and medicine generally, the rate of mortality among British garrisons abroad was very high. During the year succeeding the first siege of Gibraltar "more than half of the garrison was disabled through disease brought on by exposure, while even in 1711 the men were obliged to burn their own miserable quarters for want of fuel." (Fortescue). Furthermore, posting to the West Indies was practically a death sentence, for yellow fever quickly decimated any body of troops sent there.

To return to the home front, the Peace of Utrecht, agreed to in 1713, brought about the usual plans for effecting economies and re-organisation. In 1716 the Board of Ordnance brought out a great scheme for the reduction of coast-defence armaments. The list, which set forth the reduction in guns to be carried out, is of great interest as it shows, under the heading "Present Armament" what must have been the complete coast-defences of England, Wales and the Channel Islands at the close of the War of Spanish Succession. Here is the list:-

Place.	Present Armament. (No. of Guns.)	To be reduced to:- (No. of Guns.)
Portsmouth: *Southsea Castle*	474	143
Gosport	125	34
Isle of Wight: Yarmouth ...	73	13
Sandown ...	26	12
Carisbrook ...	10	7
Cowes ...	20	10
Hurst Castle	34	18
Calshot Castle	25	15
Portland Castle	15	7
Plymouth	311	70
Falmouth	109	30
Scilly Islands	118	30
Berwick	76	50
Holy Island	20	12
Tynemouth	30	20
Scarborough	12	8
Hull	117	50
North Yarmouth	27	15
Landguard Fort	63	20
Sheerness	150	70
Gillingham Fort	54	40
Chatham	109	37
Tilbury Fort	161	60
Gravesend Fort	17	10
Dover	98	36
Deal Castle	29	12
Walmer Castle	17	10
Sandgate Castle	16	10
Guernsey	97	80
Jersey	147	100

The naval bases and defended ports at home were evidently strongly protected by the end of the war. Reductions and re-organisation were also carried out at Gibraltar and Minorca, the guns at these two fortresses "a great part thereof being unserviceable and of improper nature". (Board of Ordnance Minutes, 1717.)

The ordnance in general use in coast defences was:-

Piece.	Approx. extreme range in yards.
42 pdrs.	3100
32 pdrs.	2900
24 pdrs.	2700
18 pdrs.	2600
12 pdrs.	1800
9 pdrs.	1800
6 pdrs.	1500
4 pdrs.	1200

The old names had quite disappeared, and all guns were now known by the weight of shot they threw. These were of course all naval guns as issued to the warships of the Navy. "Ships of the Line" had greatly increased in size and armament since the Dutch wars. First-rates now carried 100 guns, second-rates from 84 to 90 guns, third-rates from 64 to 80 guns, and fourth-rates from 50 to 60 guns. Three deckers were becoming normal, the larger ships being up to 2,000 tons burthen. There was very little urge to advance the art and skill of naval—and therefore of coast-defence—gunnery. In battle the aim of the warship was to close the enemy and fight him at point blank range. Therefore quick and efficient gun-drill was more important than accurate laying.

In the art of fortification the methods of the great French engineer Vauban were now supreme. He was no believer in systems, saying that one does not fortify by systems but by common sense. He elaborated the "bastion-trace", introducing ravelins, tenailles, and traverses. In Britain, such new coast forts and batteries as were constructed at this period, followed as near as possible the tenets of the master. Brick was now used to face the earthworks, and bastions were of diamond shape and thrown well forward, being not only built at the corners but also at intervals along the ramparts. Forts were normally of the closed lunette type, that is to say a fortified work of more than four sides with parapet and ditch all round. The guns were mounted on platforms out in the open on the ramparts and bastions, firing through embrasures and covered by breast works and parapets. Quarters, barracks, stores, etc. were built inside the hollow square of the fort, being protected by the height of the ramparts.

The small detachments of permanent gunners—now usually referred to as garrison gunners to distinguish them from the gunners of the artillery trains—each with its master gunner in charge, still maintained the armament of the coast defences. They were as in the past far too few actually to man the guns—and unfortunately only too often too old and decrepit as well—and these had to be served in action either by "extraordinary or additional gunners" taken on specially for the period of hostilities or by soldiers detailed from the infantry or militia garrisons. The permanent detachments at the more important places had also to maintain mobile armament which was held by them in store, ready for issue to any artillery-train being formed for operations in the field.

All armament, ammunition, weapons, stores etc. were issued by and were the property of the Board of Ordnance. In 1716, at the time of the great reduction, the following instructions were issued by the Board to the store-keepers of all fortresses and garrisons under their care:-

(1) *Ordnance.* As in regulations for the Sea Service some are too long for the breadth of the ships as well as others are too short, the longest are always preferable for garrison service and are only to be supplied from the shortest out of necessity.

(2) *Carriages.* It is designed for the future to supply the garrisons with standing carriages only, in the meanwhile to make use of what is in a place whether standing, travelling, or ships carriages as far as they can be repaired and not to repair or demand more than for the essential guns when reduced, the number of each nature fixed on, and the dimensions to be taken in order to make proper carriages, and the damaged travelling and old long standing carriages to be converted to short standing carriages when they will hold out from $7\frac{1}{2}$ to 8 feet long according to their natures.

(3) *Beds.* One to each gun.

(4) *Coins.* Three to each gun as the carriages are now made require but one to each.

(5) *Copper ladles with rammer heads.* For the small places half the number of guns: for the greater places $\frac{2}{3}$ the number of guns.

(6) *Sponges with Rammer Heads.* The same proportion.

(7) *Round shot.* From 50 to 100 rounds for each gun according to the bigness and consequence of the place besides what is allowed for marching trains.

N.B.—It is not intended to bring away any shot of those natures of guns that are to be left except they exceed very much and even in that case to receive directions.

(8) *Tampeons.* Two to each gun by reason of their being subject to break, and those guns kept for constant salutes are to have tampeons hung by tarred marlin, all the rest whether mounted or dismounted are to have tampeons well fixed to keep out dirt and wet particularly those next to the sea for want of which they have by long time been utterly spoiled.

(9) *Spare Ladle Staves.* To be allowed only in such places as are remote from garrisons.

(10) *Wadhooks.* One to every four guns.

(11) *Aprons of Lead.* One to each gun when reduced and only allowed to be fixed on such guns as must be always ready for warning pieces or salutes or in such places as are directed to fire watch guns and give and answer salutes over and above triumph days, all other guns whether mounted or dismounted to be fidded with cork or tow and no aprons.

(12) *Powder.* As most of the great places can be easily supplied from the great magazines and that most of the powder houses are very good, it is thought fit to allow about 20 or 30 rounds of the guns at each place which will be sufficient for the triumph days or extraordinary salutes and any emergency excepting such places as are at a great distance or others that cannot be readily supplied.

(13) *Formers.* One to each nature at the small places, and two for each nature at the greatest places.

(14) *Tacks for sponges.* About 20 to each sponge counting one with another.

(15) *Copper Nails.* One to each ladle.

(16) *Sling Carts.* Are only necessary where there are many guns dismantled and laid on skids.

(17) *Truck carriages.* This expense should be avoided in little places.

(18) *Gynns, Complete.* Are generally wanting, but in small places, where there are but few guns and not exceeding 12 pdrs., are not absolutely necessary but in great places they must be proportioned according to the number of batteries and to the distance or difficulty of access to them.

(19) *White Rope.* Five fathoms of 5 inch is for one sling, eleven fathoms of 3½ or 3 inches for one gyn rope should be carefully laid up in store: that of 6, 7 or 8 inches are only for such places as have cranes.

(20) *Tarred Rope.* From 1 to 2 inches is only wanted on extraordinary occasions.

(21) *Padlocks.* May be necessary even where there are other locks for the preservation of the stores and a particular demand of both should be made.

(22) *Lanthorns. Muscovy and Dark.* A few of these are sufficient to go amongst other stores.

Ordinary. A greater proportion of these may be necessary in case of night alarms or other extraordinary service.

(23) *Men's Harness.* In many places horses cannot be used in removing guns and carriages.

(24) *Grindstones with Troughs.* Are generally useful if well taken care of.

(25) *Intrenching tools.* A few of these will be necessary everywhere particularly at the chief places of the remote divisions.

(26) *Timber.* In most of the places in England the platforms are paved with gun-stones which may be husbandry for such guns as are left constantly mounted for salutes and alarms but for all the rest that are dismounted it will be most proper to have plank and joist left in store with spikes and weight nails therefore a computation must be made for the several platforms required at or about 12 feet deep from the breastwork.

There must have been many "triumph days" to celebrate by the firing of salutes when the great Duke of Marlborough was Master General of the Ordnance (1702-1722) and Commander-in-Chief of the Allied Armies on the Continent. It is therefore only fitting to close this chapter by giving in full the warrant of appointment of one James Hubbard to be master-gunner at *Landguard Fort* (Harwich) in 1703. The duties and position of master-gunner do not seem to have altered very much down the years.

"By virtue of the authority to me by your Queen's most Excellent Majestie on this behalf given upon ye good testimony and assurance which I have received of your Loyalty Integrity and Ability I doe hereby nominate constitute and appoint you the same James Hubbard, to be Master Gunner of Her Majestie's Garrison of *Landguard Fort* from henceforth during so long as you shall behave yourself Loyally Faithfully and Diligently in Her Majestie's Service, you are therefore to take upon you the care and charge of all ye Ordnance and Stores of War that shall be committed to you for ye service of ye garrison hereby strictly chargeing and requiring you to use your utmost skill and diligence in ye performance of your duty in ye said place and to observe and to obey the orders and directions of ye Governour and Lieut.-Governour of ye said garrison and such other your superior officers and persons as shall and may from time to time call upon you for ye performance of your duty as a Master Gunner in her Majestie's service and likewise to observe and follow such orders and directions as you shall from time to time receive from me or ye Master General of ye Ordnance for ye time being or ye Lieut.-Genll and Principal Officers of ye same, for ye better keeping and preserving of ye gunns, carriages, ammunition and other stores of War under your care and charge from decay and perishing and for rendering a true and just account of ye expenditure thereof and for ye care and diligence herein you are to have and receive ye daile Vaile or fee as usuall and is appointed by her Majestie's Establishment to be payd you by the treasr of Her Majestie's Land Forces Forts Castles and Garrisons in England. Given at ye Office of Ordnance under my hand and seal this 31st day of March 1703 in ye second year of her Majestie's Reign."(signed)   *"Marlborough".*

CHAPTER IV

1716 to 1748.

IN May 1716 the Royal Regiment of Artillery was constituted under the Master General and Board of Ordnance, two "marching companies" being formed and and stationed at Woolwich. This did not at first affect in any way the permanent detachments of artillerymen quartered in the coast-defences at home and abroad. They continued as before to maintain the guns and stores under their master-gunners, being under the Board of Ordnance for all purposes except pay which was still issued direct by the Exchequer. The first indication that the formation of the Regiment was going to affect the organisation of coast-artillery was that the detachments—or Trains as they had been previously called—at Gibraltar and Minorca were referred to as "Companies" from 1719 onwards, but they were still not part of the Royal Artillery. It was not until 1722 that these two companies were incorporated into the Regiment as Lieutenant Stepkin's Company (Gibraltar) and Captain Hughes' Company (Minorca). (Until 1825, every unit of the Royal Artillery was designated by the name of its Commanding Officer which makes it very difficult indeed for the historian, as commanders were always changing.)

From 1722 onwards any coast artillery required for the manning of the defences of naval bases and defended ports abroad was found by or from companies of the Royal Artillery, but it was not until 1749 that the detachments of coast-gunners in Newfoundland and Nova Scotia were taken into the Regiment as Captain Ord's Company with headquarters at St. John's Newfoundland.

At home a dual system prevailed until 1771. The small parties or "district establishments" continued to exist in the coast-defences quite separate from the Royal Artillery, being borne on the establishment of "Guards and Garrisons", but the "marching companies" very soon began to infiltrate into coast defence in the more important places. As early as 1721 we find Captain Pattison's Company with headquarters at Portsmouth and detachments at Plymouth and Chatham. It is probable that the "district establishment" remained responsible for the care and maintenance of the equipment while the 'marching company" trained on the guns, both static and mobile, and was ready to man either in case of need. However, during the eighteenth century, war was almost continuous, and it was very rare that a "marching company" could be spared for the coast-defences at home, so the old method went on, a small permanent detachment for "keeping and preserving ye gunns, carriages, ammunition, and other stores of war" with "extraordinary or additional gunners" found from seamen landed from the fleet, the infantry garrison, the

Place.	Garrison.		Armament.								
	Mr. Gnrs.	Gnrs.	42 pdrs.	32 pdrs.	24 pdrs.	18 pdrs.	12 pdrs.	9 pdrs.	6 pdrs.	4 pdrs.	3 pdrs.
Berwick	1	6			8		6	16		22	
Calshot	1	2							13		
Chatham	3	14	14					10	8	2	
Dartmouth	1	2					6		6		6
Deal	1	2						11			
Dover	2	8				17		9	21		
Falmouth	2	3			6	14		23	15		
Gillingham	1	2		36						4	
Gosport	1	1				21		8			
Gravesend and Tilbury	2	12	30					36			
Guernsey	1	4				48			58		
Landguard Fort ...	1	3			15				5		
Hull	1	6				20		14			20
Jersey	1	8			60				80		
Plymouth	2	18	16			16		20	10	7	
Portland	1	2			3	5					
Portsmouth & S'thsea	2	17	23	30	5	12	44	32	36		
Sandgate	1	1				10					
Scarborough	1	1				12			8		
Sheerness	1	16	30			22					
Tynemouth	1	2				19		9			
Walmer	1	2						18			
Yarmouth North ...	1	2			10				5		
Hurst Castle	1	4				14			8		
Isle of Wight, Cowes	1	5						10			
Carrisbrook ...	1	3								7	
Sandown ...	1	3						20			
Yarmouth ...	1	2				12			6		

county militia, or even the local inhabitants brought in to man and fight the guns in time of war. In February 1771 eight "Invalid Companies" Royal Artillery were formed, the personnel of the permanent detachments of coast-artillery at home being absorbed into these companies, and from that date onwards coast-artillery became an integral and intrinsic part of the Regiment both at home and abroad.

However, to return to the aftermath of the War of the Spanish Succession. By 1725 most of the reductions and re-organisation had been carried out, and about that year the chief coast-defences at home stood as the table on previous page.

Placentia (Newfoundland) was armed with 16-18 pdrs. and 4 - 9 pdrs., and Annapolis (Nova Scotia) with 6 - 24 pdrs., 4 - 12 pdrs., and 16 - 4 pdrs. From examination of this list of armaments it can be seen that 18 pdrs., 9 pdrs. and 6 pdrs. seem to have been the most favoured coast defence guns at this time.

Due to the aggressive policy of Cardinal Alberoni, the chief minister of the King of Spain, war broke out between Britain and Spain in 1718. Command of the Mediterranean was assured to Britain by Admiral Byng's defeat of the Spanish Fleet off *Cape Passaro* in that year, so it was not until 1727 that the Spaniards were able to start once again on their favourite venture, the recapture of Gibraltar. The garrison of the *Rock* at the beginning of that year consisted of Captain Holman's Company of Royal Artillery (strength about 3 officers and 55 men) with Egerton's (20th Foot) Pearce's (5th Fusiliers) and Lord Mark Kerr's (13th Foot) Regiments of Foot, in all about 1,700 men, under the command of Colonel Jasper Clayton, the Deputy Governor. The armament of the *Rock* in 1727 was:-

```
        52 — 32 pdrs.
        62 — 24 pdrs.
        85 — 18 pdrs.
         5 — 16 pdrs.
        52 — 12 pdrs.
        17 —  9 pdrs.
        22 —  6 pdrs.
        40 —  4 pdrs.
         8 —  3 pdrs.
        56 — mortars
        12 — howitzers
```

Grand total 411 pieces of ordnance

As soon as it became obvious that the Spaniards were about to try once again to retake Gibraltar, reinforcements were sent out from England; a party of 2 officers and 50 other ranks of the Royal Artillery, which included Lieut.-Colonel Watson, the Lieut.-Colonel of the Regiment, to take over command of the Royal Artillery on the *Rock*, and Newton's (39th Foot) Hayes' (34th Foot) Anstruther's (26th Foot) and Disney's (29th Foot) Regiments of Foot, bringing the garrison up to a strength of about 6,000 all ranks.

The siege lasted four months, from February to June, the Spaniards concentrating a force of some 15,000 men with 150 guns and mortars, under the Conde de las Torres, on the landward side. It is, however, probably wrong to describe these operations as a siege for the Spaniards were never able to cut off the fortress from the seaside nor bring their warships into Gibraltar Bay, for the British fleet had complete command of the Straits, and the Admiral in command was even able to send vessels to enfilade the Spanish trenches on the isthmus. Nevertheless, both sides kept up a tremendous bombardment, the British firing some 80,000 shot and shell during the four months, the Spanish reply being even greater. The enemy never made any really serious infantry assault on the landward defences of the *Rock*, their lack of command of the sea handicapping their efforts at every turn. The Spanish attack petered out in June, and hostilities came to an end. The British casualties were small, about 150 killed or died of disease, and 200 wounded: unfortunately, the former included Captain Holman, the commander of the Company of Royal Artillery.

By 1742 Britain was at war again, this time with France and Spain, being engaged in the War of Austrian Succession supporting the cause of the Austrian Maria Theresa. Ever since the early years of the seventeenth century, when both Britain and France had first founded settlements in North America, there had been bitter rivalry between the New England provinces and the French colony of Canada spread along the banks of the *St. Lawrence*. When the mother-countries were at war, the daughter provinces were generally only too eager to join in. We have already seen how, during the Spanish Succession War, Britain had taken Nova Scotia and Newfoundland from the French, and, now that the two countries were once again at war, the French seized the opportunity to make an attempt to retake the former. Early in the spring of 1744 the Governor of Louisburg, the French fortress and naval base on Cape Breton Island, decided to send an expedition to Nova Scotia with the object of capturing Annapolis, the British headquarters, and so regaining control over the last province. The permanent garrison of Annapolis at this time was the detachment of coast-artillery, under Lieut.-Fireworker Dyson who had but lately been commissioned from the rank of sergeant, of 3 N.C.O.s and 13 gunners and matrosses, and a weak company of Philips' (40th Foot) Regiment about 50 strong, all under the command of the Governor, Lieut.-Colonel Mascarene, the armament of the small fortress being 6—24 pdrs, 4—12 pdrs. and 16—4 pdrs.

As soon as Mascarene heard of the French preparations at Louisburg he sent urgently to Governor Shirley of New England, asking that energetic and far-seeing official to send him reinforcements as he fully realised the weakness of his garrison and his probable inability to resist successfully a serious attempt by the French to retake the province. Governor Shirley did not fail nor hesitate and at once raised four companies—about 200 all told on arrival at Annapolis—of provincial militia for special service in Nova Scotia. As soon as Mascarene learnt that these most welcome reinforcements were on their way, he sent forward his company of regular infantry under Captain Heron to hold Canseau, the small port opposite Cape Breton Island at the northern end of the province.

Meanwhile the French expedition had sailed from Louisburg. It consisted of 2 sloops-of-war, each of 8 guns, with about 75 soldiers embarked in them, under the command of Captain Duvivier of the Colonial Marine. On 11th May, this small flotilla arrived before Canseau where, having disembarked his troops, Duvivier was joined by about 200 friendly *Micmac* Indians from the backwoods. Duvivier then set about formally summoning Capt. Heron to surrender which most surprisingly that officer promptly did for "he surrendered himself, the same day, prisoner of war, with his whole garrison, as well as women, for one year, with all the arms, artillery, and military stores" (*Terms of Capitulation:* Rolt Vol. III Pt. V.) After this somewhat astonishing success, Duvivier, having despatched his prisoners and booty to Louisburg, re-embarked his force and set sail for Annapolis before which place he arrived at the beginning of June.

The basin of Annapolis lies on the south side of the bay of Fundy about 40 miles from the entrance. It is surrounded by precipitous and densely wooded hills, but at the top end there is a gap where the *Cornwallis River* flows in. Here lay the fort and township of Annapolis, the headquarters of British Government in Nova Scotia. It was a situation of great beauty and importance and had been occupied by either French or British since 1604. The fort was constructed of earth ramparts with a bastion at each corner, having barracks, storehouses etc. built of wood inside the fortifications, and, by landing guns from ships, the armament had now been increased to 44 pieces of ordnance. It was here that Captain Duvivier disembarked his small force on 22nd June and was soon joined by an even larger mass of friendly Indians than had been with him at Canseau, probably not less than 500 this time. On this occasion his summons to surrender was refused with scorn, the answer being that the fort would be defended to the last extremity so he set about bombarding the defences with his ships and exhausting the garrison with perpetual alarms from the landside. These efforts, however, had little effect on Mascarene, his gunners, and New Englanders, so Duvivier decided early in July on a general assault, preparing scaling ladders and offering 500 *livres* to every Indian who would mount them. However, the Indians did not look with much enthusiasm upon this sort of straight forward warfare, and Duvivier could not prevail on them to attempt such a frontal attack, so, considering his small company of Marine Infantry too weak to attempt to storm the place on its own, he was forced to re-embark his troops and return with his ships to Louisburg. Thus Annapolis survived, thanks to the resourcefulness of Governor Shirley and the bravery of his New Englanders, and was left to remain in peace in its beautiful surroundings for the rest of its existence.

This success merely whetted the appetite of Governor Shirley who in the following year raised an expeditionary force of some 4,000 volunteers from the New England colonists and despatched them, under the command of William Pepperell, a prosperous New England trader, to capture Louisburg, the great French base on Cape Breton Island. Escorted by a squadron under Commodore Warren, the convoy reached Cape Breton Island safely and on 30th April (1745) landed the New England troops in Gabaron Bay about 4 miles from Louisburg. The fortress was still incomplete, and the French were taken utterly by surprise, so after a short but energetic siege of 6 weeks, this great fortress, with a garrison of 1,200 men and mounting some 180 guns, surrendered to the victorious but somewhat amateur army from New England.

In the early spring of the following year 3 battalions of regular infantry were sent to Louisburg to take over the duties of garrison from the New Englanders, and this force was accompanied by Captain Rogers' Company of Royal Artillery which had come from Woolwich by way of the West Indies. This Company remained at Louisburg manning the guns of the coast-fortress until the end of the war in 1748 when this valuable stronghold was returned to France in exchange for Madras which had been captured by Dupleix from the East India Company in 1746. Captain Rogers' Company then (1749) moved to Nova Scotia where it set to assisting in the construction of the new seat of government (Halifax) of the province on Chebacto Harbour, and at Halifax it remained to man the guns of the defences. During the same year, 1749, Captain T. Ord, R.A. was sent from Woolwich to St. John's Newfoundland, where he formed a new company of Royal Artillery with headquarters at St. John's, by absorbing the coast-artillery detachments at that place, Placentia and Annapolis.

Meanwhile, the home-country had suffered the attempt of Prince Charles Edward the Young Pretender to regain the throne for the Stuarts. The series of surprising successes won by the *Jacobites* had greatly alarmed George II, his government in London, and the people of England, but command of the *Channel* and *North Sea*, held firmly by the British fleet, effectively prevented the French from following them up with an invasion, although a force of some 12,000 men was concentrated at the *Channel* ports with this possibility in view. A squadron under Admiral Vernon was stationed in the *Downs* to watch Dunkirk and Calais, a second squadron under Admiral Martin cruised in the *Channel*, while a third under Admiral Byng went northwards and patrolled the coasts of Scotland. It is doubtful whether the French really ever seriously meant to invade, their *Channel* fleet being greatly inferior to the British both in numbers and quality, but the threat of the concentration of troops at the French ports was sufficient to cause great alarm in the south-east and southern counties of England which were far from the immediate effects of the *Jacobite* rising.

Field Marshal Lord Stair was appointed Commander-in-chief of the forces in Southern Britain, the lords-lieutenants were instructed to call out the county militias, and the coast-defences were alerted. The coast artillery was as usual but a small squad of a master gunner and a few gunners at each fort or defended port whose business was care, maintenance and preservation only: they relied upon "additional gunners", seamen from the fleet, or the infantry garrison to provide gun detachments in case of need. During the War of Spanish Succession the War Office had discovered a method by which the Army was able to provide these infantry garrisons for the forts and defences without keeping at home regular troops who were trained and fit to fight overseas, and as the result "Independent Companies of Invalids" were formed, "Invalids" being soldiers who were either too old or not physically fit to stand the strains and stresses of active service: they must have indeed been suitable companions for the coast-artillery detachments, most of whom were in like condition. By the time of the '45 rising, there were infantry Invalid Companies in garrison at the coast-defences of Plymouth (4 companies), Hull (4 companies), Sheerness (2 companies), Gravesend and Tilbury (2 companies), Tynemouth (1 company), *Landguard Fort* (1 company) and Falmouth (1 company), Portsmouth, Dover and Chatham being still garrisoned by regular battalions. However, there was evidently a shortage of trained manpower for the gun-detachments because Rolt, writing in 1750, tells us that 200 gunners, bombardiers, and matrosses were sent from Woolwich to the forts and castles in the south and west of England "so that the whole Kingdoms of Great Britain and Ireland were properly protected from any invasion, either from the coast of Britany, Normandy, or Picardy." (Rolt, Vol. IV Pt. VII.) These Gunner re-inforcements probably came from the companies of Captains Goodyear and Belford which were at that time stationed at Woolwich. However "Bonny Prince Charlie" failed in his gallant endeavour to win the throne for his dynasty, the French never attempted invasion, the gunners, bombardiers, and matrosses returned to Woolwich, their proper home, and the War of Austrian Succession at last came to its end in 1748.

CHAPTER V

1748 to 1763.

WAR, the Seven Years War, however, broke out again between Britain and France in 1756, and the movement of French troops towards the *Channel* ports early in that year, as if in preparation for an invasion, produced a panic among the population of the south and east of England. It was indeed the intention of the French to threaten invasion, and they concentrated some 50,000 troops between Dunkirk and Cherbourg under Marshal Belleisle in February and March, but, without command of the *Channel*, it would have been a very hazardous operation. The population of the south and east of England was not alone in being alarmed, for the government, besides detaining in home waters a large fleet and thus leaving the western *Atlantic* and the *Mediterranean* insufficiently guarded, issued a scaremongering proclamation that, in case of a French landing, all horses, cattle and sheep, together with provisions and forage should be removed inland for at least 20 miles. A military report of this period gives the opinion that the most probable areas for enemy landings were (1) around Sandwich and Deal (2) between Hythe and Rye (3) between Pevensey Bay and Eastbourne, and (4) between Brighton and Chichester. (W.O.S.S./1848). The writer of the report did not consider the coast defences at Chatham, which he stated "seem to have been neglected", Deal, Walmer, Dover, Sandgate, and Portsmouth sufficient to cover this long stretch of coast and made recommendations concerning the positions a defending army should occupy in case the French landed.

The Board of Ordnance, stirred by the threat of invasion, issued instructions that nine "sea-batteries"—which today would be be called "emergency batteries"—should be erected along the unprotected coast at:-

Folkestone	6— 9 pdrs.
Hythe...	6—18 pdrs.
Rye	10—24 pdrs. and 5 of different natures.
Hastings	11—12 pdrs.
Seaford (two batteries)	5—24 pdrs.
	5—12 pdrs.
Newhaven	5—12 pdrs.
Brighton	12—24 pdrs.
Arundel Haven ...	7—18 pdrs.

"To oppose an enemy landing at the most exposed places on the sea-coast of Sussex and Kent." One master gunner and one gunner were to be allowed for each battery who were to "teach such of the inhabitants as shall be inclinable to learn how to load, point and fire the guns placed there for their defence." (S.P. 41/38). This gives an illuminating insight into how "additional gunners" were obtained and trained and a direct indication of the embryo from which local volunteer artillery eventually grew.

The Board also ordered an inspection of the coast-defences of the east coast, and the report on *Cliffords Fort*, the main defence of Tynemouth, informed the Board that the fort mounted 36—18 pdrs., in three batteries, and that it possessed "a very good magazine in which is now 388 barrels of cannon powder and 93 half barrels of fine powder. Over this magazine and round it are barracks for 40 men, a black hole, three rooms for officers, a master-gunner's room, cooperage, store-room and stables." There was also "a cistern for 1,152 gallons of rain water but there is no spring nearer than Shields." In its summing up the report stated that "notwithstanding the many disadvantages of *Cliffords Fort*, considered as a fort, yet as a battery it is much the best situation on the coast" but added that "a small battery of 6 guns should be erected at *Tinmouth Castle* (*sic.*) for the defence of ships at anchor in the roads". (S.P. 41/38).

But the French did not invade Great Britain, indeed it is doubtful if they ever intended to do so. Their object was to distract the attention of the British Government and Admiralty from the *Mediterranean* where the main blow was to fall, and very successful they were in doing so. The British concentrated their fleet and resources for the defence of the home-country and left the *Mediterranean* almost unprotected. The French therefore secretly prepared an expedition for the recovery of Minorca, and on 8th April (1756) the convoy, consisting of 138 transports, having on board some 16,000 troops commanded by Marshal the Duke de Richelieu, sailed from Toulon escorted by a squadron of 12 ships-of-the-line under the Marquis de Galissonnière, the finest admiral in the French King's service.

The British Government had meanwhile got news of what was going on and, still needlessly fearing invasion, despatched only a weak squadron of 10 line-of-battleships under Admiral John Byng which, sailing from *Spithead* on 6th April with orders to proceed to Minorca, reached Gibraltar on 22nd May. There Byng learnt that the French had landed in Minorca on 15th April and had at once set about besieging Port Mahon, the chief port and defensive post of the island. Byng, with his ships-of-the-line now reinforced to 12, left Gibraltar on 8th May and on the 20th gained contact off Minorca with the French fleet. The two squadrons were very even, each of 12 line-of-battleships with an average of

70 guns to a ship, but the French were better handled and, after two hours fighting in the afternoon, Byng was forced to draw off his squadron. In fact he continued his withdrawal until he reached Gibraltar, and so abandoned Port Mahon and its garrison to its fate, Galissonière being left in complete control of the sea approaches to the island. (*See map on page 64*).

Minorca was the smaller of the two significant *Balearic Islands* but contained the most important harbour and fortress, Port Mahon. Its garrison in April 1756 consisted of Captain Flight's Company of Royal Artillery (seven officers and 100 men) and the 4th, 23rd, 24th and 34th Regiments of Foot, in all about 3,000 men, under the command of the Lieutenant-Governor, General William Blakeney, who had reached the ripe old age of 82. Added to the garrison were five Officers and 122 ratings, Royal Navy, and four Officers and 160 men, Royal Marines, who had been landed from the small naval squadron based on Port Mahon before it sailed to join Byng. The armament of *Fort St. Philip*, the fort which commanded the town and harbour of Port Mahon, was at about this time:-

```
        24—32 pdrs.
        50—18 pdrs.
        40—12 pdrs.
        40— 9 pdrs.
        36— 8 pdrs.
        10— 6 pdrs.
        ─────
Total   200 guns.
        ─────
```

besides an unknown number of mortars. It is obvious that as usual the number of gunners was quite insufficient to provide detachments for the guns so the whole of the naval landing party, plus 220 men from the infantry battalions, were attached to the Royal Artillery to serve as "additional gunners" to man the guns. Captain Flight was appointed acting C.R.A. as Major Mace, the appointed C.R.A. had been held up at Gibraltar and could get no further on account of Byng's defeat. Flight's Company was the direct descendant of the original detachment—or train—of permanent artillerymen sent to Minorca in 1709.

As soon as the French began to land their army on 18th April, Blakeney withdrew all his forces into *Fort St. Philip*. This fortress was immensely strong, being fortified according to the school of Vauban and strengthened by innumerable galleries driven into the solid rock on which the stronghold was based. The fighting part of the fort was divided into five sectors with an officer of the Royal Artillery in command of the guns in each sector:-

No. 1 Sector—Lieut. Webdell.
No. 2 Sector—Lieut. Charleton.
No. 3 Sector—2/Lieut. Forman.
No. 4 Sector—Lieut. Inglis.
No. 5 Sector—2/Lieut. Day.

Captain-Lieutenant Gregory having general command over the guns of the inner defences (No. 4 and 5 Sectors). The French took some days to complete their landing, haul their guns overland to within range of the fort, and construct their batteries, of which they raised nine, and it was not until 30th April that their warships closed in on the entrance to the harbour, and ships and land batteries opened fire. The bombardment of *Fort Philip* had begun.

From the French account of Poncet La Grove (1799) it would appear that the Duke de Richelieu had in position 84—24 pdrs. and 22 mortars, besides several smaller pieces, with which to engage the landside of the fort. Throughout May the bombardment continued while the French infantry and sappers advanced their trenches slowly but steadily to within assaulting distance. The British guns and mortars never ceased to reply, countering the enemy batteries, destroying their trenches, and on occasion, by establishing a listening-post on the glacis and using an F.O.O. on the West Lunette, bringing mortar-fire to bear on their working-parties at night. The French concentrated their fire on the north-western corner of the defences—Argyll, Anstruther and Queen's Redoubts —and by the middle of June these started to crumble and collapse, most of the guns mounted on them being already out of action. Soon a breach began to appear in Queen's Redoubt, the hostile trenches were already within musket shot of the ramparts, and the French hail of shot, shell and bullet became so hot that it was impossible to serve any longer the few remaining guns covering the threatened sector. A sally was attempted to spike the guns of the nearest and most dangerous battery but this failed badly, the whole party being captured. On June 25th the enemy bombardment reached its greatest volume, and continued so for the next two days, the British guns and gunners being shattered and broken as they stood exposed on the now all but bare ramparts. By the morning of the 27th *Fort St. Philip* had been almost silenced.

As darkness fell that evening, the French launched their grand assault. It was directed mainly at the north-west front of the fort, with a land attack on Anstruther and Queen's Redoubts and an amphibious attack from small boats on Royal Battery and Argyll Redoubt. There were also holding attacks on the eastern front. After a tremendous struggle the French infantry stormed the breach in Queen's Redoubt, and by dawn on the 28th were holding Kane's, Queen's, Anstruther and Argyll Redoubts and Royal Battery.

During the day the French consolidated their position and prepared for the next stage of the assault. The British garrison, however, was utterly exhausted and in no state to stand a further battering so, after holding a council-of-war, General Blakeney offered to surrender, and at 6 o'clock in the evening of the 29th, the garrison hauled down the British flag from the fort and marched out with full military honours, drums beating and colours flying. By the terms of the capitulation, the British garrison was repatriated to Britain by way of Gibraltar, an indulgence granted by the French as a token of respect for their gallant and brave defence. The garrison suffered a total of 480 casualties of which 30 came from the company of Royal Artillery, 43 from the naval landing-party, and 80 from the infantry "additional gunners". It is pleasing to know that the Board of Ordnance granted to the men of the Royal Artillery company a half-day's additional pay for each day of the siege as a reward for their bravery and endurance. As General Blakeney said in recommending this award "The Company of Royal Artillery here most justly deserve encouragement and consideration of their great fatigue and the spirit they show at this seasonable time for his Majesty's service."

The Seven Years War continued on its way, and, under the bold but wise direction of the elder Pitt, Britain succeeded in winning the great contest with France for world dominion. Canada, great tracts in India, West Indian Islands, Florida, Manila, and trading posts on the west coast of Africa were taken and held by the victorious British. After the opening disasters of the defeat of Byng's squadron and the fall of Minorca in 1756—brought about entirely by the fear of the government and people of Britain of the threat of invasion simulated by the French—the British Navy generally held the whip hand on the seas beyond Europe, and the Coast Artillery serving abroad was rarely called upon to defend the ports which it was its duty to protect. In 1759 we find two companies of Royal Artillery at Gibraltar, one company in Nova Scotia (Halifax and Annapolis) one company in Newfoundland (St. John's, Placentia, and lesser harbours) and a detachment at Antigua, West Indies. *Fort Louis* at the mouth of the *Senegal River* and Goree Island, off Dakar, were captured from the French by small expeditions during 1758, and detachments of Royal Artillery (Senegal one officer, 36 O.R.s: Goree one officer, 23 O.R.s) were left to maintain the armament and form the core of the gun's crews. Quebec (two companies) Louisburg (two companies) Guadeloupe (one company) Martinique (one company) Manila (one company) Pensacola Bay, Florida (one company) Havana (one company) and Belleisle (detachment), all captured from the French and Spanish, were garrisoned by British troops, with the coast armament manned by Royal Artillery, and remained so until the Peace of Paris in February 1763. There were also two companies at *Fort St. George*, Madras, but it is doubtful whether they were employed in coast-defence duties.

However, some seven months before the signing of the Peace of Paris which brought the war to an end, Captain-Lieutenant Rogers of Captain Dover's Company Royal Artillery, suffered the shock of his lifetime. Dover's Company was stationed in Newfoundland, headquarters at Placentia, with detachments at St. John's and three small outstations. Captain Rogers commanded the detachment at St. John's consisting of six gunners and nine matrosses, which was responsible for the armament of *Fort William*, the main defence work, and there could have been no more peaceful place, far from the hazards, alarms, and excursions of war. The fort itself, built of earth, was in a state of poor repair, its effective armament at this period being:-

    20—24 pdrs.
    12—18 pdrs.
     9— 6 pdrs.
and 2—mortars.

France had suffered terribly during the contest with Britain and had lost most of her overseas possessions, but Choiseul, King Louis' minister-of-war, hoped to make some riposte before his country was forced to come to terms with her enemy. For this nuisance raid—for with Britain's command of the sea it could be little else—Choiseul chose as its objective Newfoundland, chiefly because it was isolated and a large proportion of its inhabitants were known to be Irish and disaffected towards the British. The leader of the expedition was Captain Ternay d'Arsac of the French Navy who commanded a flotilla of two ships-of-the-line (74 and 64 guns) and three frigates (each 26 guns) with embarked in them some 800 Marine Infantry under Colonel d'Haussonville. This small expedition slipped out of Brest on 8th May 1762 and, having successfully evaded the British blockade during a spell of thick weather, made off across the *Atlantic*. Early in June Ternay arrived at the Newfoundland fishing banks and caused great confusion and damage, destroying and sinking all the fishing-boats he could lay his hands on. He then moved in shore and landed d'Haussonville with his troops who proceeded to burn the fishing villages, drive off the cattle, and lay waste the surrounding country. Meanwhile, the news of the French expedition had reached St. John's which was garrisoned by Captain-Lieutenant Roger's detachment of Royal Artillery and a company of the 40th Foot (two officers and 74 men) all under the command of Captain Ross of the same regiment: also at anchor in the harbour was the frigate *Gramont* (16 guns, Captain Mouat).

Having sent off "express" to Placentia and Halifax the alarming information that a French expedition was on the coast, Captains Ross and Mouat made ready to defend their charge. A party of seamen was landed to assist in serving the guns, the *Gramont* was cleared for action, and all preparations were made for a desperate resistance. On 27th June Ternay's flotilla appeared off the entrance to the harbour, and d'Hausonville's troops, after a 25 mile cross-country march, arrived on the landwardside of the fort. At the sight of the French, Ross' courage began to ooze out of him. The Marine Infatry advanced steadily towards a rise about 300 yards from the fort. Rogers then asked Ross if he should open fire. Ross replied "You may if you please." Rogers at once started firing with his 24 pdrs. but only had time for a few rounds before the French took cover behind the rise. In accordance with the custom of the age, the French then sent forward an officer under a flag of truce to demand the surrender of the fort. He was received by a council-of-war consisting of Ross, Mouat, and Rogers. Ross by now had lost all his courage. Help was far away, and the French, in overwhelming numbers, were horribly near, so, in spite of the protests of Mouat and Rogers, he decided to capitulate. The white flag was hoisted, the British troops laid down their arms, and the French took over the fort. Mouat just had time to send orders to his first-lieutenant to burn and scuttle the *Gramont* before he himself was made prisoner. Ternay's warships then sailed into the harbour, and St. John's was in the hands of the French.

This was a shameful surrender and reflected no credit on anyone concerned. It was not until September that an expedition, despatched from New York and Halifax, under the command of Lieut.-Colonel William Amherst, reached Newfoundland and made a landing to the west of St. John's. D'Haussonville resisted fiercely, firstly in the country surrounding the fort and then from the fort itself when driven back into it, but he was finally forced to surrender on 13th September, Ternay and his ships managing to escape safely from the harbour during the previous night. A considerable body of Royal Artillery accompanied Amherst's force, a party from which remained in *Fort William* until relieved by a detachment of Captain Dover's Company from Placentia. This detachment consisted mostly of locally enlisted men, taken on to replace those so unhappily lost under Captain Rogers when the fort surrendered to the French in June. To give some idea of the conditions under which Coast Artillery abroad functioned at that time, here is an extract from the despatch to the Admiralty of Admiral Lord Colville, the naval Commander-in-Chief in the *North Atlantic*, reporting these operations in Newfoundland in 1762. He is describing the capture of Carbonera Island—one of Captain Dover's smaller outstations—by

1748 to 1763

a landing-party from Ternay's warships:- "The Island of Carbonera in Conception Bay has had no other garrison for many years but a few old men of the Artillery to take care of the guns and Ordnance stores. Had some of the inhabitants of the adjacent coast taken part here, they might easily have defended it against any force as the island is inaccessible on all sides, except one narrow landing place, and no safe road in the neighbourhood for great ships. But the enemy landed in boats and destroyed the whole without resistance". The chief gunner and his assistants at Carbonera Island had evidently failed "to teach such of the inhabitants as shall be inclinable to learn how to load, point and fire the guns placed there for their defence."

Meanwhile at home the coast-defences had been little troubled since the invasion scare of 1756. Lord Hawke's great victory in Quiberon Bay on 20th November 1759 settled the question of command of the sea in European waters until the end of the war, and French naval activities were reduced to small raiding squadrons and privateering. We are now approaching the time when the coast-artillery gunners, on the establishment of "Guards and Garrisons", were absorbed into the Royal Artillery, so let us take a last look at them as they were during the period of the Seven Years War:- (*vide table on next page*).

During the war the "marching companies" of the Royal Artillery were on occasions brought in to assist the coast-defences. In 1758 we find Captain Gregory's Company at Sheerness, Captain Pattison's at Scarborough, Captain James' at Portsmouth, and Captain Charleton's at Plymouth, this last company remaining at Plymouth until the end of the war.

The Seven Years War was ended in February 1762 by the Peace of Paris. France and Spain ceded to Britain, Canada, Florida, and the West Indian islands of Grenada, St. Vincent, and Dominica and returned Minorca. Britain handed back to France, Martinique, Guadeloupe, Belleisle, St. Lucia and the West African posts; and to Spain Havana and Manila. Louisburg, the great French fortress on Cape Breton Island—now British territory—was levelled to the ground and left to the cry of the seagulls and the roar of the waves.

Place	Mr. Gnrs.	Gunners
Arundel Haven	1	1
Berwick	1	1
Brighton	1	1
Calshot	1	1
Chatham	3	13
Dartmouth	1	2
Deal	1	1
Dover	2	4
Dunbarton	—	1
Falmouth	2	3
Folkestone	1	1
Gillingham	1	1
Gosport	2	1
Gravesend & Tilbury	2	12
Guernsey	1	4
Hastings	1	1
Landguard Fort	1	3
Holy Island	1	1
Hull	1	6
Hythe	1	1
Jersey	1	8
Newhaven	1	1
Plymouth	2	8
Portland	1	2
Portsmouth & Southsea	3	3
Rye	1	1
Sandgate	1	1
Scarborough	1	1
Scilly Islands	1	6
Seaford	2	2
Sheerness	1	16
Walmer	1	1
Tynemouth	1	2
Yarmouth North	1	2
Hurst Castle	1	4
Isle of Wight, Cowes	1	5
Carisbrook	1	3
Sandown	1	3
Yarmouth	1	5

CHAPTER VI
1763 to 1774.

WITH the end of the Seven Years War, the Royal Artillery returned to peace conditions and stations. The coast-artillery abroad was provided from the "marching companies".

Gibraltar:	3 Companies	1764-66
	4 Companies	1766-70
	5 Companies	1770-74
Minorca:	3 Companies	1764-69
	4 Companies	1769-70
	2 Companies	1770-72
	3 Companies	1772-74
New York:	1 Company	1764
	3 Companies	1764-66
	2 Companies	1766-72
	3 Companies	1772-74
Pensacola Bay: (Florida)	2 Companies	1764-70
	1 Company	1770-74
Quebec	1 Company	1764-74
Halifax (Nova Scotia)	1 Company	1764-74
Newfoundland:	1 Company	1764-74

Detachments found for outlying coast-forts at defended ports along their coasts.

In India the coast-defences were manned by the artillery of the East Indian Company. At home, the "marching companies" withdrew altogether from the role of coast-artillery—Captain Carter's Company leaving Plymouth in March 1765—and the defences were left once again to the Master Gunners of the "Guards and Garrisons". However, in 1771 the Board of Ordnance determined that this dual system must end and, deciding to come into line with the War Office, authorised on 18th February 1771 the formation of eight Invalid Companies, Royal Artillery. The main object of these companies was to relieve the "marching companies" at home from having to provide men for the many, many employments, both "on command" and at the chief Gunner stations (Woolwich, Greenwich, Chatham), which could be carried out by men not in their first youth or partially disabled. In the eighteenth century a man enlisted in the Royal Artillery for life—or as long as the Board of Ordnance wished to keep him—and many gunners grew old in the service, lost their youthful vigour, contracted rheumatism, and such like, and were most certainly quite unfit to stand up to the rigours of active service—most of the coast-artillery personnel of the "Guards and Garrisons" were usually of this category. So the eight Invalid Companies were formed from such men who were serving at home in the "marching companies," from pensioners who were considered fit for garrison duties, and from the coast-artillery detachments who were absorbed from the establishment of "Guards and Garrisons."

So the soldiers of the coast-artillery at home were now at last an integral part of the Royal Artillery after having been for so many years something apart. The headquarters of all the Invalid Companies were at first at Woolwich (or Greenwich), but of course their men were scattered far and wide over Great Britain, in coast defences and in employments, but very soon it began to be realised that their primary business was—as with the Infantry Invalid Companies—to provide garrisons for the coast ports and defences. Before 1771 was out, two companies had been moved to Portsmouth, and in 1775, on the outbreak of the American War of Independence, one company complete went overseas to take over the coast defences of Newfoundland from Captain Buchanan's Company and so release that company for active service in New England. This, however, was not a success, the climate of Newfoundland proving too much for most of the veterans, as a result the wastage was too heavy, and in 1778 the Company was fetched back home where it settled down into the coast-defences of Plymouth. In 1779 two more companies were added, bringing the total up to 10, and they were collected together for purposes of administration into an Invalid Battalion, Royal Artillery. In this year two companies were at Portsmouth, one company at Plymouth, one company in Scotland (H.Q. Fort George) and the remaining six had their headquarters at Woolwich. The officers allowed for battalion headquarters at Woolwich were one Lieut.-Colonel commanding, one Major second-in-command, and an adjutant. The establishment of each company was one Captain, one Lieutenant, one Second Lieutenant, one Sergeant, one Corporal, three Bombardiers, one Drummer, six Gunners and 36 Mattrosses. In the following year 1780, two Companies were sent from Woolwich to the Channel Islands one to Jersey and one to Guernsey.

In the summer of that year, in the middle of the War of American Independence, the detachments from the ten Invalid Companies were scattered among the coast-defences at home as follows:-

1763 to 1774

Place	Carleton's H.Q. Company Jersey		Hind's H.Q. Company Woolwich		Webdell's H.Q. Company Portsmouth		Dover's H.Q. Company Woolwich		Anderson's H.Q. Company Plymouth		Toriano's H.Q. Company Guernsey		Buchanan's H.Q. Company Woolwich		Whitmore's H.Q. Company Portsmouth		Winter's H.Q. Company Woolwich		Macleod's H.Q. Company Fort George		Totals	
	off.	o.r.	off.	o.r.	off.	o.r.	off.	o.r.	off.	o.r.	off.	o.r.	off.	o.r.	off.	o.r.	off.	o.r.	off.	o.r.	off.	o.r.
Berwick				4																		4
Tynemouth				3																		3
Scarborough				1																		1
Hull			1	6																		7
N. Yarmouth					1																	1
Harwich				1										2								3
Sheerness				1										10								11
Tilbury								5						1				1				7
Gravesend								4										1				5
Chatham								7						7				1				15
Gillingham								2														2
Hythe								1						2				1				4
Rye				1																		1
Seaford				1																		1
Newhaven		1																				1
Hastings																		2				2
Brighton																		1				1
Arundel Haven								1										1				2
Portsmouth					1	45		1	1	8				1	2	6		1		4		62
Sandown														1				1				2
Hurst														1		5						6
Carisbrook				1																		1
Yarmouth				3																		3
Cowes								2														2
Calshot						2																2
Gosport						1								3								4
Portland								2														2
Dartmouth		1						1										1				3
Plymouth		1						2	2	45		3	1	10	1	28		7		4		96
Falmouth				1												1		2				4
Scilly Islands																6						6
Jersey	3	41						1		1								4		3		47
Guernsey								3		1	3	44								3		48
Dumbarton																				2		2
Totals	3	45		23	1	49		32	3	55	3	47	1	38	3	41		29	2	14		361

Dover, Deal, Walmer and Sandgate were garrisoned by a special *Cinque Ports* detachment. It is to be noted that Master Gunners were not included in the strength of the Invalid Companies. There must have been Master Gunners in charge at all of the coast defences but their distribution is not known.

As these Invalid Companies formed the core of coast-artillery at home throughout the War of American Independence (1775-83) and the great war against revolutionary France and the Emperor Napoleon (1793-1815), it is necessary to take a closer look at them. The Company commanders were normally officers who had served for many years with distinction but who had become in course of time much too old for active—and sometimes for any kind of—service. We hear of Invalid Captains with 69, 62, and 57 years of service, and of one in the Channel Islands in 1797 who was "between 70 and 80 years of age and had been confined to his room for three years mostly bedridden" The subalterns were usually sergeants who had been promoted from the ranks as a reward either for bravery in action or long and good service, and they did most of the work, keeping the books and accounts, and carring out the training if any. However, they also were only too often allowed to remain in the service far beyond their years of usefulness, and a medical report on one stated "That of late years he has been incapable of attending his duty on account of violent fits of the stone, a considerable number of which he has been passing daily. At present his habit is debilited as also on account of his age, being about 70, to be unfit for his duty of the garrison." The rank and file were very little, if any, better. The Invalid Companies provided a much needed shelter for many old gunners who might otherwise have found it extremely difficult to exist outside the Army and once in that shelter were very loth to leave it. They remained on as long as they were allowed to do so, many being quite unfit for any sort of duty, much less to serve the guns in action, and a large number were usually sick, lame, or absent through disability.

In 1760 the Royal Irish Regiment of Artillery was formed under the Master General and Board of Ordnance in Ireland. As in Britain, the coast-artillery had previously been in the hands of small parties of permanent gunners scattered amongst the forts at the principal ports, and in 1687 there was a gunner or gunner's mate acting as caretaker or storekeeper at the following coast-defences:-

        Cork   Carrickfergus
        Charles Fort (Kinsale)   Passage
        Limerick   Fort Duncannon
        Galloway   and Waterford.

1763 to 1774

By the time of the War of Spanish Succession, gunners and matrosses had been added so that by 1734 the coast-artillery in Ireland stood as follows:-

Place	Master-Gunners	Store-keepers	Gunners	Matrosses
Cork		1	2	2
Kinsale (Charles Fort)	1	1	8	4
Limerick		1	4	2
Galloway		1	3	2
Carrickfergus		1	1	1
Waterford and Duncannon	1		2	2

Nothing is known of the armaments except that in 1763:-

Duncannon Fort (Waterford) mounted 10—24 pdrs.
                                                                  15—18 pdrs.
Cork ... ... 8—24 pdrs.
Charles Fort... ... ... 11—24 pdrs.

In 1776 the Royal Irish Artillery formed an Invalid Company from "such men who now receive their pensions or men discharged and not provided for who may be entitled to pensions as shall appear to be fit for garrison duty" (*Dublin Gazette* 24th February 1776) and into this the coast-artillery garrisons were to be absorbed, but this was found to be quite impracticable. Although the establishment of the Invalid Company was 53 all ranks, very few of its members were ever fit for any sort of duty, and we find the Colonel of the Royal Irish Artillery writing in 1783 that "There are at present 27 Invalids who are worn out or disabled in the Regiment and now paid a bounty of sixpence and threepence per day by the Ordnance". So the coast-artillery garrisons had to be formed from the "marching-companies" of which by 1783 there were six, and in that year (the last year of the War of American Independence) we find no less than 400 all ranks on detachment in the coast-defences of Ireland, disrtibuted as follows:-

Place	Officers	N.C.O.s	Gunners	Mattrosses	Total
Cork—Cove	2	2	1	5	
Spike Island	1	5	1	31	
Ramhead	1	5	3	30	
Carlisle Fort	2	11	5	57	
Roche's Tower	1	3	—	31	
Cork Garrison	2	9	2	30	240
Charles Fort (Kinsale)	2	9	8	48	67
Clugheen	—	2	2	10	14
Waterford—Passage	1	2	—	24	
Duncannon	2	6	—	25	60
Tarbert Island	1	2	—	19	22

And the armaments were:-
Cork—Spike Island ... 18—24 pdrs.
    Ramhead ... ... 8—24 pdrs.
        5—18 pdrs.
        6—12 pdrs.
    Carlisle Fort ... 30—24 pdrs.
        10—12 pdrs.
        30— 6 pdrs.
    Roche's Tower ... 12—24 pdrs.
    Charles Fort (Kinsale) ... 11—24 pdrs.
    Waterford—Passage ... 5—24 pdrs.
        3—12 pdrs.
    Duncannon ... 10—24 pdrs.
        10—18 pdrs.
    Tarbert Island ... ... 9—24 pdrs.

while Dublin's harbour was protected by two floating batteries, H.M.S. Britannia (18—12 pdrs.) and Hibernia (16—9 pdrs.) each specially converted for the purpose and carrying a detachment of one N.C.O. and nine gunners and matrosses from the Royal Irish Artillery. It is to be noted that Cork had evidently displaced Kinsale as the most important naval base and defended port in southern Ireland.

By 1770 the great sailing wooden-walled warship was rapidly reaching its culmination. No ship of less than 50 guns was considered a ship-of-the-line, and a First Rate carried 100 guns or more. The typical first rate of this period mounted 30—42 pdrs., 28—24 pdrs. 30—12 pdrs. and 12—6 pdrs. and was of about 2,200 tons burthen, so the necessity for coast defences to be armed with 42 and 32 pounders is obvious. The guns themselves had not changed and were still mounted, both afloat and ashore, on solid wooden carriages with small trucks but in coast-defences "these in turn were placed on a short length of railway so that they could be run back for reloading or to take up the recoil. Then the whole of the 'railway' was slightly elevated and pivoted at the front and wheeled at the rear end so that it could manœuvre on an arc. Thus the gun could be pointed over a wide traverse" (*A History of Firearms* by W. Y. Carman.) Although elevation by the horizontal screw was introduced generally for field-guns about the middle of the century, heavy fortress guns still retained the old wedge system.

A new type of gun was brought into service during this period for both warships and coast-defences. This was the carronade, invented by General Melville and Mr. Gascoign and made at the ironworks of the Carron Company in Scotland. This gun had a much shorter and lighter piece than the normal heavy weapons of the mid-eighteenth century, but its calibre was larger for the weight of shot fired. It was easy to handle, could fire either solid shot or hollow shell filled with explosive, outranged the more normal types of gun, and, having a greater muzzle velocity, had such destructive effect on ships' timber that it was nicknamed "*the smasher.*" These carronades were gradually taken into use by the Royal Navy between 1775 and 1790, being usually mounted on the forecastle or quarter-deck so as to engage a fleeing or following enemy at long range, but no record can be found of them being introduced into coast-defence before the French Revolutionary Wars. Also about this this time Captain Sir Charles Douglas of the Royal Navy invented the gun-flint-lock with tin firing tube which replaced the use of priming powder and match. This lock was a simple flint and steel mechanism which, being fastened to the gun, was operated by pulling a lanyard and so ignited the tube fixed in the vent. There was no change made in the Vauban system of coast-fortification, in fact it remained practically unaltered until long-range, high velocity, breech-loading artillery came into use in the mid-nineteenth century.

CHAPTER VII
1774 to 1783.

BY the summer of 1774 the British government was forced to recognise that very shortly it was going to have on its hands a serious rebellion, the revolt of its North American Colonies. Discontent and open sedition was at its worst among the New England colonists who began to collect weapons and ammunition, to assemble for drilling, and to prepare in every way for an armed struggle. In October they openly defied both the British Government and the commander of the troops at Boston, General Gage, by siezing the small coast-forts at Rhode Island and Pistcataqua New Hampshire. In April next year there occurred the affair at Lexington and on 17th June was fought the battle of Bunkers Hill outside Boston. The War of American Independence had now started in earnest. The rebellious colonies possessed no navy so that it was a war on land only that faced Britain, the Government's main problem being the supply and nourishment of the campaign across the great expanse of the Atlantic. This handicap made it almost impossible to concentrate sufficient force in America to crush the rebellion, and the errors of both government in London and the commanders on the spot did not help matters, with the result that the contest did not go well for Britain. On 17th October 1777 General Burgoyne's force, advancing down the Hudson to New York from Canada, was defeated by the New England rebels and compelled to surrender with some 5,000 men.

This disaster awakened the hopes and ambitions of Britain's rivals on the continent of Europe, for here was a chance too good to be missed of being revenged on the old enemy, and so France declared war in February 1778, to be followed in June 1779 by Spain, and in December 1780 by Holland, while Russia, Sweden and Denmark banded together to form an "Armed Neutrality" hostile to Britain. This combination of the maritime powers completely altered Britain's situation. Besides being faced with an enemy fleet superior to her own in home waters, the lines of communication across the Atlantic, the very life line of the campaign in America, were also threatened, so she had to concentrate her naval forces in the Channel, the Western Approaches and off the coasts of the rebellious colonies which meant abandoning the Mediterranean and the Caribbean Seas. In 1779 the French and Spanish opened proceedings in Europe by attempting to seize command of the Channel. By early June it was quite obvious that some such venture was in preparation. A combined Franco-Spanish fleet of 66 sail had been concentrated at Ferrol and Vigo, and it was reported that a French force of some 50,000 men was being massed in Normandy and Brittany with 400 vessels to transport them over to England. All that Britain had to oppose this formidable array was the Channel Fleet of no more than 35 ships under Admiral Sir Charles Hardy which was cruising between the Scilly Isles and Ushant.

It can be well imagined that this powerful threat to their security greatly alarmed the government and people of Britain, especially those living in the south. As usual, nothing was ready, and the defences of the south coast were in a lamentable condition. Most of the regular army was in North America, and arms, ammunition, and warlike stores for home defence were in very short supply. However, again as usual, the British people rose to the occasion. The *Militia* was embodied, a great number of nobles and gentlemen offered to raise regiments, both horse and foot, of *Fencibles* (regulars for home service only) and, for the first time, Parliament authorised the enrolling of volunteer companies, average strength three officers and 50 men per company, to be added to the *militia* regiments. For purposes of defence, the threatened coasts were divided into five districts, each under a general-officer, No. 1 District: Lincoln, Norfolk and Suffolk. No. 2 District: Essex. No. 3 District: Kent and Sussex. No. 4 District: Hampshire. No. 5 District: Dorset, Devon and Cornwall, and reserves were concentrated at Coxheath—six infantry brigades—and at Warley—four infantry brigades—these reserves being mostly *fencible* and *militia* regiments.

The Board of Ordnance was not behind-hand in preparing for action the defences for which they were responsible. At the beginning of June the Master General warned the commanders of all coast-defences on the south and east coasts to be ready to resist an attack by superior forces and to oppose an attempt at invasion. He also instructed them to let the Board know what equipment, stores, supplies etc. they required to bring their defences up to a state of efficiency. This produced such a flood of demands and complaints that inspectors were hurriedly sent off to the south-coast to report direct to the Board on the condition of some of the more isolated defences. The information they brought back with them is very enlightening. The inspector who went to the batteries at Hastings, Seaford, Newhaven and Brighton, stated that a great many of the wooden carriages had rotted and were quite unserviceable, in fact in some cases the iron pieces were lying on the platforms, that the iron shot had become considerably scaled, and that the powder was suspect. Inspector No. 2, who went further west, presented a more detailed report:-

1. *Calshot Castle*   Defences of this castle are much out of
     9—18 pdrs.   repair. The gun carriages quite rotten and
     4— 6 pdrs.   so much that many of them are broken down under the weight of the guns.
2. *Arundel haven*   Guns and carriages in good order and
     7—18 pdrs.   complete.
3. *Hurst Castle*   Masonry in tolerable order. 6 pdr. carriages are decayed but the 18 pdrs., which are mounted in casemated batteries, are in general pretty good.
4. *Cowes Castle*   Is being put in a state of defence by G.O.C.
     10— 9 pdrs.   Portsmouth.
5. *Carisbrook Castle*   Dismantled and being used as a hospital.
6. *Sandown Fort*   Guns mounted on sea-service truck carriages
     20— 9 pdrs.   from Portsmouth. The fort is being repaired.
7. *Portland Castle*   12 pdrs. guns and carriages, being mounted
     5—12 pdrs.   in casemated batteries, are in partly good
     3—18 pdrs.   condition. The 18 pdr. guns are scaled
     3—18 pdrs.   and their carriages decayed.
8. *Dartmouth Castle*   The platforms of the 6 pdr. battery want
     5—12 pdrs.   new laying. The carriages are in moderate
     5—12 pdrs.   order. The bores of the guns are much
     6— 6 pdrs.   damaged from having been exposed to the
     3— 3 pdrs.   sea-air and spray for 20 years past, the Master Gunner in general being negligent in securing them properly with tampions.

The Board of Ordnance immediately set about making great efforts to get things put right. Large quantities of stores were sent out from the Tower, and arrangements made for obtaining locally material for making palisades, fascines, and pickets for defending the batteries against land attack. Instructions were also issued for the construction of new batteries at:-

     Caister Heights ... ... 3—24 pdrs.
     Gorleston Hill ... ... 6—24 pdrs.
     Lowestoft ... ... 4—32 pdrs.
                        2— 9 pdrs.
     South Lowestoft ... ... 10—32 pdrs.
                        3—18 pdrs.
     Parkefield Battery ... 4—32 pdrs.
     Harwich Battery ... 4— 9 pdrs.
     New Tavern (Gravesend) 16—32 pdrs.

Margate	...	4—24 pdrs.
		6—18 pdrs.
Broadstairs ...	...	4—12 pdrs.
New Work (Dover)	...	6—12 pdrs.
Blackington ...	...	5—24 pdrs.
Weymouth ...	...	5— 9 pdrs.
		2— 6 pdrs.
Swanage	...	7— 9 pdrs.
Poole (4 batteries)	...	12— 9 pdrs.
		6— 6 pdrs.
Lyme Regis ...	...	10—18 pdrs.
Torbay (4 batteries)	...	25—24 pdrs.
Mevagisey ...	...	6—18 pdrs.
Looe	...	14— 6 pdrs.
Fowey (3 batteries)	...	2—18 pdrs.
		5—12 pdrs.
		2— 9 pdrs.
Guernsey (4 redoubts)	...	6—24 pdrs.
		22— 6 pdrs.

A certain number of these batteries, such as those at Weymouth and Lyme Regis, were demanded, paid for, and manned, by the townships concerned, the mayor and corporation of each town contracting with the Board of Ordnance to do so, the guns and stores being supplied by the Board. The *"Town Guard"* who manned these municipal coast-defence batteries must have been true forerunners of the Home Guard of the 1939-45 War.

There was as usual a great shortage of men to serve the guns, the small detachments of Royal Artillery Invalids in the permanent coast-defences being totally insufficient for this purpose. Many expedients were resorted to. At Portsmouth two men from each company of infantry in the garrison were detailed for the guns and "practised in the Artillery Exercise". At Plymouth the dockyard workers were fetched in and taught how to serve the guns. At many of the other coast-defences whole companies of the County *Militia* were assigned for duty as coast-artillery. At Dover the gun-detachments were found by the *Cinque Ports* Volunteers, an Invalid sergeant being specially detailed to instruct them. And this brings us to the "Volunteers of 1779", probably the first forerunners of all the various volunteer forces which have so willingly and bravely served Great Britain in her times of trial. During this year of 1779 Parliament passed an act authorising the raising of 150 companies of volunteers, companies to be attached to the county *militia* regiments. This Act stated that the volunteer officers were to be commissioned by the King but were only to be entitled to rank and pay when called out on active service. The private men were not to be bound by any engagement beyond attendance for exercise at the places appointed for that purpose until summoned for actual service when they would be required to enter into an engagement to serve for a term not exceeding six months in their own county and would be entitled to pay on the same scale as the regular forces. There was no difficulty in completing the 150 companies to establishment.

None of these "Volunteer Companies "were formed definitely as artillery, most were infantry, a few were light-horse, but, as can be seen from the case of the *Cinque Ports* Volunteers, some of them were very soon drawn into the service of coast-artillery. It is not known which groups of volunteer companies—except for the *Cinque Ports* Volunteers—were so employed, but it is possible to make a shrewd guess from the places at which some of them were formed or mustered, for the volunteer-companies at Sussex, Devon and Cornwall were formed "for the defence of the coast-line" and were mustered at Hastings and Newhaven, Plymouth and Falmouth respectively. It is therefore probable they were used to reinforce the slender manpower of the coast artillery. In Ireland the inhabitants of the principal sea-ports formed themselves into armed companies for defence of the coast.

We left the combined fleets of France and Spain in the month of June 1779 concentrated at the north western Spanish ports, while well-founded reports of a French expeditionary corps collecting at the ports of Normandy and Brittany continued to come across the Channel. Early in July it was established that the landing—if attempted at all—would be made near Plymouth, Cawsand Bay being the most popular choice for the actual disembarkation. Lieut-General Sir David Lindsay was G.O.C. of the Western District —Dorset, Devon and Cornwall—and was responsible for the defence of Plymouth, and the Board of Ordnance sent specially one of their own engineers—Lieut.-Colonel Dixon, Royal Engineers—to assist him. Lindsay was rather overwhelmed by the task before him. His troops were few—almost entirely *militia*—and practically untrained, and the coast-defences were in poor repair.

The *Citadel* was armed with some 20 heavy guns only, being supported by:-

Mount Wise	8—32 pdrs.
Western King	10—18 pdrs.
Drake's Island	21—32 pdrs.
	2—18 pdrs.

Under the direction of Dixon new works were begun to strengthen the defences, especially at Cawsand Bay the expected place of enemy landing. Here it was decided to construct three redoubts so sited as to be able to cover the foreshore by firing grape. To equip these redoubts the Board of Ordnance sent 16—12 pdrs. from their gun-park at the Tower of London. Dixon also had new batteries constructed at:-

Eastern King	4—18 pdrs.
West Hoe	6—18 pdrs.
Passage Point	4—18 pdrs.
Stonehouse Bridge	6—12 pdrs.
Cavalier Battery	6— 9 pdrs.
Lower Mount Wise...	6—24 pdrs.

and a Mr. McCormick raised 2,000 Cornish tin-miners to serve the guns and defend the new works, the Board authorising the issue of muskets and long pikes from their store at Plymouth to arm them.

In spite of these efforts, Lindsay complained continually to the War Office that he did not think it possible to hold Plymouth against a determined enemy attack. He reported that he was having difficulties with the admiral in command of the station, with the master-gunner at the Citadel, and even with Dixon who was forced to ask the Board whether he was to obey their orders or Lindsay's. The Board very properly replied that normally he would carry out their instructions "but in time of danger must obey the orders of the G.O.C." Early in August it was learnt that the combined Franco-Spanish fleet had left port and was making for the *Channel*, on the 14th it was sighted off the Lizard, and on the 16th it appeared off Plymouth. On the next day it was seen by the look-out at Dartmouth Castle, and the master-gunner ordered two rounds to be fired to give the alarm—an unauthorised expenditure of ammunition which he had the greatest difficulty in justifying to the Board when called upon to do so. On the close approach of the enemy, Lindsay let out another howl to the War Office, stating "in my opinion Plymouth if attacked must unavoidably fall." Lord Amherst the Commander-in-Chief, very wisely promptly relieved him of his command, sending Lieut.-General Haisland to replace him. However, Plymouth was not attacked nor was England invaded. An easterly gale came which, blowing hard for several days, drove the combined fleet out of the Channel. D'Orvilliers, the French Admiral in command, then held a council-of-war which, on account of the shortage of supplies and the mounting sickness among the crews, decided to make for Brest at which port the combined fleet anchored on September 14th. Once again Britain had survived the threat of invasion, and the coast-artillery were able to relax for a while.

However, the temporary superiority of the enemy in the Channel and Western Approaches had exposed parts of the United Kingdom to hostile enterprises. Jersey was twice attacked. In May 1779 a

French squadron of three ships of 44, 28 and 20 guns respectively, appeared off St. Helier and opened fire on Elizabeth Castle (23—24 pdrs; 8—6 pdrs.) the main defence-work of the port. The fort at once replied with its 24 pdrs. which were manned by the *Jersey Militia*. Unfortunately, one of the guns immediately burst, killing and wounding several men, but, in spite of this initial set-back, the coast-artillery kept up such a hot fire that the French ships were forced to draw off without effecting anything. As the result of this raid Charleton's Invalid Company, Royal Artillery, was despatched from Woolwich to Jersey in May 1780 to reinforce the *Jersey Militia* coast-artillery and two regiments of Foot—the 78th Highlanders and the 95th—were sent to strengthen the garrison.

In January 1781 the French made a much more determined attempt to seize the island. A force of about 900 men of the Regiment of Luxembourg Volunteers under Baron de Rullecourt was collected at Cherbourg and, embarking in 26 small boats escorted by three warships, set out for Jersey of which Rullecourt in the previous year had made a thorough reconnaisance while disguised as a grain merchant. Late in the evening of January 5th the 26 small boats, loaded with troops, crept into the cove of La Roque which Rullecourt had discovered was left unguarded except for a small picquet. Overpowering this picquet which was found to be asleep, the French force pushed on rapidly to St. Helier, entering the market-place without resistance about midnight. The Lieutenant-Governor, Moyse Corbet, like his island was taken completely by surprise and, being hauled out of his bed, was forced to sign a capitulation for the whole island before he fully realised what was happening. However, except for the St. Helier Town Battalion of *Jersey Militia*, the main-body of the garrison was cantoned outside the capital, and by dawn Major Francis Pearson of the 95th Regiment, the senior military officer on the island, had concentrated a force of about 1,600 men on the rising ground less than three miles from St. Helier.

Rullecourt now began to realise that his task was by no means finished, and that the capitulation he had forced so roughly from the Lieutenant-Governor was not going to be accepted by the military forces. He therefore sent off Corbet with an escort under a flag of truce to Pearson to point out to him that the capitulation covered the whole island and all the troops on it, while he himself summoned Elizabeth Castle to surrender. However, the coast-artillery would have none of him, rudely firing a 24 pdr. at him and his party, and Pearson was equally firm, sending back Corbet with a message that he intended to attack as soon as the Lieutenant-Governor was safely returned to St. Helier. About noon Pearson led his troops to the assault. They quickly entered the town, penetrating to the market place where a fierce hand-to-hand struggle ensued until at last the French, completely overpowered, were forced to lay down their arms and surrender. Unfortunately, both the gallant Pearson and Rullecourt were killed during the short fight, but Jersey had been saved for Britain, and the Union-Jack still flew over Elizabeth Castle.

Jersey was not the only part of the United Kingdom to be attacked at this period. Both in 1778 and 1779 Captain Paul Jones the famous American Naval commander, was at large around the coasts of Britain. In April 1778 he appeared in the Irish Sea in his sloop *Ranger* (18 guns) and during the night 21/22 silently landed a small party in Whitehaven Harbour. Quickly overpowering the guard, Paul Jones and his men spiked all the guns of the municipal coast-battery which protected the haven, and then, as soon as it was light, burnt the many vessels that were lying at anchor in the harbour. Paul Jones then slipped across the Irish Sea and on the morning of the 24th arrived off Carrickfergus where, keeping carefully out of range of the coast-battery, he attacked and captured H.M.S. Drake (20 guns). In September of the following year (1779) he sailed into the North Sea with a Franco-American squadron of five ships with embarked about 120 French marines. His plan was to ravage the eastern coast of Great Britain from the Forth to the Thames. On September 13th, he was off the Firth of Forth, his intention being to sail up to Leith, which had no coast-defences, and destroy both town and shipping, but contrary winds effectually preventing him from doing so. He then cruised southwards down the coast, capturing and destroying such merchant-ships as he met on the way, and causing the wildest alarm along all the shores of east Britain. On September 13th off Flamborough Head he fell in with the Baltic convoy escorted by H.M. Ships Serapis (44 guns) and Countess of Scarborough (20 guns). A tremendous fight ensued by the end of which both British ships had been captured, but Jones' own ship, the *Bonhomme Richard*, sunk shortly after the close of the action. Owing to the most vigorous resistance of the British ships, the convoy escaped into Hull without any loss, and Jones' squadron was so heavily damaged that he was forced to take refuge in the *Texel*. This was the end of his raiding, and he did not again appear in the narrow seas.

Britain's naval supremacy was also challenged in the Caribbean Sea, and between February 1778 and April 1782 the French fleet under d'Estaing and de Grasse more often than not had the whip hand in West Indian waters, their most notable victory being the defeat of the British fleet under Admiral Byron off Grenada on July 6th 1779. It was not until Rodney overwhelmed de Grasse on 12th April 1782 at the battle of *"The Saints"* that the command of the Caribbean

—and of the ocean generally—was finally settled. This temporary superiority at sea enabled the French to despatch expeditions against some of the British held islands which brought into action the detachments of British coast-artillery stationed on them. In 1778 the Royal Artillery in the West Indies consisted of a number of small parties, detached from the "marching companies" serving in north America, scattered among the various islands and manning the coast-defences at Barbados, St. Kitts, Grenada, Dominica, Antigua etc. As usual the strengths of these parties were quite insufficient to serve the guns at the several defended ports so the actual gun detachments and ammunition numbers were provided from the island inhabitants: in some of the smaller islands the *militia* manned the guns alone without a stiffening of regular gunners. It was indeed advisable to keep the number of men of the Royal Artillery serving in the West Indies down to a minimum, yellow-fever being rampant and endemic, and the mortality rate among British troops extremely high.

The French made the first move in September 1778 by embarking a force of some 2,000 men under de Bouillé at Martinique and descending upon British held Dominica without any warning. The garrison, which consisted of a R.A. detachment—one officer and 27 men—and a weak half-battalion of 48th Foot, was taken completely by surprise, and the commander, being only too pleased to accept the very favourable terms offered him by de Bouillé, surrendered, so that on September 8th Dominica fell into French hands without any fighting. The British, however, were not long in launching their counter-stroke. A force of 10 battalions of infantry and two companies of Royal Artillery (those of Captains Standish and Williamson) was concentrated under General Grant at Barbados in December and, escorted by a squadron of 10 ships commanded by Admiral Barrington, set sail on December 12th in 59 transports for St. Lucia. The troops were landed that same evening, and within 24 hours Port Castries, the main French naval-base and capital of the island, had been successfully captured after very little resistance.

The British, however, had scarcely taken over the fortifications and raised their flag when warning was recieved of the approach from the north of d'Estaing with a fleet of 12 ships-of-the-line having on board 7,000 troops. Barrington drew his ships in line across the entrance to the bay under cover of the guns of the fortress, which had already been manned by the Royal Artillery, while the infantry took up position on the heights covering the landside. On December 15th d'Estaing made two attacks to break Barrington's line but the combined fire of the ships and coast-guns was too much for him, and he had to abandon the attempt. So he sailed up the

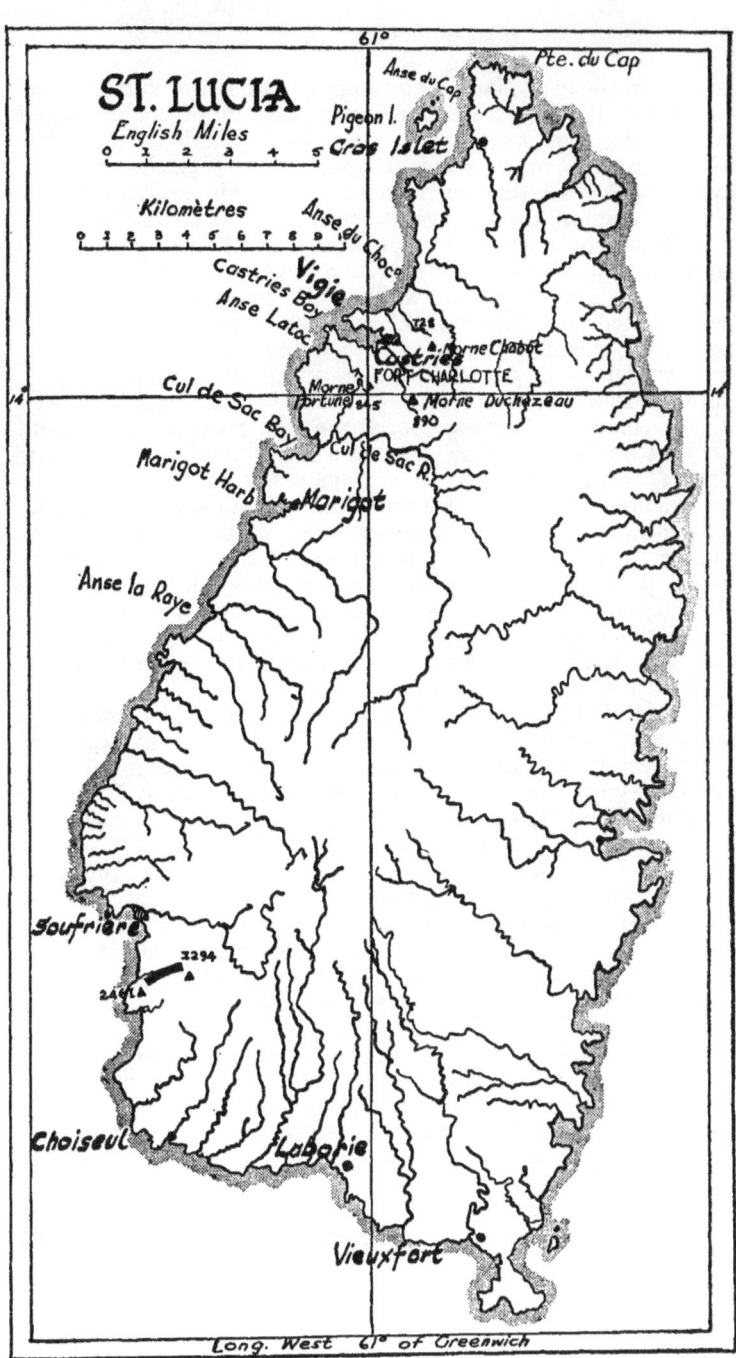

coast of the island and landed his troops to the north of Port Castries. On the 18th the French launched an assault with the main body of their infantry, supported both by the guns of their fleet and the few field guns they had managed to land, against Vigie, the position protecting Port Castries on the north side. This was defended by 5th Foot, reinforced by the grenadier and light companies of seven other battalions, covered by four 3 pdrs. and two 12 pdrs. which had been captured from the French in position on Vigie and were now manned by the Royal Artillery, all under the command of General Medows. The French made a very determined effort to capture Vigie and drive through to the port and harbour in rear. Soon after dawn they developed their attack on the main position, pushing forward their light troops while the main-body advanced in solid columns. Medows opened fire with his guns which swept away the leading enemy files but nevertheless the French still came forward. A desperate fight ensued, and it was not until the British had exhausted almost all their ammunition that the French gave up and "the whole body of the enemy faced about and retired, with no indecent haste, but with evident disinclination to advance any more". (Fortescue). The French had lost very heavily, suffering about 1,500 casualties. This was enough for d'Estaing who, re-embarking the remnants of his troops, sailed away for Martinique, leaving St. Lucia, which was considered to be the key to the Windward Islands, in the hands of the British.

The R.A. companies of Standish and Williamson remained to man the defences of St. Lucia, but in July 1779 their headquarters were moved to St. Kitts, and they became generally responsible for the artillery defence of the islands with detachments at St. Lucia, Barbados, Grenada, Antigua etc. Meanwhile, d'Estaing, now having command of the Caribbean, attacked St. Vincent and Grenada, capturing both these islands with very little trouble, the guns defending the former being manned by *militia* only, but at the latter a detachment of one officer and 24 men of the Royal Artillery fell into the hands of the enemy. These set-backs and the entry of Spain into the war determined the Board of Ordnance to send Royal Artillery reinforcements to the West Indies, and during 1780 drafts of about 100 in all were sent out from home to the two companies stationed there, including a detachment of 60 for the defences of Port Royal, Jamaica, this detachment apparently being the first British coast-artillery to carry out duty on that island. In February 1781 this detachment became a special "Jamaica Company" of the Royal Artillery under the command of Captain J. D. Gall.

In 1781 the French fleet of 25 sail in the Caribbean was under the command of Count de Grasse, and, in spite of all the efforts of the British squadrons, now totalling 18 ships under Admiral Rodney, the French retained their superiority. In May de Grasse sent a small expedition of 1,200 troops, escorted by a squadron of his ships, to capture Tobago, the garrison of which consisted of a half-battalion of 86th Foot, and about 400 colonial *militia*, those of the latter manning the guns being trained and encouraged by a bombardier and one gunner of the Royal Artillery. After a stout defence, the governor was forced to surrender the island, de Grasse having effectively prevented an attempt at relief by a squadron of six ships under Admiral Drake sent from Barbados. Later in the year the French swept down upon the small islands of St. Eustatius and St. Martins, and on November 27th captured them without difficulty, the governor, Lieut.-Colonel Cockburn, surrendering the island with their garrison of 800 men, which included seven men of the Royal Artillery employed in the coast-defences, without a fight. He was subsequently brought before a court-martial and sentenced to be cashiered.

In January of the next year (1782) de Grasse determined to attack St. Kitts, and on the 11th landed 6,000 troops under de Bouillé at Basseterre, the capital of the island. The garrison of St. Kitts was about 800 men, consisting of one officer and 31 men of the Royal Artillery and 1st and 15th Foot, under command of General Fraser who, realising the impossibility of holding Basseterre, abandoned the defences and withdrew his troops to Brimstone Hill, a fortified height at the north-west of the island. Here he was beseiged by the French but was able to repulse all their efforts.

Meanwhile, Admiral Hood with 22 ships of the line and 69th Foot on board, sailed from Barbados and, having eluded the French fleet, arrived at Basseterre roadstead where he most successfully resisted all attempts by de Grasse to drive him out. Unfortunately, his small force of troops was insufficient to relieve Brimstone Hill, although he landed 69th Foot and about 500 Royal Marines from his ships under the command of General Prescott in the hope of doing so. The French meanwhile had opened a heavy bombardment against the defences of Brimstone Hill and on February 12th "the little garrison having lost over 150 killed and wounded, besides many men from sickness, and being quite worn out by excessive fatigue" (Fortescue) Fraser was forced to capitulate. Having re-embarked his troops, Hood slipped out of the roadstead by night and safely withdrew his fleet right under the nose of de Grasse to Antigua where he was joined on February 25th by Rodney with 12 ships.

De Grasse now planned to capture Britain's largest island in the Caribbean, Jamaica, and set sail from Martinique on April 8th with 35 ships-of-the-line and a great convoy of transports and store ships. This time, however, he had met his match. As soon as Rodney learnt that the French fleet had left harbour, he gave chase with his whole force (36 ships-of-the-line) and, coming up with the French on April 12th off the north of Dominica near the group of small islands called *The Saints*, immediately set about bringing them to action. The French admiral, realising he could not avoid battle, ordered the convoy to make for shelter in Guadeloupe and made ready to engage. The battle lasted from 9.45 a.m. until about 2 p.m. ending in complete victory for Rodney and the British fleet, de Grasse himself being captured, so that the question of command of the Caribbean was settled finally once and for all, and with this resounding success operations in the West Indies came to a finish. The conclusion of the war in 1783 found three companies of Royal Artillery in garrison in the West Indies, Parry's with headquarters at Barbados, finding the detachments for the Windward Isles, Downman's with headquarters at St. Lucia providing the detachments for the Leeward Islands, and Gall's at Jamaica.

Coast Artillery took little part in the tremendous struggle for the independence of the American colonies on the continent of North America. Detachments—and sometimes whole companies—had to be found to man the coast-defences of the British main bases, such as New York, Charleston and Savannah, and of several other lesser defended ports along the eastern coast. In July 1778 Rhode Island was attacked by the French fleet under d'Estaing and an American army under General Sullivan. The island, which mounted, besides field-guns, 72 pieces of various calibres, was held by a British force of some 5,000 troops, which included John Innes' Company of Royal Artillery, under the command of General Pigott. D'Estaing arrived with his fleet before Rhode Island on July 29th, and, having destroyed seven small British warships, set about bombarding the defences. In this the French were soon joined by the guns of General Sullivan who had by now closely invested the island on the landside. In spite of this heavy fire Innes' Company manfully stuck to their guns, losing 35 men killed and wounded, and were specially commended by Pigott for "the severe labour and exposure cheerfully undergone." It was not until 8th August that the allied commanders were ready to launch a joint assault, but this never took place for the very next day Admiral Howe appeared with a British squadron, and d'Estaing was forced to put to sea. Sullivan remained blockading Pigott's entrenched camp until the end of August when he abandoned the whole operation.

The port of Savannah in Georgia was the main base in 1778-79 for the British campaign in the southern colonies, and in the summer of 1779 the Americans decided to capture it and so bring to an end the British operations in the south. Savannah was held by General Prevost with a garrison of some 2,500 men, which included a detachment of Royal Artillery of three officers and 53 men, supported by a small naval squadron of two frigates and four armed vessels. At the beginning of September d'Estaing appeared off the mouth of the Savannah River with 22 ships-of-the-line and, having sunk the two British Frigates, sent his smaller ships up the river towards the town. He then disembarked his French troops and some sixty heavy guns from his ships, and, joining with an American force under General Lincoln, set about investing Savannah. Two sides of the town were covered by the river and marshes, and the other two had been considerably strengthened by the British with redoubts and entrenchments. By 4th October the besiegers had their batteries ready and at dawn on that day opened a heavy bombardment which lasted four days. On the 9th d'Estaing and Lincoln launched their main assault. Their plan was to attack the redoubt, Springhill Redoubt, at the south-west angle of the town's defences with two columns of French and American regulars, in all about 4,000 men, while the remainder of the Americans made demonstrations against the other parts. But the bombardment had not been very effective, and, as soon as the enemy columns could be seen advancing in the light of early dawn, the British guns opened a tremendous fire at them and tore terrible gaps in the foremost ranks. Nevertheless, the French and Americans managed to scale the ramparts and enter the redoubt where a desperate hand-to-hand struggle with the small garrison of gunners and 60th Foot took place. A counter-attack by Royal Marines and the grenadier-company of the 60th under Colonel Maitland arrived in the nick of time and, charging the attackers, tumbled them back in confusion to their own lines. This repulse much discouraged d'Estaing and Lincoln, and, after a good deal of mutual recrimination, they decided to raise the seige, d'Estaing re-embarking his troops and sailing to the West Indies, while Lincoln and his force returned to South Carolina.

We have seen how in the period between the two wars—1763 to 1774—a company of Royal Artillery was normally stationed in West Florida, headquarters Pensacola, to provide the coast-artillery detachments for the small defended ports along the southern coast. With the entry of Spain into the war in June 1779 the garrisons of these ports found themselves completely isolated in their fortifications. In December 1779 the Spaniards from New Orleans began to mop them up in a leisurely fashion. They first dealt with the fortified posts that covered the main stream of the Mississippi, collecting eight men of the Royal Artillery, besides infantry, from them, and in March 1780 appeared before Mobile with a small sea-borne expedition. Fort Charlotte was held by a garrison of 300 men, including an officer and 10 men of the Royal Artillery, but unfortunately they were very short of supplies and stores. The Spaniards invested the place both by sea and land, and by the 14th of the month the British were forced to haul down their flag and surrender.

In the following year (1781) the Spaniards decided to deal with the centre of British power in West Florida, the defended port of Pensacola. On 9th March a Spanish squadron appeared before the harbour and during the next two weeks disembarked a force of some 8,000 troops. Pensacola was held by about 900 men under Brigadier Campbell and with him were the headquarters and main-body of Captain W. Johnston's Company of Royal Artillery. The Spaniards set about besieging the place in a regular manner, digging trenches and throwing up batteries, and on 22nd May opened a bombardment on the fortifications. This had little effect until 8th May when a mortar-shell penetrated into the principal magazine, the resulting explosion destroying the main redoubt. The Spaniards immediately leapt to the assault, and the defenders, having suffered terrible losses and being half stunned by the explosion, made what resistance they could among the ruins of their fortifications. This however did not last long, and Campbell was soon forced to capitulate. This left the small defended port of St. Augustine, the garrison of which included one officer and nine men of the Royal Artillery, the sole remaining British held post in all Florida. On the other side of the Atlantic the British had been somewhat more successful, for in 1778 they had mopped up the French posts in Senegal, and in May 1719 a detachment of two officers and 26 men of the Royal Artillery arrived at Goree to man the fort.

This startling series of losses of coast fortresses and forts, both in the West Indies and Western Florida, during the period 1778-82 shows that, though a defended naval base—or protected harbour—may provide a safe refuge for a fleet in which to carry out repairs and take in munitions and supplies, should that fleet lose superiority in the surrounding waters for any appreciable time, the safety of that base is seriously endangered. An enemy fleet in command of the sea may transport at leisure an expeditionary force, which once landed, can set about reducing the base without fear of interruption, and the defending power, being inferior at sea, can do nothing to save it. Great Britain for most of the period between 1540 and 1956 had superiority at sea over her rivals and rarely experienced her coast fortresses and forts falling into the hands of the enemy, but in 1941-43, when she lost the command of the Indian Ocean and China Sea to Japan, she was to see her great coast fortresses in that part of the world go the same way as they had in the West Indies and Florida during the War of American Independence—and for the same reason.

CHAPTER VIII

1774 to 1783. (concluded)

IT was not only in the West Indies that Britain felt the adverse results of losing her superiority at sea. Having to concentrate her naval strength in home waters to retain command of the narrow seas, and in the west Atlantic to cover the vital lines of communication to North America, she was forced to withdraw her fleet from the Mediterranean and accept inferiority in the middle sea. This exposed her two naval bases in that sea, Minorca and Gibraltar, to attack and capture just as Hong Kong and Singapore were so exposed under similar circumstances more than 150 years later. Rodney, when on his way to the West Indies in January 1780, put into Gibraltar with a fleet of 22 ships-of-the-line, and, having sent on a convoy under escort to Minorca, landed reinforcements and a vast quantity of stores and supplies for the Rock garrison. On 13th February, Rodney sailed away out of Gibraltar Bay for the West Indies, and, as his ships disappeared into the west, the garrison knew they were being left to defend themselves alone and unsupported, without expectation of assistance or relief. However, they were not dismayed for they were well led, strong, and plentifully supplied, and quite confident they could resist successfully any attack launched against them.

Minorca, the other British naval base in the Mediterranean, was not so well placed. This island was to the east of the Straits of Gibraltar, and, with the French and Spanish in control of those Straits, could only be reinforced and supplied with much greater difficulty than Gibraltar. Rodney's convoy of store ships reached Port Mahon safely in January 1780 but that was the last addition to their supplies the garrison was to see. That garrison was not a strong one, consisting of three companies of Royal Artillery (about 280 all ranks), four battalions of infantry, two British—51st and 61st Foot—and two Hanoverian, all very weak in strength, and a landing-party from the Royal Navy of about 200 officers and seamen, in all some 2,300 men under the command of General James Murray, the Governor. The three companies of Royal Artillery were those of Captains J. Walton, G. Fead and B. Lambert of whom Walton was the senior, being therefore appointed C.R.A. of the fortress with the rank of major. The fort of St. Philip, still the main defence-work of Port Mahon, at this time, mounted some 230 guns and mortars of all calibres, and of course the party of seamen and a contingent from the infantry were required as "additional gunners" to complete the gun detachments.

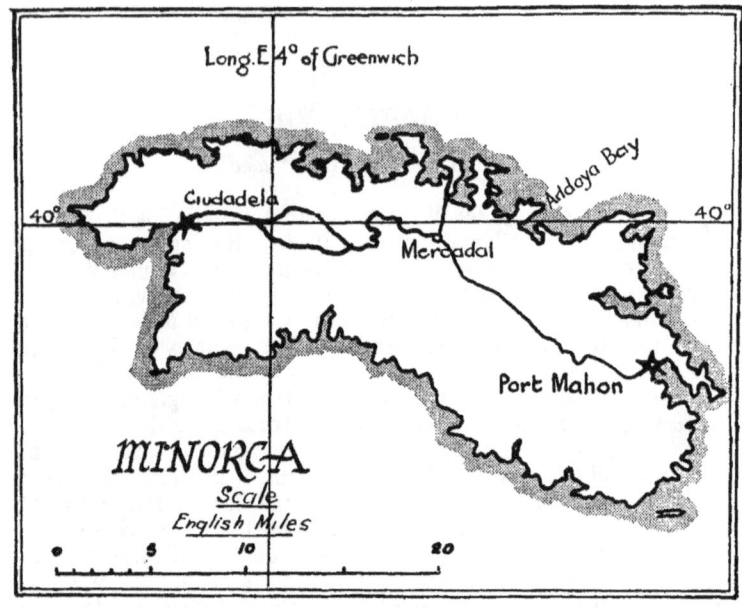

The enemy, a Franco-Spanish force of about 12,000 men under the Duke de Crillon escorted by a great fleet, landed on the island during the last two weeks of August 1781 and rapidly occupied Port Mahon without meeting with any resistance, the garrison having wisely withdrawn into Fort St. Philip. The French and Spanish then set about laying regular siege to the fort, digging parallel trenches and constructing a great tier of batteries in which they mounted some 200 guns and 20 mortars. These batteries opened fire on 15th September but could make very little impression on the fort and garrison, and towards the end of the month Murray was able to slow up considerably the efforts of the enemy by launching a fierce sortie which destroyed several of the Spanish batteries and returned with a hundred prisoners. The British garrison however was soon being attacked by two enemies much more dangerous and fatal than the French and Spanish besiegers. By December starvation and disease had fixed their deadly grip upon the defenders. Only salt rations had been available since the start of the siege with the result that scurvy broke out among the troops already weakened by the shortage of food. The sickness spread rapidly with great virulence, and very soon an average of 50 men a day were going down with the malady, Captain Lambert of the Royal Artillery being among the victims. It became almost impossible to find sufficient men to man the guns and defences, and even those who were ill had to take their turn of duty.

1774 to 1783 (concluded)

Meanwhile the enemy had pushed forward their trenches and batteries close to the advanced works of the fort, and on January 6th (1782) opened a tremendous bombardment. The garrison, now too weak to launch a sortie, bore this bravely, and the British guns returned as good as they received, Murray declaring he had never seen guns and mortars better served. Throughout the following weeks the enemy kept up this heavy cannonade, firing sometimes as many as 750 shot and shell into the fort in an hour. The garrison suffered severely, Captain Fead of the Royal Artillery being amongst those wounded, the scurvy continued to take terrible toll of the troops, and such men as could be found to stand at their posts were almost too weak to load and fire the guns or raise their muskets, being far too few to hold adequately the defences. By the beginning of February the fort was in ruins, a great number of the guns destroyed or dismounted, and the garrison reduced by the enemy's fire, disease, and starvation to a mere remnant. Throughout the bombardment Murray had continuously visited his men and done his utmost to keep them in good spirits, but he saw now that no more could be done. So on 4th February, he hoisted the white flag and offered to capitulate. De Crillon was generous—he and his generals had been greatly impressed by the bravery of the defenders—and on the next day a pitiful and wasted handful marched out with drums beating, colours flying, and muskets shouldered between the ranks of the waiting French and Spanish soldiers. Minorca was then handed back to the Spaniards, and the survivors of the siege were returned to Britain in a convoy of French ships. In his final general order, Murray stated that he had not words to express his admiration of his men's behaviour; and that while he lived he should be proud of calling himself father of such distinguished officers and soldiers as he had had the honour to command. It is to be noted that during the siege Murray commissioned as officers three sergeant-majors of the Royal Artillery for their gallantry.

We left the British at Gibraltar watching the last of Rodney's ships as they disappeared over the western horizon on 13th February 1780, just two years before the loss of Minorca. The garrison of Gibraltar, about 6,000 strong, consisted of 5 companies of Royal Artillery, and nine battalions of infantry—39th, 56th, 58th, 72nd, 2/73rd and 97th Foot and three Hanoverian battalions—commanded by General George Eliott, the governor. The five artillery companies were close on 500 strong and were commanded by Captains V. Lloyd, G. Groves, G. Lewis, J. Eyre and P. Martin, the C.R.A. being Lieut.-Colonel A. Tovey. The armament of the Rock at the time of the seige was:-

F

        77—32 pdrs.
        122—24 pdrs.
        104—18 pdrs.
        70—12 pdrs.
        16—9 pdrs.
        25—6 pdrs.
        38—lesser pieces.

        452 guns plus 48 mortars and howitzers:

so as usual the R.A. had to be assisted by "additional gunners" supplied by the infantry. A small naval force—H.M.S. Panther (60 guns) three frigates, and a sloop—under Admiral Duff lay off the New Mole. Since June 1779 Gibraltar had been blockaded by a large Spanish army, some 14,000 strong, under Lieut.-General Alvarez de Sotomayor, encamped on the mainland opposite the isthmus, and a powerful Franco-Spanish naval squadron, commanded by Admiral de Cadova, based on Cadiz, but it was not until sometime after the departure of Rodney's fleet that the blockade became an active, offensive siege, and the Rock was completely cut off by land and sea from all supplies from the outside world.

During the autumn of 1780 the Spaniards constructed two forts, St. Barbara and St. Philip, at the northern end of the isthmus, and using these as a base, set about digging a series of batteries and trenches across the isthmus, advancing steadily towards the north front of Gibraltar as they did so. By the middle of April 1781 these works were sufficiently far forward for the Spaniards to open fire with 114 guns from their most advanced batteries on the town and landward fortifications. Meanwhile the British had been busy strengthening and improving their defences. The batteries covering the land front were Willis', Green's and Queen Anne's, all three mounted on a ridge about 900 feet above sea level, Grand Battery at Land Port, Water Port Battery, and the redoubt on the end of the Old Mole, and an attempt was made to reinforce these by hoisting a 24 pdr. to the summit of the northern apex of the Rock. This was carried out successfully, and the gun—known as the Rock gun—remained there in action throughout the seige. However, during the early months of 1781, it was not the Spaniards who were causing most trouble to the British garrison but those two dread spectres, starvation and disease, and by April the supplies landed by Rodney the year before were running short, food, especially fresh food, was becoming very scarce, and scurvy had already broken out among the troops.

However, the British government was not unaware of the plight of Gibraltar, and in February (1781) had already ordered

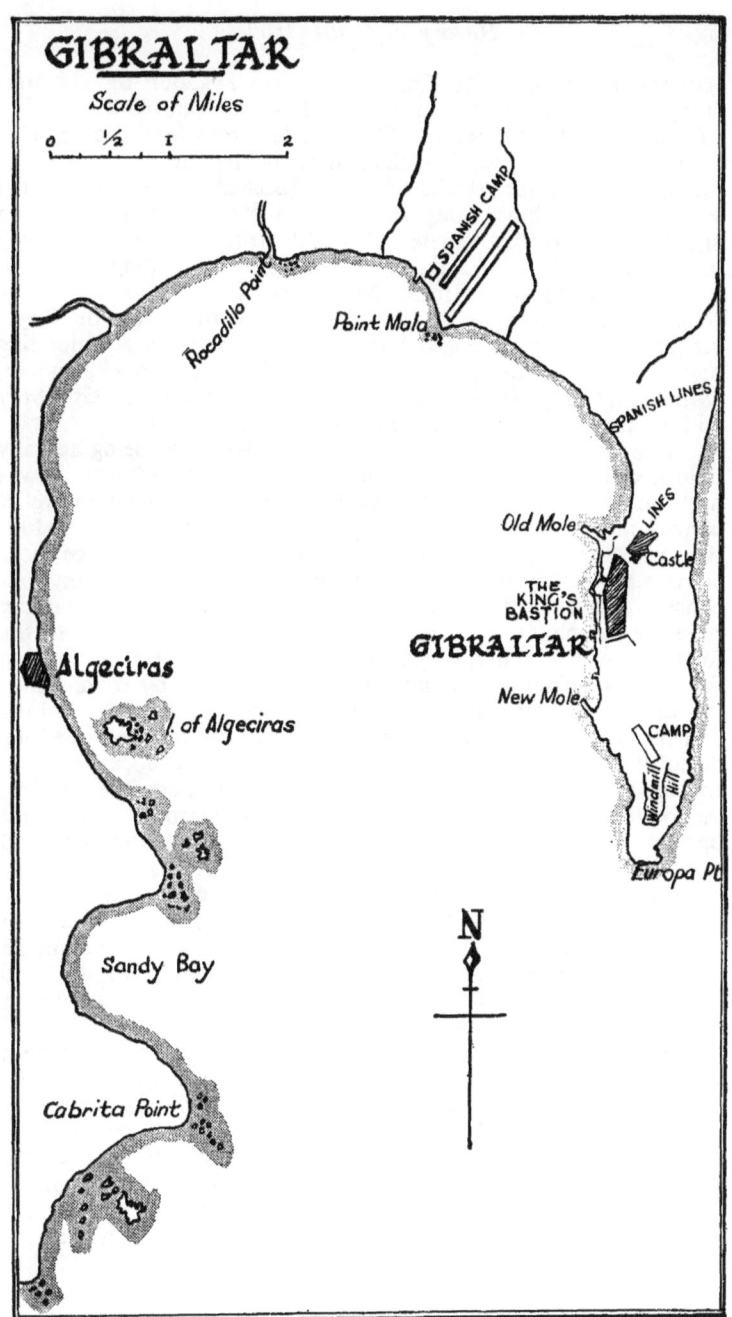

Admiral Darby with the Channel Fleet to abandon for the time being his normal watch and ward and escort from Cork a convoy of store-ships for the Rock. Darby sailed from Spithead on 13th March with 28 ships-of-the-line, and, picking up the convoy at Cork, made for the Straits which he reached on 11th April, the Franco-Spanish blockading squadron taking refuge in Cadiz on sighting his advanced frigates. On 12th April at noon the convoy anchored in Gibraltar Bay, and the work of unloading the food, ammunition, and stores from the hundred or so ships began, the reinforcements brought in them including a draft for the Royal Artillery of seven officers and 70 men. The business of unloading and stocking up the garrison was completed in a week, and on 19th April, Darby sailed for Spithead which he reached on 22nd May without molestation from the enemy.

It was while the store-ships of the convoy were being actually unloaded that the Spaniards opened their first bombardment. It was as if on seeing the ease with which Darby had brought the relief convoy into Gibraltar Bay, they realised they could never reduce the fortress by blockade and starvation alone and must take more active steps if they wished to bring about its fall. The enemy first concentrated their fire on the town of Gibraltar itself, rapidly reducing the buildings to ruins and driving the inhabitants to the fortifications for shelter, they then switched to the batteries covering the landward front, increasing the volume of fire by rowing into the Bay from Algeciras small gunboats, each armed with a long 36 pdr., which engaged the British warships lying off the Old Mole, demolished the workshops and storehouses in the dockyard, and enfiladed the landward defences. To deal with these, the guns on the batteries covering the harbour, King's, Orange, and South Bastions, were brought into use, and indeed all the batteries replied to the enemy bombardment with vigour and effect, silencing some of his most forward guns and setting fire to his camp, the Gunners being specially complimented for the steadiness and accuracy of their fire. During this period the Royal Artillery suffered but 23 casualties, one of whom unfortunately was the C.R.A. Lieut.-Colonel Tovey, who died of fever. He was succeeded by Captain Lewis who was promoted major to fill the vacancy.

It is reckoned that during the first six weeks of this bombardment the Spaniards expended 56,000 shot and 20,000 shells, but by the end of May their fire had slackened considerably. Although they had almost entirely destroyed the town of Gibraltar, the effect of their fire on the fortifications had been insignificant, and all the defending batteries were still in action. There is no doubt that the enemy were greatly disappointed by the small results their bombardment had produced, and with the approach of warm weather

1774 to 1783 (concluded)

Spanish lethargy took charge, the cannonade was reduced to occasional bursts of fire, and the siege works on the isthmus were advanced extremely slowly. By the end of the summer the most forward trenches were still more than half-a-mile from the British ramparts.

Meanwhile, Governor Eliott had not been idle. Having repaired and strengthened his defences, he determined upon a plan to arrest the enemy's slow but nevertheless inexorable progress along the isthmus and decided that a sortie was necessary. Accordingly at midnight 26th/27th November three columns were paraded under Brigadier-General Ross, consisting of 12th Foot, Hardenberg's Hanoverian battalion, four officers and 110 men of the Royal Artillery, three officers of the Royal Engineers with 40 workmen from the Artificer Company assisted by a working-party of 160 men from the infantry, and a party of naval artificers from the warships, in all about 2,200 men. Shortly before three o'clock, the three columns crossed the defending ramparts and, moving silently down the isthmus towards the Spanish lines, had to advance close on three-quarters of a mile before they ran into the Spanish outposts. The enemy were taken completely by surprise, and the British poured into and over the siege works in an irresistable torrent. The guns were spiked, gabians and platforms set on fire, trenches levelled, breast-works pulled down, and powder blown up. The Spaniards made only a poor resistance—there being about a thousand of them on duty in the trenches at the time—and then decamped at great speed to the rear. By five o'clock the business was done, the enemy's siege-works and batteries were in flames or destroyed, and the three columns were safely on their way back to the fortress of Gibraltar, having suffered only 30 casualties.

A devastating blow had been struck at the besiegers, and for sometime the Spaniards appeared to be stunned, but soon after Christmas they revived somewhat and started once again to dig their parallels and construct new batteries. On their side the British did not rest on their laurels and continued to improve their fortifications. It had been found during the enemy bombardment of the previous spring that the guns in the higher batteries often could not be depressed sufficiently to engage the enemy works on the isthmus, so an intelligent subaltern of Eyre's Company, Lieut G. F. Koehler, invented a carriage from which the guns could be fired at considerable angles of depression. Lieut. A. Witham of the same Company also produced an incendiary shell weighing some 14 lbs., which could be discharged with excellent results from a 32 pdr. During March 1782 a small convoy, evading the blockading squadron, slipped into Gibraltar harbour bringing with it welcome reinforcing drafts and 12 gunboats in sections—to be put together at Gibraltar—for strengthening the seaward defences. At this period nearly 700 men of the Garrison were absent from duty sick.

The monarchs of France and Spain had not been pleased with the Spanish failure of the previous year, and they decided that a truly stupendous effort should be made during 1782 to capture Gibraltar. Minorca having fallen in February, the Duke de Crillon, with the laurels of his victory at Port Mahon fresh upon his brow, was summoned to the Spanish mainland to take command of the operations against the *Rock*. He brought with him his victorious veterans from Minorca, 4,000 French troops and 8,000 Spanish, which raised the total of the besieging force to some 25,000 men. (The garrison opposing them could not have been much more than 6,000 strong.) De Crillon's plan for the assault on Gibraltar was as follows:-

(1)—The parallels on the isthmus were to be strenuously pushed forward covered by a vigorous bombardment from the advanced Spanish batteries, but this was to be only a feint to draw the attention of the British to their land front and away from:-

(2)—The main attack which was to be an amphibious operation launched against the waterfront between the Old Mole and Europa Point, the southernmost point of Gibraltar. A corps of about 3,300 men under the command of General Cagigal, a Spaniard, was to be embarked at Algeciras in small boats covered by large planks on hinges which on unfolding would form ramps for disembarkation. This landing was to be prepared and covered by the fire of the combined fleet under Admiral de Cordova and the floating batteries designed by the French engineer d'Arcon.

(3)—As soon as the landing force, or even part of it, had been successfully disembarked, the main army under Marshal Burghesi was to advance from the trenches on the isthmus and storm the landward fortifications.

A description must now be given of the famous floating batteries designed by Chevalier d'Arcon who claimed that they were both incombustible and insubmergible. They were constructed from warships of between 500 and 1,500 tons burthen, cut down to their main gun-decks. "Each battery was clad on its fighting side with three successive layers of squared timber, three feet in thickness; within this wall ran a body of wet sand, and within that again was a line of cork soaked in water, and calculated to prevent the effects of splinters, the whole being bound together by strong wooden bolts. To protect the crews from shells or dropping shot, a hanging roof was contrived, composed of strong rope-work netting, covered with wet hides, and shelving sufficiently to prevent the shot from lodging. Not the least remarkable part of these vessels was a plan for the prevention of combustion from red hot shot. A reservoir

1774 to 1783 (concluded)

was placed beneath the roof from which numerous pipes, like the veins of the human body, circulated through the sides of the ship, giving a constant supply of water to every part, and keeping the wood continually saturated" (Sayer—*The History of Gibraltar*) Each of these strange craft was armed with from 8 to 20 guns, manned by a crew of between 250 to 750 men, and propelled in and out of action by one large sail.

Governor Eliott was not discouraged by the tremendous preparations being made for the assault on his fortress. He had heard rumours of the floating batteries and in consultation with his Gunners, had decided to use red-hot shot, heated in furnaces placed within the principal batteries, to counter them. The safe loading of such a dangerous projectile obviously presented great difficulties. These were overcome by introducing a substantial and soaking-wet wad of cotton rags into the bore between the charge and the shot, then laying and firing the gun as quickly as possible. This method seems to have been successful, nevertheless it must have been a lively business—to say the least of it—being a loading or ramming number. With this ace up his sleeve and having done everything to augment and repair his fortifications, Governor Eliott—and his garrison—awaited the coming storm in good heart.

The Spaniards as usual had been slow in pushing forward their trenches and batteries on the isthmus, so that when it was decided that the landward bombardment should open on 8th September (1782) they were by no means ready, the parapets and breastworks being encumbered with material which had not yet been built into the works. Nevertheless, as soon as it was light on the 8th the Spaniards began the bombardment with some 170 guns and mortars. Major Lewis, Gibraltar's C.R.A., seeing the piles of inflammable material lying around the Spanish batteries, suggested to Governor Eliott that red-hot shot might be tried out on them as an effective counter-battery weapon. To this the governor agreed, and Lewis ordered Grand (Land Port) and Water Port Batteries, then manned by Martin's Company, to return the fire of the Spanish guns with red-hot shot. This they did, and the effect exceeded their most hopeful expectations. Very quickly the fascines and baulks of timber caught fire, and the flames, spreading rapidly to the batteries, put one, *Mahon*, completely out of action, whilst two others, *St. Carlos* and *St. Martin*, were temporarily silenced. Nevertheless, the Spaniards persisted with their cannonade, and on the next day were reinforced by seven Spanish and two French line-of-battleships which, moving across the Bay, engaged the batteries at King's, Orange, and South Bastions, covering the harbour and manned by Lloyd's Company.

This bombardment from land and sea continued throughout the 10th and 11th, the Spaniards on the isthums steadily bringing more and more guns and mortars into action until they numbered over 200. The British gunners returned the enemy fire with even greater energy and accuracy, but the incessant artillery dual slowly began to wear them down. They could get no sleep nor rest for the nights had to be spent in removing and replacing casualties, replenishing ammunition, and repairing damage to both guns and fortifications. Meanwhile de Crillon had decided to launch his main assault on the 14th. The bombardment was to be stepped up throughout the 12th and 13th, the floating-batteries being introduced into the battle on the latter day as the allied trump card. De Crillon reckoned that by night fall on that day the guns of the fortress would be silenced, and he would be able to carry out successfully the amphibious attack next morning at dawn. So on the 12th the combined Franco-Spanish fleet of 40 battleships, commanded by de Cordova sailed into Gibraltar Bay and added their fire to that of the ships already in action and the Spanish guns on the isthmus. On the next day at 7 o'clock in the morning the floating-batteries got under way and, sailing majestically across the Bay, took station and anchored in line fore and aft about a thousand yards from the seaward front of the fortress. There were ten of these floating batteries under the command of Admiral Buenoventura Moreno, and they were:-

Pastora (flag)	21 guns
Paula Primera	21 guns
Talla Piedra	21 guns
El Rosario	19 guns
San Christoval	18 guns
Principe Carlos	11 guns
San Juan ...	9 guns
Paula Segunda	9 guns
Santa Anna	7 guns
Los Dolores	6 guns
Total	142 guns

All these guns were mounted on one side only of each battery.

It is almost impossible to describe adequately that last 24 hours of the Franco-Spanish attempt to capture the famous *Rock of Gibraltar*. The enemy's primary object now was to blast to pieces and pulverise the British defences covering the sea-front between the Old Mole and South Bastion along which they intended to land next morning, and their principal instrument for this destruction was the floating batteries. The warships and guns on the isthmus continued to thunder, but the vital duel was between the British guns at the Old Mole, Orange Bastion, King's Bastion, and South Bastion and d'Arcon's new contrivances.

At first the floating-batteries appeared to be getting the upper hand. Red-hot shot from the 32 pdrs. and shells from the mortars seemed to have no effect on them whatsoever. The British artillery and the batteries suffered seriously, being especially troubled by the enfilade fire on their northern flank from the guns on the isthmus. Many guns were dismounted, fortifications began to crumble and collapse, casualties were heavy including the C.R.A., Major Lewis, Captains Reeves, Groves and Seward, and Lieutenants Boag and Godfrey, and death and destruction were in every quarter. The British however ignored the fire from the isthmus and concentrated all their efforts against the floating batteries. An incessant fire of red-hot shot and mortar shells was directed at them, and slowly but surely, as the long day wore on, the British began to get the upper hand. During the afternoon the fire of the battering-ships gradually slackened and suddenly the *Talla Piedra*—in which was d'Arcon himself—was seen to be ablaze. Shortly afterwards the flagship, the *Pastora*, was in evident distress, her guns silenced and flames devouring her upper works. One by one the floating-batteries ceased fire and drifted out of line, their crews feverishly trying to extinguish the flames which were consuming them. When darkness set in there was no doubt with whom the victory lay. Gibraltar Bay was littered with the burning hulks of floating-batteries which one after another blew up and sank as the fires reached their magazines, their own death pyres providing sufficient light for the British gunners to continue to harass them. When dawn came the artillery of the defences was so wearied and exhausted that seamen and Royal Marines had to be detailed to replace the gun-detachments while they rested.

This was in fact the end of the battle for Gibraltar. Neither de Crillon nor his French and Spanish admirals and generals had any heart left in them to carry on operations. The guns of the fortress had not been silenced, the decisive amphibious attack along the water-front could not be attempted, and their "secret-weapons", their sure recipes for victory, lay at the bottom of Gibraltar Bay shattered and burnt-out wrecks. On their side the British, though victorious, were near the end of their tether. Most of the garrison had been closed up within the fortifications since June 1779, there had been almost a continual shortage of food, and sickness had been rife throughout the seige causing the death of some 530 of the troops. The five companies of Royal Artillery—with their "additional gunners"—had borne the brunt of the prolonged enemy attack and were mainly responsible for the successful defence of the fortress. They suffered severely having three officers and 64 men killed or died of disease, and six officers and 121 men wounded, the officers who were dead being Lieut.-Colonel Tovey and Captains Reeves and Grumley, and those wounded Major

Lewis, Captain Seward, and Lieutenants Boag, Willington, Godfrey and Cuppage. They fired more than 200,000 shot and shell from their guns and mortars of which, it is said, no more than 96 were still left in action after that terrible battering on 13th September. On the conclusion of the siege the Master General of the Ordnance (Duke of Richmond) wrote a letter to Major Lewis in which he said "His Majesty has seen with great satisfaction such effectual proofs of the bravery, zeal and skill by which you and the Royal Regiment of Artillery under your command at Gibraltar have so eminently distinguished yourselves during the siege; and particularly in setting fire to and destroying all the floating-batteries of the combined forces of France and Spain on the 13th September last".

After that 13th September the Franco-Spanish operations ceased to have any vigour or resolution in them. They maintained the blockade by land and sea and from time to time carried out a desultory bombardment from their forward batteries on the isthmus, but it was quite evident they no longer had any hope of capturing Gibraltar. Rodney's victory over de Grasse in the West Indies on 12th April had quite altered the situation at sea, and by the summer of 1782 the Admiralty had been able to concentrate in home waters a fleet of 34 ships-of-the-line under the command of Howe. On 11th September this fleet set sail from Spithead with a convoy of 31 transports and store-ships carrying reinforcements and supplies for Gibraltar. On 11th October Howe's fleet entered the Straits, and the combined Franco-Spanish fleet of 48 sail emerged from Algeciras Bay but, being badly handled, made off eastwards into the Mediterranean and was thus unable to interfere with Howe's operations. Between 16th and 18th October the ships of the convoy safely entered Gibraltar Bay and reinforcements, supplies and stores were soon landed, Gibraltar being now stocked up for another year. On the 19th Howe put out to sea, had a brush with the enemy towards sunset on the next day, and was safely back at Spithead by 14th November.

In January 1783 the war came to an end, and peace was concluded with France, Spain, Holland and the rebellious American colonies. These last obtained their independence as the United States of America, France recovered St. Lucia, and her posts in Senegal, Spain retained Minorca, West Florida, and Tobago, and received East Florida: Britain regained such West Indian islands—except Tobago—as she had lost to the allies. It was not a pleasant peace for Britain who was forced to pay the penalty for losing her naval superiority and command of the seas once the foremost European maritime powers came in against her.

CHAPTER IX
1784 to 1793.

BRITAIN had but ten years to recover her breath between the close of the war against the European maritime powers in alliance with her rebellious colonies in North America and the outbreak of the greatest struggle for existence she had yet had since her invasion and conquest by William the Norman. The war against the French republic and empire broke out in 1793 and continued without interruption—except for short breaks in 1802-03 and 1814-15—until 1815, a period of 22 years, during the whole of which Britain was the mainstay of the opposition against the Revolution and the Emperor Napoleon. Throughout this war Britain had command of the seas. The admirable and efficient navy of Louis XVI of France never recovered from the effects of the Revolution. The senior officers were either murdered, executed, or driven into exile, and those, who in due course were promoted to fill their places, never attained the skill, knowledge, or abilities of their predecessors, while the junior officers and crews were wanting in sea-going and fighting experience and above all lacking in discipline, that essential military accomplishment having been utterly wrecked by the principles and teachings of the Revolution. The French ships were well built and armed, and their fleets and squadrons efficiently organised and equipped but they were badly handled and fought, and their officers and men were never able to compete with their opponents in seamanship, steadiness, and above all discipline when it came to a fight.

Coast artillery therefore did not have the busy time it had had during the previous war when Britain's supremacy on the seas had been so effectively challenged. Its duties were mainly defending overseas bases which had been captured or occupied by British amphibious expeditions. The military superiority of France on the continent of Europe brought about three threats of invasion, 1797-98, 1801, and 1803-05, but all three proved ineffective because of her weakness at sea. Nevertheless, they caused great alarm in Britain, tremendous activity along the threatened coasts, and the raising of large numbers of volunteer corps among whom for the first time we find definite artillery units being raised to man coast-defences. The warships did not differ from those of the War of American Independence. The great wooden walls of 2,500 tons burthen, carrying more than 100 guns, were lords of the ocean, and frigates of between 500 and 1,000 tons, carrying up to 50 guns, their most efficient eyes. The guns had not altered either, 42 pdrs, 32 pdrs., 24pdrs., and 18 pdrs., being the principal ones in the large ships, but carronades, 68 pdrs., 32 pdrs., 24 pdrs. and 18 pdrs. were being increasingly employed. Nor had the guns ashore changed, either in pattern or in the actual pieces themselves, the same weapons, which had guarded the coasts so well throughout the century, continuing to do so, although carronades were being slowly introduced.

The coast-artillery, with its forts, defence-works, guns, stores etc., both at home and abroad, was the responsibility of the Board of Ordnance which consisted of the Master-General, the Lieutenant-General, the Surveyor-General, the Clerk of the Ordnance, the Clerk of Deliveries, and the Principal Store-Keeper. Only the first two of these were soldiers, the other four being civilians. The Board was in no way under the War Office, being directly responsible to the Cabinet of which the Master-General was a member. The appointment of Master-General was therefore a political one, and, although it was normally given to an eminent general, it was also sometimes awarded to a great political lord who had done very little soldiering. The Lieutenant-General on the other hand was always a distinguished soldier but never a Gunner, so that the Board's knowledge of the technical side—one might almost say of any side—of the work of the artillery was practically nil. In fact the Board was a civilian organisation ruled by its civil-servants who were much more interested in economy than efficiency; indeed it was intolerably stingy and parsimonious, spending its time trying to save twopence on paint whilst the gun-carriages rotted. The general result of course was disastrous.

This system of civilian control was carried on right down to the very coast forts and defences themselves. At any coast-artillery station, at home and abroad, there were two civilian subordinates of the Board, a Clerk of the Cheque and a Storekeeper, who, together with the senior Gunner and Sapper officers, formed a local station-board which alone could authorise any expenditure of money or stores, demands for supplies, or repairs to the works and equipment. In fact the two civilian members would very rarely agree to any such expenditure and spent most of their time struggling with the parent Board in London to obtain their own—and their civilian underlings—proper pay and allowances which the Board would only disgorge after a determined resistance. The coast-defences with their equipment therefore slowly but surely rotted away. We read continuously:- "the platfroms are in a very bad order and the carriages want repair", or "the fort is in an exceeding bad state, the weather dripping thro' in many places", or again "the magazine in this castle is a very dangerous and inconvenient one, it being under the building in which the master and quarter-gunner reside, and is no otherwise secured than by a common trap-door" or again "the platforms here are mostly too old to be taken up without great injury". These are all quotations from records of reports made by the Inspector of the Royal Artillery to the Board of Ordnance between 1783 and 1793, and are but a few typical quotations from many, many similar ones.

The coast-defences and equipment were maintained in Great Britain and the Channel Islands by detachments from the Invalid Companies R.A. although complete Marching Companies were stationed at the larger places, usually in order to make full use of the available barracks. For instance in 1789, the year of the fall of the *Bastille* and four years before the outbreak of war with France, there were four Marching Companies at Chatham, two at Portsmouth, and two at Plymouth, but these companies only trained on the fixed and movable armaments of the fortress, the whole business of care, preservation and maintenance being in the hands of the master-gunner and his squad of Invalid gunners. At each coast-defence station or work there was one of these master-gunners who was not yet a soldier of the Royal Artillery but still a quasi military/civilian subordinate of the Board of Ordnance and responsible for the upkeep, storage etc. of the guns, ammunition, equipment and stores. The Invalid Companies had not changed, except perhaps to become even more aged and decrepit. There were still ten of them, seven with their headquarters at Woolwich, one at Jersey, one at Guernsey and one in Scotland. In 1789 they were distributed among the coast defences of Great Britain and the Channel Islands as shown on following page. Dover, Sandgate, Walmer and Deal were still garrisoned by a special *Cinque Ports* detachment, responsible to the Lord Warden. It is to be noted that Jersey and Guernsey each had a full company, whilst the remainder of the garrisons only had small detachments, the individuals of which were gathered in from many companies.

The list of places in the left column gives a complete catalogue of the coast-defence stations around Great Britain which were the responsibility of the Board of Ordnance. To these must be added the *Cinque Ports* and such minor ports as had paid for and now owned their own coast-defence batteries. King Henry VIII's coast-forts still formed the basis of the chain of defences and, except for the fact that their batteries were now outside on ramparts, had altered very little. In Scotland the entrances to the Firths of Forth and Clyde were alone protected, the former by Blackness and the latter by Dumbarton, both mediæval castles adapted for mounting guns, but Paul Jones' threatened attack in 1779 upon Leith, which was then quite defenceless, had persuaded the Govern-

Place	COMPANY H.Q. Jersey	Godwin's Woolwich	Van's Woolwich	Gostling's Woolwich	Whitmore's Woolwich	Tiffin's Scotland Ft. George	Westman's Woolwich	Anderson's Guernsey	Toriano's Woolwich	Fairlamb's Woolwich	Standish's Woolwich	Totals
Berwick		4				1	1				2	8
Holy Island								1				1
Tynemouth			1	3								4
Whitby			2								1	3
Scarborough			1	1	1			1				4
Hull			2			2			1		2	7
N. Yarmouth				2		1						3
Lowestoft										1		1
Landguard Fort				1	1	1						3
Tilbury						3		1		1		5
Gravesend				1		1				1	1	4
Sheerness			2	2				2		4	5	15
Chatham Harbour						1					2	3
Upnor Castle			1		1	2				2	1	7
Gillingham Fort			1			1						2
Camber Fort								1				1
Hythe				1	1							2
Hastings				1								1
Blatchington										1		1
Seaford					1							1
Newhaven										1		1
Brighton					1							1
Arundel Haven					1							1
Portsmouth Harbour					1	2	1	1		3		8
Eastney Battery								1				1
Lumps Fort										1		1
Blockhouse Fort										1		1
Yarmouth I.O.W.			2									2
Sandown Fort				1						1		2
Cowes Castle								1				1
Calshot Castle										1		1
Hurst Castle			1							1		2
Dartmouth					2							2
Portland								1				1
Plymouth			1	2	9	1		5				18
Falmouth: Pendennis			1		1			2		1		5
St. Maws						1						1
Scilly Islands				4						1		6
Liverpool										1		1
Blockness								1				1
Dumbarton						1						1
Jersey	36*											36
Guernsey								42*				42
Totals	36	19	15	24	18	2	19	42	17	20		212

* includes 2 officers.

ment to construct coast-defences there, and in that very year Leith Fort was started upon. However, with the peace of 1783 the building work languished, but, with the threat of war once again hanging over the country, construction was hurried forward with the result that the fort was ready for occupation in the summer of 1793, armed with 5—24 pdrs. and 4—15 pdrs, garrisoned by a marching company, and became the main base of the Royal Artillery in Scotland throughout the war.

In Ireland the one Invalid Company of the Royal Irish Artillery was even too "invalid" to provide the care and maintenance detachments for the coast-defences, the majority of its personnel being quite beyond any form of duty. This is not surprising when we learn that in June 1786 a gunner, who had been discharged from the R.I.A. for insanity, was admitted into the Invalid Company, and there must have been many more like him. In 1800, of the total strength of 50, only 12 were serving in the coast-defences—at Duncannan Fort—and no less than 21 were "totally unfit for any duty". So the coast-artillery detachments had to be found from the marching companies, and in 1795, two years after the outbreak of war, were distributed as follows:-

	Officers	O.R.s	Total
Carrickfergus	1	11	22
Drogheda		7	7
Dundalk		7	7
Charles Fort, Kinsale	1	29	30
Cork Harbour	1	28	29
Galway Bay	1	23	24
Tarbert Island	1	24	25
Dublin Harbour		8	8
Duncannan Fort		5	5
Total ...	5	152	157

In 1793 three companies of the Royal Irish Artillery went to the West Indies, and these were followed in 1795 by two more. They took part in several operations, serving both as mobile artillery when attacking an enemy island or stronghold, and as coast-artillery when holding or actively defending it afterwards. Detachments from these companies served at Martinique, St. Lucia, Guadeloupe, St. Vincent, Grenada, Demerara, Berbice, Trinidad, Surinam and the Virgin Islands. In March 1801, as the result of the union of Great Britain and Ireland, the Royal Irish Artillery was abolished and incorporated in the Royal Artillery.

During the ten years between the two wars the coast-defences abroad were garrisoned by Marching Companies. In 1789 they were distributed as follows:-

Gibraltar	6 companies
Newfoundland	1 company
Quebec	4 companies
Halifax (Nova Scotia)	1 company
St. John (New Brunswick)	1 company
West Indies:	
Jamaica	3 companies
Dominica	1 company
St. Vincent	1 company
Grenada	2 companies

It is to be noted that a new defended port, St. John New Brusnwick, had been established in Canada. This was to provide a protected base on the Bay of Fundy when the St. Lawrence was frozen up, Annapolis having been abandoned. It must also be pointed out that, although companies are shown as stationed at definite places, this only means that their headquarters were there. Detachments were sent out from the parent company to man all the coast-defences and forts within the area. This was especially true of the West Indies where the location of the headquarters of a company might move from island to island, while the detachments remained in the various coast defences. Likewise the artillery garrisons of all the fortified posts on the St. Lawrence and the Great Lakes were found from the companies with their headquarters at Quebec.

The general system of command and organisation of coast forts and fortresses at this period must now be considered. Each "strong place" had a governor, lieutenant-governor, and fort-major. It varied very much as to who did the actual commanding. In a great place such as Gibraltar, the governor would be a general officer who actually commanded the fortress, and the troops holding it, with the lieutenant-governor acting as his second-in-command. In some places at home the post of governor might be entirely honorary, and the lieutenant-governor—or the fort-major if, as in some cases, the lieutenant-governor was also an "absentee landlord" —would be the executive commander. In some very small places all three appointments would be held by the same person who would be the senior officer present in the garrison. However, no matter what his title, the actual executive command in any coast fortress or fort lay with the senior military officer. The garrisons normally consisted of artillery, artificers, and infantry. At home the infantry garrisons were found from the Independent Companies of Invalid Infantry. In 1789 these were distributed as follows:-

1784 to 1793

Berwick	3 companies
Hull	2 companies
Landguard	1 company
Tilbury and Gravesend	1 company
Sheerness	2 companies
Dover	1 company
Plymouth	6 companies
Falmouth	1 company
Scilly Islands	1 company
Jersey	6 companies
Guernsey	6 companies.

Again, detachments from these companies would be provided for the smaller forts, and Portsmouth and Chatham were garrisoned by regular battalions. In time of war battalions or companies of militia were brought in to reinforce these infantry. Abroad, the garrisons were found by regular battalions, either complete in a large place or distributed in companies among lesser defended ports. In all cases "additional gunners" for manning the "great guns" had to be found either from the infantry garrison or from some other source, suitable or otherwise.

This survey of the 10 years between the two wars cannot be concluded without considering what "foreign service" meant to the coast-artilleryman at this period. Even at the most healthy stations such as Gibraltar, Quebec, or Halifax, it meant privation and suffering as the Board of Ordnance was much too mean ever to provide proper clothing, decent accommodation, or sufficient money to purchase adequate rations. The correspondence from stations abroad to the Board of Ordnance at this time is one long, continuous plea for more money, more equipment, more stores to furnish the bare necessities of existence. Posting to the West Indies was almost a sentence of death. Yellow fever was the killer, and it slew without distinction or mercy. Medical science had not yet identified the mosquito as the assassin so that no precautions were taken against its attack, and the garrisons in the West Indies islands were continuously decimated by the fatal disease. Two years service in the Caribbean was sufficient to destroy a unit without it suffering any battle casualties whatsoever. Foreign service in those days was in the very truth "no picnic".

CHAPTER X

1793 to 1815

BRITAIN went to war with Revolutionary France on 7th February 1793, and an expeditionary force was sent across to Flanders to assist her allies, Austria and Holland. There was no alarm at home, Britain was supreme at sea, the French fleet being disorganised and enfeebled by the inroads of the Revolution. This was underlined by Lord Howe's victory over the Brest fleet under Villaret Joyeuse on 1st June 1794. The coast-defences around the British Isles were put on a war footing, that is to say efforts were made to obtain and train "additional gunners" to serve and fire the guns in case of necessity. Marching companies were sent from their home stations to Leith Fort, Landguard Fort, and Guernsey while there were companies at Portsmouth, Plymouth, and Chatham. An extra Invalid Company was raised at Woolwich to provide an artillery-garrison for Bermuda and reached that island in June 1794. This was only the second time that an Invalid Company had been sent abroad. Once arrived at Bermuda it was treated in the best Ordnance Board manner. It was sent no drafts, money, clothing, stores, nor supplies and was apparently completely forgotten. In July 1797 it was reported as being "in a terrible state, having no officer". Still the Board took no action, and by 1807, when the company was at last brought home after 13 years service in Bermuda, there were only seven of the old men left alive. However, the Board without doubt must have saved a lot of money by neglecting the poor veterans.

The war appeared to be progressing well for Britain and her allies as 1793 developed. Several cities in the south of France revolted against the Republican government and declared for the Royalist cause, amongst them being the great naval base and arsenal of Toulon. In August the city council invited Lord Hood, who commanded the British Mediterranean Fleet, to occupy the town which he proceeded to do with a mixed force of British, Spanish, Neapolitan and Piedmontese troops. This force, about 8,000 strong under the command of the British general, Lord Mulgrave, at the beginning of September held a long line of forts, redoubts and field works covering the landside of the town. There were many guns in these works but no trained gunners to serve them, so a hurried request was sent off to Sir Robert Boyd, Governor of Gibraltar, for "as many artillerymen as can be spared". Boyd responded nobly and sent off Captain E. Stephen's Company (four officers and 85 men) which disembarked at Toulon on 27th October, and later a detachment, composed of drafts from the remaining companies at Gibraltar, of three officers and 49 men under the command of Captain J. Wilks which reached the beleaguered port on 15th December. Captain Stephens was appointed C.R.A. and the Royal Artillery were scattered in small packets among the various defence works.

Meanwhile, the French Republican army had closed in on Toulon and, urged on by the political representatives of the terrible Jacobin Committee of Public Safety, was determined to recapture the city and extirpate once and for all the nest of royalist vipers, the French efforts being greatly assisted by the fact that their artillery was commanded by Colonel Napoleon Bonaparte who had just arrived with a convoy of guns from the coast-defences along the Riviera. The French opened their main assault on 15th November, attempting to storm Fort Mulgrave which dominated the entrance to the main harbour. This attack was successfully repelled with heavy loss to the enemy, General O'Hara, who had now taken over chief command from Lord Mulgrave, reporting that the Royal Artillery detachment under Lieut. J. Lemoine, who was wounded in this engagement, "were the principal means of repulsing the enemy and securing that important post". On 29th November the Allies made a desperate sortie to destroy the French batteries creeping forward on the west side of the town, but unfortunately this ended in disaster, the advancing columns encountering far superior numbers and being driven back into Fort Mulbousquet, leaving behind them a great number of killed, wounded, and prisoners which included O'Hara, the commander of the allied forces, Stephens the C.R.A., one officer and six men of the Royal Artillery, and 19 "additional gunners" enrolled from the local French royalists. As the result of this General Dundas became the supreme commander, and Captain W. Collier succeeded as C.R.A.

On 17th December the French, now directed by Bonaparte, attacked Fort Mulgrave once again and, in spite of a desperate resistance from the mixed garrison of British and Spaniards, this time successfully, the Royal Artillery detachment, which was commanded by Lieut. J. Duncan, suffering 25 casualties in the struggle. The French followed this up immediately by storming *Faron* and *Grasse* Heights which were the key to the landward fortifications. The enemy had now penetrated the main defences, and, by their possession of Fort Mulgrave, threatened the works which covered the entrance to the harbour. It was now obvious that the Allies could remain in Toulon no longer, and the various

allied contingents began to retire from the inner ring of forts to their ships with or without orders as they felt inclined. It was therefore decided that a general evacuation of the town and remaining defences should be set in action at once, and throughout the 19th a flood of troops and terrified refugees poured down to the quays, seeking embarkation and refuge in the warships and transports laying at anchor in the harbour. Confusion and disorder reigned supreme, only to be increased by the explosion of the magazines and the spreading flames of the large ammunition stores which had been set on fire by British seamen as the troops retired.

Amidst all this chaos the Royal Artillery detachments withdrew steadily towards the waterfront, manning their mobile 6 pdrs., holding off the enemy advanced guards as they pushed forward into the town with grape and canister. Most of these guns had to be abandoned at the water's-edge, but nine were successfully got away into barges. Meanwhile, Major Koehler, R.A., who had been specially sent out from Gibraltar as a technical expert, was busy spiking the guns of the fixed armament in the waterside forts and is said to have been the last soldier to leave the shore. During the afternoon, covered by a great pall of smoke and flame, the allied fleet sailed out of the harbour, leaving most of the wretched inhabitants of Toulon to the revengeful fury of their republican fellow-countrymen. Not for another century and a half were British coast-artillery to watch such scenes, not until in fact they saw the crowds of refugees on the quays at Singapore clamouring for ships to take them away to safety. In the defence of Toulon the Royal Artillery suffered about 40 casualties, most of which came from Captain Stephen's company.

If 1793 ended badly for the Allies with the loss of Toulon, 1794 was a year of unmitigated disaster for them on the continent of Europe. When that year reached its close, the Allies had been driven out of Belgium and across the Rhine, the victorious French were rapidly overrunning Holland, and the British Expeditionary Force, under the Duke of York, was retreating steadily through bitter winter weather towards the northern frontier of that country. By the beginning of February 1795 it had crossed the River Ems and, withdrawing into Hanoverian territory, was finally evacuated in April from Bremen to Britain, having suffered crippling losses from disease, cold, exhaustion, and undernourishment, and comparatively few from the enemy. Thus by the spring of 1795 Britain found herself opposed by a France which was holding the European coast-line from the Ems to the Pyrenees, with Antwerp like a loaded pistol pointed at her heart. This length of hostile coast-line was to be gradually increased, first by Spain deserting the Allies and joining France in 1796, then by Denmark's enmity in 1801, and finally by the occupation of north-western Germany by the French in 1803, so that by this last year the western coast of Europe from North Cape to Cadiz—except for the coast of Portugal—was in enemy hands.

Therefore from the spring of 1795 onwards Britain felt herself to be threatened by invasion. As long as there were the numerous fleets of France, Spain, Holland, and Denmark in being in the ports and harbours, and large, victorious French armies under their famous generals rampaging aggressively about the Continent, Britain did not feel secure. The fear of invasion ebbed and flowed according to the situation. It rose to its greatest heights during the years 1797-98, 1801, and 1803-05 when there were large concentrations of enemy troops along the coasts across the Channel and North Sea actually preparing for invasion, and subsided during 1798-1801 when France was too busy in Europe and Egypt to consider such an undertaking, and in 1802-03 during the short peace of Amiens, but it was never really and finally layed to rest until Nelson's destruction of the French and Spanish fleets in Trafalgar Bay in October 1805. Of course the French occupation of Belgium and Holland and the withdrawal of the British Expeditionary Force from the Continent found Britain quite unready and unprepared to resist any attack on her home lands. Her forces were few, and her defences ill-armed and neglected, but she set to with tremendous energy and enthusiasm, not unmixed with apprehension, to remedy all those defects.

It was the attempted invasion of Ireland by the French in December 1796 which really first opened the eyes of the British to their danger. Ireland was seething with discontent and rebellion, and the French felt that, if they could once land a strong force there, the country would rise to support them, and the hated English thrown out without much difficulty. A force of 15,000 men under the famous French general, Lazare Hoche, was therefore concentrated at Brest and, having been embarked on board a fleet of 17 ships-of-the-line, 19 frigates, and 17 transports, set sail on 15th December for Bantry Bay at the southwestern corner of Ireland where it arrived, after a severe buffeting from stormy seas, on the 21st. Bantry Bay was completely undefended, not a gun nor a coast-gunner nearer than Kinsale, the only possible opposition being a force of about 500 Irish Militia which had occupied Bantry at the head of the Bay. For 14 days the French fleet rode at anchor in Bantry Bay, but the senior French general would not land a man until his commander, Hoche, arrived, and unfortunately for the French, the ship carrying that general and the naval commander-in-chief had become separated from the main fleet and been blown far out into the Atlantic. After waiting a fortnight for their superiors, the senior French commanders on the spot decided they could delay no longer, so on the 25th they issued orders to raise anchors and return to Brest at which port they arrived, not without some loss to their ships from bad weather, towards the middle of January. The ship carrying Hoche and the French admiral at last struggled into Bantry Bay on 29th December, only to find their fleet had departed for France four days before and the whole expedition been abandoned.

Britain was greatly shaken by this attempted invasion of Ireland which had failed only because of misfortune and bad weather, especially when it became known that an enemy fleet of 43 ships had sailed to and fro' Bantry Bay and been at large for close on a month without even being sighted by the Channel Fleet which, under Lord Birdport, had been vainly searching for it. This exposed the fact that, in spite of Britain's naval superiority, it was possible for a French fleet to remain at sea undiscovered for sufficient time to transport and land a considerable force on the coasts of the British Isles. And this was followed almost immediately by the strange affair of the raid on Fishguard. Hoche had originally planned that this operation should take place simultaneously with his invasion of Ireland, hoping it would act as a diversion and distract the British from sending reinforcements across St. George's Channel. However, for a variety of reasons, this secondary expedition was not ready when Hoche sailed for Bantry Bay, but the French government would not abandon it altogether and instructed the naval commander at Brest to despatch it as soon as the little squadron, which was to carry it, was prepared for sea.

It was indeed a strange expedition. It consisted of a force of some 1,300 infantry of whom about one half were convicts specially released from French gaols, and the other half prisoners-of-war of various nationalities—mostly Irish—who preferred to fight for their late enemies rather than rot in a prison-camp. Their commander was William Tate, an American soldier of fortune who had served in the South Carolina Artillery against Britain in the War of American Independence, now aged about 70, and their officers were mostly Irish adventurers or renegade French aristocrats. This force was called the 2nd Free Legion, but was known popularly as "The Black Legion" from the captured British uniforms, dyed dark brown, which had been issued out to them. Very rarely can such a collection of scoundrels and rogues been collected together in one body, and it is to be suspected that the French government was hoping to kill two birds with one stone, to harass the British and at the same time rid themselves of a large number of their undesirables. Hoche's orders to Tate were that he was to sail into the Bristol Channel, land his force at Bristol, destroy that important port and city by explosives and fire, and then, under cover of the smoke, flames, and general confusion, march his force away northwards through the Welsh border country to Chester and Liverpool which were to be treated in the same way. As an alternative to this plan, Hoche told Tate that, should he find it impossible to enter the Severn Estuary, he was to sail round to Cardigan Bay, land his force at a suitable place there, and march rapidly through Wales to Chester and Liverpool. It was thought the Welsh might help him and rise against their brutal English oppressors.

Tate and his Black Legion sailed from Brest on 16th February 1797, embarked in a squadron of four French warships, two frigates, *Vengeance* and *Resistance*, each of 40 guns, a corvette, *Constance* 24 guns, and a lugger, *Vantour* 14 guns, all under the command of Commodore Castagnier who flew his broad pennant in the *Vengeance*. On the afternoon of the 19th the enemy squadron was sighted for the first time off Lundy Island by the Revenue look-outs at Ilfracombe, who at once gave the alarm, thus causing an awful pother in North Devon, the local regiment of Volunteers making a forced march of 15 miles to Bideford on arrival at which place, so their colonel repeated later to Lord Portland, he was extremely proud and pleased "at seeing the men so willing to defend their king and country, at the same time as silent, orderly, and sober as might be expected at a morning parade of an old regiment". In the meantime, the enemy squadron was attempting to beat up the Bristol Channel, but, a persistent easterly wind effectively preventing this, Castagnier and Tate were forced to put into operation the alternative plan and make for Cardigan Bay. They were seen making for St. Anne's Head by a coaster on the 21st, were sighted from the shore off St. David's Head on the morning of the 22nd, and appeared off the entrance to Fishguard Bay during the afternoon of the same day.

Meanwhile the alarm had been spreading far and wide. There were not many troops stationed in Pembrokeshire. Around Fishguard itself were the *Fishguard Fencibles*, about 200 strong, under Lieut.-Colonel Thomas Knox, at Milford a company of the *Pembroke Fencibles* (100), at Pembroke a company of *Cardiganshire Militia* (100) guarding prisoners-of-war, at Castlemartin Lord Cawdor's Troop of *Pembrokeshire Yeomanry* (50), and, at Milford Haven two revenue cutters (150 seamen with 2 mobile 9 pdrs.). At the entrance to Fishguard harbour there was also a coast-fort (8—9 pdrs.) garrisoned by a corporal and two gunners of the Royal Artillery Invalids and 30 "additional gunners" from the *Fishguard Fencibles* under Ensign David Bowen. The fort was on a rocky point overlooking the entrance to the harbour, well sited to carry out its role, but was unfortunately very short of ammunition—typical Ordnance Board parsimony—there being only 16 cartridges in the magazine. The Lieutenant-Governor of the fort was one Colonel Gwynne Vaughan, a local landowner.

When the look-out on the fort sighted the French squadron off the entrance to Fishguard Bay during the afternoon of February 22nd a round of blank was at once fired from one of the 9 pdrs. to give the alarm. This evidently warned Castagnier that the harbour was defended, for the squadron sheered off, and moving further along the coast, finally anchored off a small, rocky cove some two and a half miles to the west of Fishguard. Here during the evening and throughout the night Tate and his 1,300 men of the Black Legion. with their arms, ammunition, and stores, were landed safely from small boats. At dawn next morning the French squadron, having successfully accomplished its task, weighed anchor and sailed away, leaving the Black Legion, now busy carrying its impedimenta inland over the rocks, to its own devices. Tate had already pushed forward patrols—supported by an advanced-guard—to the low line of scrub-covered sandhills which straddled the route to Fishguard about a mile to the east, and, having done this, settled himself comfortably into a farm-house, summoned his officers, and set about making plans for the advance of his force.

During the night Knox had collected his *Fencibles* into Fishguard, making his headquarters in the fort, and at dawn sent out an officer's patrol to reconnoitre the French. This soon returned with the information that the enemy appeared to be between 1,200 and 1,400 strong and had already occupied the *Carn Wnda* hills. Now poor Knox had less than 200 men with him, he was quite isolated, so he decided he must abandon Fishguard and retire on Haverfordwest where he knew Lord Milford, the Lord-Lieutenant of the County, was concentrating all the available forces. So during the morning of the 23rd Knox and his *Fencibles* marched out of Fishguard, Ensign Bowen having been ordered to spike the guns of the fort before withdrawing his men. This, however, he was unable to do because "the Woolwich Bombardiers refused to let the guns be spiked" (*West Wales Historical Records*) and the Invalid gunners gallantly remained behind at their post. When Knox had retired about 9 miles, at about half past one in the afternoon, he met Lord Cawdor advancing from Haverfordwest with the relieving force, and, having described to him the situation at Fishguard, joined his *Fencibles* to the mainbody. Lord Cawdor then pushed into Fishguard as fast as his men could march, his vanguard entering the town, which to their surprise was still empty of the enemy, about 5 o'clock of the evening. His little force consisted of:-

  Castlemartin Troop, *Pembroke Yeomanry* ... 50
  Pembroke Fencibles 100
  Fishguard Fencibles 200

Cardiganshire Militia	100
Seamen from Revenue Cutters with 2—9 pdrs.	150
Total	600

a small enough number indeed to drive the 1,300 of the Black Legion back into the sea.

All, however, was not well with the enemy. Tate had sent his best disciplined contingent forward to act as advanced-guard and occupy the *Cam Wnda* heights. He was now left with some 900 ruffians who, for the first time since they had been let out of gaol, were at last really free and were all thoroughly determined to enjoy themselves while they could. A regular orgy of looting and drinking set in. Every farmhouse and cottage in the small area occupied by the Legion was broken into and plundered. There was no shortage of "hard liquor", for this part of the Welsh coast was a well-known smuggling area, and every dwelling house had its cask of spirits. By the afternoon the invasion expedition had deteriorated into a mad, intoxicated debauch and such discipline as there ever had been had long since disappeared. When Tate or any of his officers tried to interfere, their gallant troops threatened to shoot them, and it was soon quite obvious that no attempt to advance on Fishguard could be made that day—or any day if it came to that. Before evening set in Tate learnt that his advanced-guard, holding the *Carn Wnda* heights, was in touch with the van of a great enemy force coming towards him from the direction of Fishguard and, after conferring with his officers he determined to capitulate as the only possible course. In fact the "great enemy force" was only Lord Cawdor with his 600 men, but the large number of Welsh women, who had climbed all the surrounding low hills wearing their long scarlet cloaks all agog to see what was happening, added to the total of red-coats as seen by the French pickets.

During the night two French officers were brought into Fishguard by the wife of a local farmer and, having arrived at the *Royal Oak Inn*, presented a letter from Tate offering surrender on terms to be discussed between the two parties. This was considered by a Council of War, presided over by Lords Milford and Cawdor, and an answer was sent back to Tate demanding unconditional surrender by 10 o'clock that morning. Ignorant of the small numbers of the force opposed to him, deceived by Cawdor's bluff, Tate was forced to accede to this demand, and during the afternoon of the 24th the French force, the dreaded Black Legion, marched—or rather staggered—down to the Gatwick Sands outside Fishguard and laid down its arms before Lord Cawder's little army drawn up to receive them. So ended with very few casualties on either side the famous raid on Fishguard.

CHAPTER XI

1793 to 1815 (*continued*)

THE British Government and people had not only attempted invasions of Ireland and raids on Fishguard to worry them.

By the autumn of 1797 it was obvious that preparations were being made across the Channel for the direct invasion of Britain. General Bonaparte, with the laurels of his great victories in Italy fresh upon his brow, was given command of the "Army of England", and all the harbours from Antwerp to Cherbourg were busy with the construction of gun-vessels and flat-bottomed boats. Britain had already taken steps to increase her forces at home. She had close on 60,000 Regulars for the defence of Great Britain, about 70,000 embodied *Militia* and *Fencibles*—these latter being regulars enlisted for home service and for duration of the war only—and was rapidly organising an army of Volunteers. In April 1794 an act had been passed by Paliament authorising "the raising of Volunteer Corps and Companies for the defence of counties, towns and coasts, such corps to be subject to military discipline and to be entitled to pay in the event of being called out for service". The coast-defence situation was gloomy. There was indeed the chain of coast fortresses, forts and batteries stretching from Berwick to the Scilly Isles but they were far apart, almost unmanned, many of the guns unmounted and unservicealbe and their fortifications in a state of disrepair. The War Office had to consider the defence against invasion of the whole United Kingdom, so they divided the coasts into twelve different districts.

(1) Northern, the coasts of Northumberland and Durham.
(2) Yorkshire, the coasts of Yorkshire and Lincolnshire.
(3) Eastern, the coasts of Norfolk, Suffolk and Essex.
(4) Southern, the coasts of Kent and Sussex.
(5) South Western, the coasts of Hampshire and Dorset.
(6) Western, the coasts of Devon, Cornwall, and part of Somerset.
(7) Severn, the coasts of part of Somerset, Gloucester, and South Wales.
(8) North Western, the coasts of North Wales, Cheshire, Lancashire and Cumberland.
(9) Scotland.
(10) Ireland.
(11) Jersey.
(12) Guernsey and Alderney.

A general officer was put in command of each district and instructed to carry out—with the help of a senior officer of the Royal Engineers nominated by the Board of Ordnance—a detailed reconnaisance of his stretch of coast, reports with recommendations to be submitted as soon as possible.

Now it must be remembered that, in the course of the long war against the French Republic and Empire, there were three periods during which Britain appeared to be in danger of direct invasion by the French across the Channel, 1797-98, 1801, and 1803-05. The first invasion plan was abandoned because Bonaparte came to the conclusion that, without even temporary command of the narrow seas, it would be "the boldest and most difficult operation ever attempted", so he turned away from Britain and made for Egypt and the Levant instead. In the summer of 1801, Napoleon, now First Consult of the French Republic, again concentrated an army of invasion along the Channel shores. He needed peace with Britain, and this threatened invasion was a form of blackmail to force the British government to come to terms with him. He was successful in this, and the Peace of Amiens, 25th March 1802, was the result. However, Napoleon wanted peace only for his own schemes and purposes, and the British Government, quickly realising this, declared war on France once again in May 1803. Britain was now to go through her period of greatest peril. Napoleon, soon to become Emperor, was well aware that Britain was his strongest, most steadfast, and most dangerous enemy, and moreover, at the moment, his only one. Here then was his opportunity to deal with her once and for all while she was isolated, for he knew that the only way to defeat her quickly and decisively was by direct invasion. So from 1803 to 1805 the Grand Army was distributed along the western ports, from Hamburg to Bordeaux, and on the immediate "invasion front", from Flushing to Havre, he massed 80,000 veterans with over 2,000 barges, pinnaces, and gun-vessels, ready to slip across the Channel when the opportunity came. But the opportunity never did come because Napoleon still realised that he could not launch his invasion flotilla into the sea unless and until he had at least temporary command of the Channel, and that, in spite of all his deep laid schemes, he was never able to obtain. So in August 1805 he abandoned the whole project, and, turning his back on the narrow seas, marched into Southern Germany and Austria and to the victories of Ulm and Austerlitz. In October, Trafalgar put an end for ever to all his hopes for the successful invasion of Britain, and Napoleon had to seek for other means to defeat her.

Turning back to 1797-98, it must be realised that the recommendations and efforts made then to defend the coasts of Britain were gradually put into effect and greatly expanded during the subsequent years until they reached their culmination in 1805. Of course, the progress was not regular but proceeded in fits and starts. When it was discovered in 1798 that General Bonaparte was no longer inspecting the invasion forces in Belgium and north western France but was actually on his way to Egypt from Toulon, Britain heaved a sigh of relief, and all defence works were slowed down. In 1802, at the Peace of Amiens, defence work was given up altogether, the guns being removed from the emergency batteries, the *Fencibles* and Volunteers disbanded, and all thought of possible invasions dismissed, so that in 1803, when war broke out once more and Napoleon massed the Grand Army along the shores of the narrow seas, everything had to be done all over again, general-officers had to reconnoitre and report on their stretches of coast-line once more, and the Master General of the Ordnance had to issue orders "Immediately to proceed upon placing the batteries round the coast in the same state of defence as they were at the close of the last war and to report the steps taken in consequence of such orders and to transmit a particular statement of the number and nature of the ordnance which shall be mounted and the names of the batteries on which the same are placed" (Board of Ordnance Minutes, 1803.) Nevertheless, the general idea remained the same throughout, at any rate as far as coast-defence was concerned. That was to surround England, Wales, and Scotland with a ring of batteries from the Firth of Forth to Liverpool which would cover every likely anchorage or landing place, the Channel Islands also being defended at every point. (Ireland was treated quite separately as it presented a totally different problem.)

The final design of defending the coast with batteries from the Firth of Forth to Liverpool was, of course, never reached but great efforts were made to attain it. It was decided that the enemy's main objective after landing would be to capture London, and it was therefore concluded that the shores of Britain from (North) Yarmouth to Selsey Bill were the most threatened, so work was concentrated to begin with on that stretch of coastline, but even certain portions of this were considered unsuitable for landing troops, and likely places of assault were finally narrowed down to:-

(1) On the East Coast: Hollesley Bay; the beaches between the Deben and the Orwell ; Harwick; Clacton.

(2) On the South coast: Eastware Bay, Hythe Bay, the beaches around Hastings, the beaches between Pevensey and Beachy Head.

There were difficulties right from the first as to who should be responsible for the erection of the necessary defence works. If they were classified as permanent fortifications, it was the responsibility of the Board of Ordnance to construct them, but if they were considered to be field-works, the onus was on the War Office.

The Board even claimed that field-works, which were likely to develop later into permanent fortifications, were also their business and watched most jealously over their prerogative. Nevertheless, the Board, having interferred most officiously in the War Office's programme of construction of first line defences, were most indolent themselves in carrying out their own work and wasted a lot of time devising reasons why the department, in order to save money, should not meet the demands made upon it. The Board also continued with its system of forcing cities, towns, and small ports to pay for the construction of their own defending batteries. Dunbar was told to build a fort protecting its approaches at its own expense when the G.O.C. Scotland pointed out that such a defence-work was necessary, while one of the King's sons, the Duke of Cumberland, was sent specially to Liverpool to persuade that city to subscribe sufficient funds to erect the extra batteries which were considered essential for the defence of the Mersey entrance.

Nevertheless, the work continued as the years went by, the general policy laid down being that fixed defences on the coast should be confined to anchorages of local importance, to points where a landing threatened the essential interests of the county, or to parts of the coast where, by judiciously strengthening the few places which were accessible, a line of considerable extent might be rendered secure. The coast-line Yarmouth—Selsey Bill was first covered by batteries, and then the works were extended slowly northwards and westwards. The Board experienced difficulties in the west-country where the G.O.C. Western District had no faith in isolated fixed batteries and demanded that either all the guns should be mobile or placed in strong redoubts defended by at least 200 infantry. He even suggested that two light mobile-guns should be kept at each naval signal-station along the coast so that, in case of a landing in the vicinity, the guns could be run out and served by the signal-lieutenant and his men. Needless to say the Board took very little notice of such views and went on with the work. From Berwick the line of batteries crept into Scotland, the stretch of coast from Dunbar to Musselborough being given priority. On the western side of Britain the Severn District was next in order, and the approaches to Bristol were somewhat inadequately defended, but there the work more or less came to an end, except for isolated defences at Milford Haven and such places like Liverpool, Whitehaven and Swansea which were prepared to pay for the construction and upkeep of their own batteries. However, by August 1805, when Napoleon finally gave up his plan of invasion, the coast-defences around Great Britain and the Channel Islands, according to the Ordnance records, were as given below. The list is probably not complete but must be nearly so: where no armament is given it is not known:-

1793 to 1815 (continued)

District and Place	Pounder Guns											Pounder Carronades						
	42	36	32	24	20	18	12	9	6	4	3	68	32	24	18	12	9	6
1.—Northern and Yorkshire																		
Berwick...				6			5			16								
Bamborough Castle		1				2		3					1					
Blyth				3														
Seaton Sluice						3				6								
Tynemouth						32	8	11										
South Shields						4												
Hartlepools				8														
Whitby						14												
Scarborough						15												
Hull				5		13		4										
Hull Southend				2		2												
Pauls Cliff				6														
Stallingborough				4														
White Booth				2														
2.—Eastern																		
Cromer...				4														
Yarmouth			14	6		9	1	4										
Lowestoft			15			6												
Southwold						6												
Gorleston Cliff				8														
Aldeborough				2														
Hollesley Bay and R. Deben				60														
Landguard Fort	11		11												20			
Harwich				8														
R. Colne Mouths				72														
Stroud...						2												
Fingenhoe						2												
3.—Southern																		
Sheerness Fortress																		
Half Moon Bty.	10																	
Cavaliers Bty.			3															
Medway Bty.						4												
Flagstaff Bty.			2			2												
Craigs Bty.	7																	
Queensborough						4												
Minister	11					6												
Chatham Fortress																		
Fort Pitt						10									18			
Lower Cornwallis				2									2					
Upper Cornwallis				3									2					
Belvedere				2									6					
Spur Bty.				13									6			6		
Prince of Wales' Bty.				1		8							5					
Broken Curtain				5									2					
St. Mary's Bty.				4												2		
King's Bastion				1			3						5			2		
Prince Edward's Bty.				1			2						4					
Prince Henry's Bty.				1			1						2					

History of Coast Artillery

District and Place	Pounder Guns											Pounder Carronades						
	42	36	32	24	20	18	12	9	6	4	3	68	32	24	18	12	9	6
Upnor Castle				4														
Gillingham Fort						2												
Tilbury Fort																		
West Gun Line			14															
East Gun Line	14																	
Six Gun Bty.			6															
South Bastion							2											
S.E. Bastion							11											
N.E. Bastion							6											
N.W. Bastion							6											
S.W. Bastion							10											
Coal House Bty.				4														
Gravesend Fort																		
Blockhouse Bty.			19															
New Tavern Bty.			2	14			1											
Thornmead Bty.				4														
Hope Point				4														
Westbrooke				2														
Margate				3	1													
Broadstairs				2		2												
Ramsgate				5														
Pegwell Bay				2														
Fort No. 1.		6																
Fort No. 2.		6																
Deal Castle		9																
Walmer Castle		8				1												
Sandwich		9					2											
St. Margaret's Bay			3															
Dover Fortress																		
Dover Castle			20	8	64	6						9	11	4	3			
Drop Redoubt				13														
Guildford Bty.			4											2		1		
Moat Bulwark			3															
Amherst Bty.			4															
Townshend Bty.				2														
Archcliff Fort			5			6								2				
Folkestone					4													
Eastware				4														
Sandgate Castle				8														
Shorncliff				10														
Twiss Bty.				6												2		
Sutherland Bty.				8														
Moncrieff Bty.				8														
Saltwood Hts.				2														
Dungeness Btys.				20														
Dungeness Rdbt.				11														
Rye				13														
Hastings				6														
Langleys East				6														
Langleys West				6														
Eastbourne				11														

District and Place	Pounder Guns											Pounder Carronades						
	42	36	32	24	20	18	12	9	6	4	3	68	32	24	18	12	9	6
Seaford				4														
Blatchington				6														
Newhaven				8														
Brighton	6																	
Arundel Haven ...	5																	
4.—*SouthWestern*																		
Portsmouth Fortress																		
Portsmouth Harbour		19				18	33	30	22				4					
Portsea Dockyard		28	14			43	4											
Southsea Castle ...			8						5									
Lumps Fort ...			3															
Eastney Fort ...			3															
Fort Cumberland				8		41	25		7									
Fort Blockhouse ...		15				15												
Gosport Works ...				19		36												
Priddys Hard ...						14												
Fort Monkton ...		24				23	12											
Yarmouth I.O.W. ...								4	8									
Freshwater						3												
Cowes Castle								10										
Sandown Fort ...						20												
Calshot Castle ...						9			4									
Hurst Castle				6			18											
Weymouth				6														
Portland Castle ...				6			6	2										
Swanage								3										
Bridport							3											
5.—*Western*																		
Seaton				4														
Berry Head	12			16	12													
Dartmouth						10				2								
Plymouth Fortress																		
Citadel	11			8	20	17		13	3						3			
Dockyard					49	38									35	8		
Western King ...						10												
Drake's Island ...				22		6												
Staddon Hts. ...							10											
Maker Hts. ...				4	6	32												
Looe						4												
Fowey						2	5	2	1									
Crinnis Cliff						4												
Mevagissy						6												
Falmouth Fortress																		
St. Maws				10														
St. Anthony's ...				4														
Pendennis Castle				22		14								4	8			
Mounts Bay				10														
Scilly Islands			14	4		11		2	2						18			
St. Ives						6	6											
Portreath							4											
Ilfracombe						5												

District and Place	Pounder Guns											Pounder Carronades						
	42	36	32	24	20	18	12	9	6	4	3	68	32	24	18	12	9	6
6.—*Severn*																		
Bridgewater Bay																		
Axemouth																		
Portishead Point				4														
Avonmouth				4														
Barry Islands																		
Swansea							2											
Milford Haven																		
Fishguard								6										
7.—*North Western*																		
Liverpool																		
North Batteries						6		12		2								
Red Noses				5														
Rock Channel						3												
Whitehaven	6		8	8		7												
8.—*Scotland*																		
Firth of Forth																		
Blackness Castle								5										
Leith Fort				5		4												
Queensferry				8														
Inchgoln				7														
Inchgarvie						4												
Dunbar																		
Dundee																		
Aberdeen																		
Peterhead																		
Fraserburgh																		
Dumbarton Castle						3	13	1	4									
Orkney Islands				10														
Shetland Islands						8												
9.—*Jersey*																		
Elizabeth Castle				26		6			1				5	14	1	12		
Mount Orgueil				3			6											
St. Alban's Fort						7									3	3		
Fort Henry				4														
Archirondelle						4												
La Rocca				5														
Prince William's Rdbt.				4														
Nourmont						2												
Seymour				2														
New Fort (Regent)				38		19	22							32				
10.—*Guernsey*																		
Fort George				31	1	11	19		10				2	9		27		
Donne Bty.							1		2									
Jerbourg Bty.				2														
Fermain Bay				6										1	3			
Moulin La Huette				3														
Saints Bay				3		1										1		
Soumilleuse Bty.				2														
La Tuelle						2												
Narrow Port								1										

1793 to 1815 (continued)

District and Place	Pounder Guns											Pounder Carronades						
	42	36	32	24	20	18	12	9	6	4	3	69	32	24	18	12	9	6
Plein Mont						1												
Pessery Bty.						3												
Rocquain Bay				6		3										1		
Fort Saumarez				3												1		
Perelle Bay					4													
Fort Richmond					4													
Le Crocq Point					3													
Vazon Bay					14													
Houmette Bty.				7														
Fort Burton					4													
Cabbo Bay					2													
Grand Rocq Bty.				3														
Rouse Bty.				3												1		
Pickerie Bty.						2												
Mont Cuette				5												1		
Le Ree Point						2												
Lancresse Bay				3	9		1									4		
Fort Le Marchant				6														
Fort Doyle						3												
Bouset Point				2														
Vale Castle				2														
Mont Crevet				5												1		
Delaney Bty.						2												
Belle Grove						4												
Houge a la Pierre				7			2									1		
Fort Perrin					4													
Mont Arrive						3												
Sallerie					4													
Fort Amherst					2													
Castle Cornet				10		12	17					3	2	9				
Alderney				4	25	34	2	1				3	4				4	4
Sark							6											

To complete this long line of coast-defence fortresses, forts and batteries, and to fill in the gaps, it was suggested in the summer of 1803 that fortified towers should be built. Such towers were common along the shores of the Mediterranean, and several had already been constructed by British engineers in Minorca and Guernsey. The War Office was enthusiastic about the idea, but the Board of Ordnance as usual was much less eager, the Master General (Lord Chatham) writing to the Secretary of State for War in February 1804 that "as it is entirely novel in principle and connected with an infinite variety of details, I should be unwilling to occupy the time of the (Engineer) Committee on this point unless your Lordship continued to feel a desire to receive a report on the subject, and that the idea was generally approved by His Majesty's Confidential Servants". (W.O. 1/783). This was followed by a long discussion between the two departments concerned, and it was not finally decided until October when the Privy Council, at a meeting specially convened at Rochester to consider the whole system of coast-defence, pronounced definitely in favour of towers. So instructions were issued to the Board of Ordnance for the immediate construction of these fortifications, or *"Martello Towers"* as they were called from Mortella Bay in Corsica where there had been a similar tower on the cape at the north-end of the bay when the British landed there in 1793. The work of erecting the towers began during the early summer of 1805 and was not completed until 1812, there being 74 built along the south-coast—27 between Folkestone and Dymchurch and 46 between Rye and Eastbourne with an isolated one at Seaford—and 29 along the east coast—11 from St. Osyth to Walton-on-the-Naze and 18 from Shotley to Aldeburgh. The towers were made of brick covered with stucco, were of elliptical design, the longer axis being 40 feet at the top and 45½ feet at the base, and were 34 feet high with walls nine feet thick. Each tower had two floors and a roof surrounded by a parapet which carried the armament, usually a long 24 pdr. gun and two 24 pdr. carronades, while the first floor housed the garrison—about 25 men—arms and stores, and the ground floor the magazine and water-cistern. The object of these towers was either to give artillery defence to possible landing places not already covered by batteries, to provide support to a battery already in position, or to protect specific vulnerable points. Similar towers were also built in Ireland (14) and along the coasts of the Channel Islands.

The Board of Ordnance had never faced up to the problem of providing gun-detachments for their coast-defences. Always these had to be found from sources other than the Royal Artillery, but with the continuous threat of invasion, and the resultant tremendous expansion between 1797 and 1805 of the coast-defences of Great Britain, these sources were no longer able to produce the necessary numbers to man all the new batteries. Even at the great coast-fortresses such as Portsmouth, Plymouth, Sheerness, and Leith Fort, which had been accustomed to getting their "additional gunners" from either the fleet, infantry garrison, or county *militia* regiment, it was no longer an easy matter, the men of these formations being urgently required for their own proper spheres of action. Marching Companies had to be retained in the coast fortresses, and detachments from other companies sent to assist the Invalid gunners at the lesser forts. Marching Companies were stationed as follows:—

	1798	1801	1804
Portsmouth	2 Companies	1 Company	2 Companies
Plymouth	2 Companies	2 Companies	2 Companies
Tynemouth	1 Company	1 Company	2 Companies

1793 to 1815 (continued)

Leith Fort	1 Company	1 Company	1 Company
Dover		1 Company	1 Company
Hull			2 Companies
Cork			1 Company
Waterford			1 Company
Jersey			1 Company

whilst as early as 1795 detachments from other Marching Companies were serving at:-

Plymouth	3 Officers	100 O.R.s.
Portsmouth	3 Officers	100 O.R.s.
Chatham	1 Officer	22 O.R.s.
Dover	1 Officer	22 O.R.s.
Landguard Fort	1 N.C.O.	10 O.R.s.
Tynemouth	1 Officer	11 O.R.s.
Berwick	1 N.C.O.	10 O.R.s.
Sunderland	1 N.C.O.	10 O.R.s.
Dungeness	1 Officer	22 O.R.s.

and these tended to increase and to spread more widely as the war went on.

But the difficulty of finding gun detachments was not insuperable at the great fortresses and forts: if the worst came there was always the *militia* who could be called upon if all other sources failed, for example the Royal Pembrokeshire Militia Regiment provided the "additional gunners" for Landguard Fort almost throughout the war. However, when it came to manning the guns of both the old and new sea—or emergency—batteries, the situation was very different. It will be remembered how in 1756, when invasion threatened, nine new sea-batteries were built along the south-coast, the master-gunners being instructed to find the gun detachments by teaching "such of the inhabitants as shall be inclinable to learn how to load, point, and fire the guns", but with General Bonaparte, First Consul Bonaparte, or the Emperor Napoleon just across the Channel at Boulogne with 80,000 or so veteran troops it was obvious that "inclinable inhabitants" were going to be neither sufficient nor proficient enough to serve the coast defence guns. It was at first hoped that the detachments could be provided by the "*Sea Fencibles*," a force raised by the Royal Navy in 1797 from coasting seamen, fishermen, longshoremen etc. for close-in-shore coastal defence, but the Admiralty very quickly objected to this, needing the men themselves to man the gun-boats, gun-hulks, and armed vessels required to guard the small havens and anchorages. We have already seen how in April 1794 Parliament passed an act authorising "the raising of Volunteer Corps and Companies for the defence of counties, towns, and coasts," and, although the natural tendency at first was to form only yeomanry cavalry and infantry corps, it was very quickly realised that here lay the solution to the problem of finding the coast-defence gun detachments.

The Government having refused permission for the forming of Volunteer mobile artillery corps, many cities and towns situated on the coast very soon set about raising bodies of volunteers to be trained in the use of the great guns in the coast-batteries which were notably deficient in gunners. Such corps were only supplied with muskets to the extent of one third of their total strength, the remaining two thirds being available for training on and manning the guns. Indeed, after war broke out again in 1803, the Government laid it down that the primary object of Artillery Volunteers was to aid in the most efficient manner in the manning of the batteries erected for the protection of the coast-towns. As early as May 1794 Gravesend and Milton raised a "Stationary Company of Artillery" for the defence of Gravesend and Tilbury forts and to exercise the great guns in them. It was this unit, as the "Gravesend Volunteer Artillery" which in June 1797 forced a ship-of-the-line, in the hands of the naval mutineers during the Nore mutiny, to strike her colours and surrender whilst trying to run past Gravesend Fort. During 1796-97 many such corps were raised, especially in the larger coast-towns. It was in 1797 that the town of Falmouth formed the Pendennis Artillery as the defences of the port at that time were only manned by a few Invalid gunners and an Invalid Infantry Company not more than 40 strong. It is said that Volunteers flocked to Pendennis Castle with great enthusiasm to learn the use of the guns, their numbers increasing at such a rate that later it was possible to send detachments to man the guns at other defended places in Cornwall. Most of these Volunteer Corps were disbanded at the Peace of Amiens in 1802 but reformed, with others, when war broke out again in 1803. It was in this latter year that *Trinity House* formed the Royal Trinity House Volunteer Artillery from officers and seamen of the Merchant Marine. This corps was employed in manning 10 armed hulks moored across the Thames from Tilbury to Gravesend.

The general conditions of service of these various Volunteer corps were that, on being called out "on actual service," officers and men were paid the same as the Regular Army. When not so called out, the officers were paid for two days each week when assembled for exercise, and Other Ranks two shillings weekly for two whole days exercise of six hours each or a number of hours in the week equivalent thereto. Clothing was supplied by the Government (Home Office) and arms and accoutrements by the Board of Ordnance. In 1803 the War Office began the system of posting a permanent staff from the Regulars to Volunteer Corps. In that year

1793 to 1815 (continued)

the Sunderland Corps of Volunteer Artillery received an adjutant, a sergeant-major, and four additional sergeants, who had been recommended by the C.R.A. of the Northern District, and this staff was to be "on constant pay". In 1804 Field Officers were also appointed from the Regular Army to Volunteer Corps "to afford commandants of corps the benefit of the assistance and advice of officers not their superiors, but from experience competent to aid them" (Sebag-Montefiore). Below is given a list of the Volunteer Corps which, as far as can be ascertained, were providing gunners for the coast-artillery by the end of 1797. Next to it is a list of those which had been added by 1801, and again, next to that, those which had been added by 1805. Now it must be remembered that very few of these corps began life as artillery units. They were mostly raised as infantry battalions or companies, a proportion of which were sent as "additional gunners" to the local coast-batteries. As time went on, these detachments serving with the guns tended to become quite separate from their infantry parents, and so formed themselves into distinctive Artillery Volunteer Corps, although some Corps remained combined units right up to 1814. On the other hand, some started as artillery corps and continued as such; some indeed were formed specially to man the local coast-defence fort or battery All these Volunteer Corps were disbanded soon after 1815.

1797	1801	1805
Aberdeen	Appledore	Aldeburgh
Beer and Seaton	Ayr	Barmouth
Berwick	Brixham	Barnstaple
Bridgewater	Chatham &	Bridlington
Bridport	Gillingham	Brighton
Bristol	Crinniss Cliff	Campbelltown
Cinque Ports	Dartmouth	Coleridge
Dover	Exmouth &	Dumbarton
Hastings	Sidmouth	Eastbourne
Hythe	Hartlepools	Farringdon
Romney	Lyme Regis	Grimsby
Rye	Maker	Haytor
Seaford	Mid-Lothian	Heaton
Sandwich	Penzance	Herne Bay
Winchelsea	St. Ives	Lancaster
Deal	Selsey	Littlehampton
Dunbar	Sheerness	Looe
Dundee	Stonehouse	Lowestoft
Falmouth	Sunderland	Mersey Island
Fishguard	Swansea	Milford
Folkestone	Whitehaven	Newhaven
Fowey		Orkney & Shetland
		Portreath

Fraserburgh
Gosport
Gravesend & Tilbury
Greenock
Hull
Leith
Liverpool
Lydd
Margate
Mountsbay
Musselborough
Petershead
Plymouth
Queensferry
Ramsgate
Salcoats
Scarborough
South Shields
Tynemouth
Weymouth
Isle of Wight

Ringwould
St. Osyth
Southwold
Stannary
Workington &
    Maryport
Yarmouth

The situation in Ireland throughout the "invasion period" was quite different to that in Great Britain. The inhabitants of John Bull's other island were seething with revolt and ready for rebellion. The ideas and successes of the French Revolution had filled them with hopes and expectations that they would soon be able to throw off the hated English yoke and already most of their revolutionary leaders were in Paris concocting plans with the French government. In 1796, when Hoche made his attempted invasion at Bantry Bay, there were about 40,000 Regulars, 23,000 *Militia*, and 25,000 Volunteers for the defence of Ireland. There were permanent coast-defences at:-

    Carrickfergus
    Charles Fort (Kinsale)
    Cork
    Drogheda
    Dublin Harbour
    Dundalk
    Duncannon Fort (Waterford)
    Galway Bay
    Tarbert Island.

These were manned by detachments from the "Marching Companies" of the Royal Irish Artillery plus "additional gunners" from the Irish *militia* regiments. It is to be noted that there were no coast-

1793 to 1815 (continued)

defences whatsoever at Bantry Bay. The threat to Ireland was appreciated by the authorities at Dublin Castle as invasion by a body of enemy troops which was prepared to abandon its communications with France and rely on support and assistance from the disaffected inhabitants. It was therefore considered that invasion need not necessarily be made at or near one of the great harbours, but by a comparatively small corps landing on a wide front at small havens and coves, rapidly concentrating, and then pushing quickly inland with final objective Dublin, the seat of English dominion in Ireland. It was realised that it would be quite impossible to cover every possible anchorage and likely landing place by coast-batteries. The coast-line was too long, there were far too many possible anchorages and likely landing places, and the inhabitants were much too friendly with the enemy. The most probable areas of enemy landings were thought to be Galway Bay, Mouth of the Shannon, Bantry Bay (after 1796) Kinsale, or a direct attack on Dublin itself from Dublin Bay. So orders were issued for the construction of defences at Bantry Bay, for increasing the batteries on Tarbert Island and the raising of new ones on Scattery Island, both at the mouth of the Shannon, for strengthening the works covering Galway Bay, and for building new fortifications at Pigeon House and Ringsend, both protecting Dublin Harbour. By 1811 the main coast-defences of Ireland were:-

Cork
- Cove Fort ... 24—24 pdrs.
- Fort Westmorland 24—24 pdrs: 2—12 pdrs: 10—6 pdrs: 5—Swivels.
- Fort Carlisle ... 26—24 pdrs: 7—18 pdrs: 6—12 pdrs: 14—6 pdrs.
- Fort Camden ... 14—24 pdrs: 6—18 pdrs: 7—12 pdrs: 4—6 pdrs.
- Haulbowline ... 2—18 pdrs.

Bantry Bay
  Bere Island
- No. 1. Battery 2—24 pdrs.
- No. 2. Battery 2—24 pdrs.
- No. 3. Battery 2—24 pdrs.
- No. 4. Battery 2—24 pdrs.
- Western Redoubt 6—24 pdrs.
- 4 Martello Towers 6—24 pdrs.

  Whiddy Island
- Centre Redoubt 12—24 pdrs.
- East Redoubt 8—24 pdrs.
- West Redoubt 8—24 pdrs.

  Garnish Island
- Battery ... 3—24 pdrs.

Charles Fort, Kinsale ...	17—24 pdrs: 14—12 pdrs: 4—6 pdrs.
Duncannon Fort ...	38 guns.
Waterford	20 guns.
Galway Bay	7 guns.
Tarbert Island	13 guns.
Scattery Island	8 guns.
Dublin Bay: Pigeon House Fort	13 guns.
Ringsend ...	13 guns.
Carrickfergus	22 guns.

These were manned by detachments from the Royal Artillery "Marching Companies" stationed in Ireland, the Royal Irish Artillery having been disbanded and absorbed in April 1801. But 1811, of course, was long, long after Ireland had had her insurrection and her French invasion.

Both of these took place in 1798. The rebellion broke out in the south-east in May, but only took a real grip in County Wexford. The French had in preparation two expeditions to support this rising: one under General Humbert, consisting of 1,150 troops and four field guns, set out from Rochefort in four frigates under the command of Commodore Savory on 6th August, and the second, under General Ménage, of 3,000 troops and several field and heavy guns embarked in one ship of the line (*Hoche* 74 guns) and nine frigates under Admiral Bompart, sailed from Brest on 16th September. These two expeditions were of course far too late, the rebellion having been finally crushed—not without some difficulty—by the end of July. However, Humbert and his corps, having dodged the British fleet, landed at Killala Bay, County Mayo—where there was no coast-defence battery—on 22nd August, and, pushing rapidly southwards, met and routed a small British force at Castlebar on the 27th. The French then marched as fast as they could to the north-east and east away across Ireland, hoping to avoid the much superior British forces which under Lord Cornwallis were closing in on them. Humbert's only chance of survival now lay with the local inhabitants. If they would revolt spontaneously, harass the British forces, cut their communications, and join him in their thousands, he might still compel the British to abandon most of west and central Ireland and retire on Dublin. But the Irish would not rise, they already knew the fate of the Wexford rebellion, and on 8th September at Ballinamuck, Humbert and his troops, hopelessly outnumbered and all but surrounded, were forced to lay down their arms and surrender to Cornwallis. The second expedition, under Ménage, was even less successful. The French squadron was sighted soon after it left Brest by British frigates and kept under observation throughout the time it was on its way to Ireland.

Bompart, not knowing what had happened to Humbert, decided to land his troops at Lough Swilly, but, when off Bloody Foreland on 12th September, was intercepted and engaged by a British squadron of three ships-of-the-line and five frigates under Sir John Warren. The French were heavily defeated, losing their flagship and three of their frigates during the engagement, and three more during the subsequent chase, only two ships at last returning to Brest on 3rd October. Thus disastrously came to an end all Ireland's hopes of obtaining freedom during the long French war, and after 1798 there were really no more serious fears of a French invasion.

The Channel Islands, although that part of the United Kingdom nearest to the enemy, came through the long years between 1793 and 1815 unscathed and free of invasion. As can be seen from the list of coast defences and batteries given above, the islands were heavily defended by the time the French emperor gave up all hope of invading Britain and marched away into central Europe. Later *Martello-towers* were added in both Jersey and Guernsey, and the great new fort on the former island, *Fort Regent*, was completed. The coast-defences in Jersey and Guernsey were maintained throughout the war by the two Invalid Companies, Royal Artillery, which were stationed in them, and these were reinforced by marching-companies in 1793-94 (Guernsey), 1803-14 (Jersey) and 1808-14 (Guernsey) but the backbone of the coast-artillery was always provided by the Channel Islands *Militia*. This ancient force, whose origins go back to Norman times before the Conquest, had long been the mainstay of the defence of the islands against invasion. Every man, between the ages of sixteen and sixty, was legally bound to be a serving member of the *Militia* and to turn out whenever ordered, and from the time of James I, when the *Militia* was re-organised as the result of a report by Royal Commissioners, gunners for manning the coast-batteries were always provided from the infantry battalions or companies. Even when in the middle of the eighteenth century separate Militia Artillery Corps were formed in both islands, this custom continued as the Militia Artillery was never able to find enough men to serve all the great guns. These artillery corps were divided into two divisions; a field-artillery regiment with mobile guns and a "battery-corps" for coast-defence, but the latter always had to rely on the infantry battalions to provide a high proportion of their gunners. In 1804 the companies of the Guernsey East Militia Regiment (infantry), which had for so long been manning the coast-defence batteries covering Fermain Bay, were formed into a special coast-artillery corps, called the Fermain Corps, under Major Commandant de Lisle, which continued to exist as an independent unit until 1822. In Alderney the coast-defence batteries were always manned by the island *militia*.

CHAPTER XII

1793 to 1815. (*continued*)

THE history of Coast Artillery in the British Army during the war of 1793-1815 outside the British Isles has now to be considered, and the operations in the West Indies will first be dealt with. In 1793 there were nine companies of Royal Artillery in the West Indies, three at Jamaica, five with their headquarters at Barbados, and one at Grenada, the Barbados companies finding the artillery garrisons for the other islands, Dominica, St. Kitts, Antigua and St. Vincent. In the first year of the war, the British sent out three expeditions in the Caribbean, one to Martinique which failed, one to Tobago, and one to San Domingo. The expedition to Tobago was a complete success, the island being taken during April with practically no fighting, and a detachment of Royal Artillery of one Sergeant and eight Rank and File was left in the captured fort together with the rest of the British occupying garrison. San Domingo presented a different problem. Here the mulattos and negroes, filled with the ideals and aspirations of the French revolution and actively assisted by the Republican government in Paris, revolted against their white masters and set about an orgy of murder, arson, plunder and destruction. The white officials and planters, terrified for their lives and property, begged for British intervention and aid from Jamaica, so that in September a small force sailed from Port Royal and occupied the ports and towns along the west coast, the coast-forts being taken over by the British and garrisoned by a combination of British infantry and French royalist troops, each fort—the most important of which was Mole St. Nicholas the so-called Gibraltar of the Caribbean—having a small squad of coast-artillery from the Jamaica companies.

The situation in San Domingo was very confused. On the one side were the French landowners and planters with the royalist remnants of the French garrison and such negroes as remained faithful to them: on the other were the refractory mulattoes and negroes assisted by Republican commissioners and later Republican troops. The British were trying to help the former, the real strength of the latter lay in the revolted slaves under their two outstanding leaders Rigaud and Toussaint L'Ouverture. The most dangerous enemy to the British troops, however, was yellow fever which killed them off almost as quickly as they arrived in the country and reduced units to mere handfulls in a few months. By the end of 1794, of the four officers and 60 O.R.s of the Royal Artillery who had been despatched to San Domingo, one officer and 40 O.R.s were already dead. Drafts were poured in from Jamaica

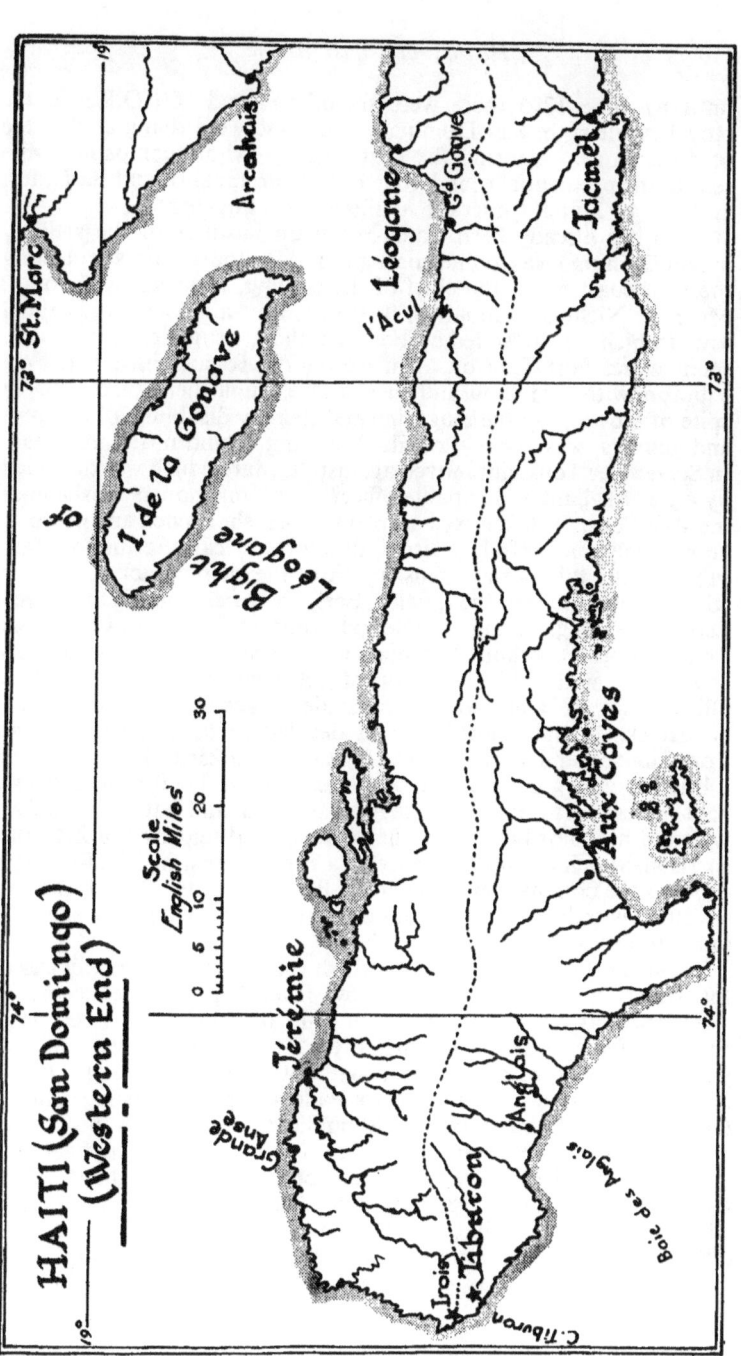

until by June 1795 there were six officers and 200 O.R.s of the
Royal Artillery in San Domingo, and these were dying at the rate
of 40 a month. During the next year six whole companies were
sent to the pestilential island, four R.A. from Jamaica and Barbados,
and two of Dutch Emigrant Artillery from England.

As has already been stated the main business of the artillery
in San Domingo was to find coast gunners for the defences protecting
the west coast ports, the chief of these being, from north to south,
Mole St. Nicholas, Gonaives, St. Marc, Port-au-Prince (the capital
town) which was defended by Fort Bizothon, Leogane, Fort L'Acul,
Jérémie and Fort Tiburon. In April 1794 Rigaud assaulted Fort
Tiburon with two thousand men and a single field piece, but, in
spite of blowing up the magazine and thereby disabling all the guns
and gunners, was severely repulsed, leaving behind nearly 200 dead.
In September Toussaint moved against St. Marc which was defended
by a most gallant and capable officer, Captain Thomas Brisbane of
the 49th Foot. With a mixed garrison of British, French and negroes,
he not only successfully resisted all efforts to capture his fort but,
subsequently taking the offensive, drove the enemy back across the
River Artibunite into the jungle. Early in December Rigaud seized
Leogane by engineering an internal rising and then pressed on to
capture Fort Bizothon, but once again was unable to secure his
prize, the small British garrison of 120 men—out of which three
officers and 22 men became casualties—successfully resisting all
his attacks. Undismayed Rigaud decided to have another go at
Fort Tiburon and arrived before that place on 23rd December with
a brig (16 guns), three armed schooners, about 3,000 men, and five
field-pieces. Setting about a regular siege Rigaud battered the fort
with red-hot shot both from the sea and land and by the 29th the
garrison, which consisted mostly of a negro corps, had had enough
of it, so the commander, Lieutenant Bradford of the 23rd Fusiliers,
stealthily withdrew his men from the fort, broke through the enemy's
lines, and retired on Jérémie.

During 1795 the arrival of British reinforcements greatly eased
the situation, but very soon yellow fever began to reduce their
numbers, and by the autumn "not only was the sea covered with
French privateers, but it was impossible for a man to move a hundred
yards outside the British lines without danger of being shot."
(Fortescue). However, the negro leaders were quarrelling among
themselves, and the arrival of Republican troops from France in
August 1796 made them hesitate as to which enemy to fight, an
uncertaintity which paralysed their operations throughout 1797.
By the beginning of 1798, they had at last decided to co-operate
with General Hédauville, the French commander, sink their own
differences, and make a united effort to drive the British out of the

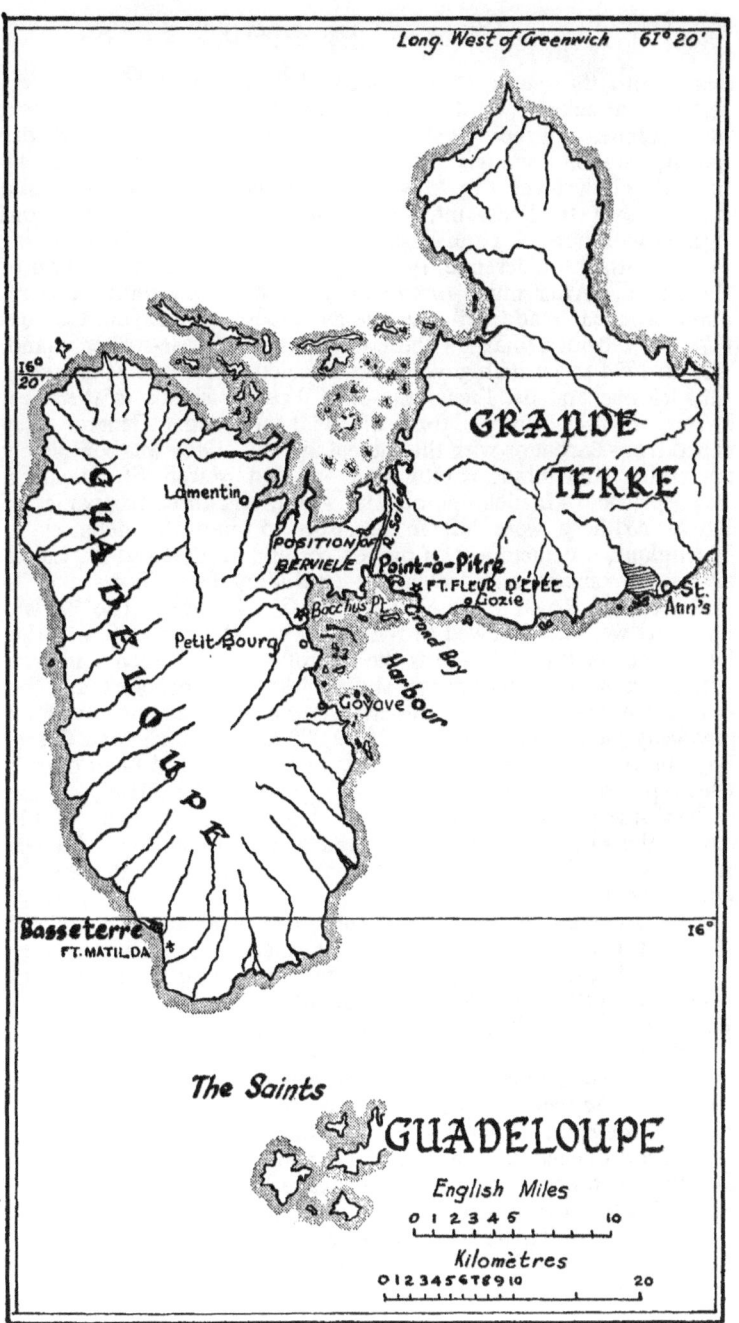

island into the sea so that January, February and March saw a series of attacks on Mole St. Nicholas, St. Marc, Port-au-Prince, and Jeremie. All these posts were successfully held, but the British had by now had enough of San Domingo and its terrible casualty lists of yellow fever. In May Port-au-Prince was abandoned and handed over to Toussaint, and during the summer negotiations opened with both him and Rigaud, who was already busy preparing another attack on Jérémie, to arrange for a peaceful evacuation of the island. After much discussion and hard bargaining, a secret agreement was made with the enemy leaders, and during October (1798) the British finally withdrew from that disease-stirken island.

In 1794 the British government decided to capture Martinique, Guadeloupe and St. Lucia, the chief French islands in the West Indies, so in February a force of 7,000 men under General Grey sailed from Barbados with that object in view. Grey was completely successful, Martinique capitulating on 23rd March, St. Lucia on 2nd April and Guadeloupe on 21st April, and three companies of Royal Artillery were left in garrison to man the defences at Martinique, a detachment of one officer and 53 O.R.s at St. Lucia, and seven officers and 109 O.R.s (including five officers and 70 O.R.s of the Royal Irish Artillery) at Guadeloupe. The British, however, were not allowed to remain in peaceful possession of this last island for long. Towards the end of April a French squadron with a convoy of transports slipped out of Rochefort and on 4th June appeared off the southern shore of Guadeloupe. Without any delay the French rapidly landed 2,000 men at Gozier and Grand Bay and on the 6th stormed the forts of St. Louis and Fleur d'Epée which protected Pointe-à-Pitre, the capital of the island, on the eastern side and which were only held by a weak garrison of 120 men of the 43rd Foot. By the 7th the French were in Pointe-à-Pitre, and the British had retired to Petit Bourg and Berville on the western side of the harbour.

Between 10th June and 22nd July Grey, who was in command of the British troops on Guadeloupe, made several attempts to recapture Pointe-a-Pitre and the forts but, failing to do so with heavy loss, fortified his position at Berville and occupied it with his remaining troops, about 1,800 men under Brigadier Graham.

Yellow fever very quickly took its deadly toll of this force and, when the French advanced against the position both from land and sea on 26th September, Graham could only raise some 550 men to man the defences. Having brought forward their heavy guns, the French opened a bombardment against the entrenched camp, and this was followed by a series of savage assaults which were only repulsed with great difficulty. By 5th October Graham was reduced to desperate straits, he had no more than 150 men left on their legs, his supplies were only sufficient to last another two days, and his ammunition was all but expended, so on the next day he capitulated with the honours of war. Among that small body of troops which surrendered were four officers and 29 O.R.s of the Royal Artillery, all that were left of the coast-artillery defending the approaches to Grandeterre and Pointe-à-Pitre. Basseterre, at the south-western end of the island was now the only place left to the British in Guadeloupe, and Fort Matilda, which protected Basseterre, was garrisoned by about 600 men, including some 20 coast-artillery gunners, under Brigadier Prescott. The French some 2,000 strong rapidly pushed on from Berville by land and sea and by 14th October had invested the fort on the land side. Prescott held out until 10th December, when the fort being greatly battered and having most of its guns dismounted, he embarked the whole garrison and slipped away to Martinique. Thus, after eight months of occupation, Guadeloupe was once again lost to the enemy.

The recapture of Guadeloupe by the French gave them a firm base in the West Indies and enabled them to send their revolutionary agitators into the British islands with the result that throughout 1795-96 Jamaica, Grenada, St. Vincent, Dominica and St. Lucia were engulfed in negro revolts which they only just managed to survive. In fact it needed the arrival of Sir Ralph Abercromby with strong reinforcements from England in the spring of 1796 to put down the risings, pacify the islands, and recapture St. Lucia which had fallen once again into French hands. The Royal Artillery in the West Indies in 1795-96 were distributed among the coast-forts of the islands which very often, due to the negro revolt, were the only places left in the hands of the British. Fort St. George in Grenada, Fort Charlotte in St. Vincent, Fort Cabril at Dominica and Vieuxfort in St. Lucia proved godsent havens of refuge to the British inhabitants and troops, surrounded as they were by bloodthirsty masses of insurgent slaves. At the beginning of 1796, the Royal Artillery garrisons in the Caribbean, exclusive of San Domingo, were as follows:-

Martinique	...	3 companies
Jamaica...	...	1 company
Barbados	...	1 company
St. Kitts...	...	1 company
St. Lucia		
Grenada		
St. Vincent	}	2 companies, plus 3 companies of Royal Irish Artillery, distributed in small detachments among these islands.
Dominica		
Antigua		
Tobago		

J

Their strengths were very weak due to the continuous ravages of yellow fever. Nevertheless, with the arrival of reinforcements under Abercromby, an expedition was sent to capture the Dutch ports of Demerara and Berbice, both of which surrendered without resistance, and small garrisons of Royal Irish Artillery and part of the 39th Foot were left to hold them. Spain having joined France and declared war on Britain in September 1796, Trinidad was captured by an expedition under Abercromby in January of the next year and was subsequently held by a strong force which consisted of a detachment of Royal Artillery—later raised to a company—and three battalions of infantry under that famous Welsh soldier, Thomas Picton, while it was found necessary to provide a party of one officer and 24 men from the Royal Artillery in Jamaica for Belize a British port on the Bay of Honduras. During 1801 the islands of St. Eustatius, Saba, St. Bartholomew, St. Croix and St. Martin were taken from the enemy, and small artillery garrisons—both from the Royal Artillery and Royal Irish Artillery—amounting to some 100 all ranks were left in them.

The Peace of Amiens came in March 1802, and all Britain's captures in the Caribbean, with the exception of Trinidad, were returned to their French, Dutch, and Spanish owners so that when war broke out once more in May 1803 the whole business had to be started over again. During 1803 and 1804 St. Lucia and Tobago were captured from the French, and Demerara, Berbice and Surniam from the Dutch, so that Royal Artillery detachments had to be found to man their defences. By the end of 1804 the Royal Artillery in the West Indies were distributed as follows:-

Jamaica	4 companies
Barbados	2 companies
Antigua	1 company
Dominica	1 company
Grenada	1 company
St. Kitts St. Vincent Trinidad St. Lucia Tobago Demerara Berbice Surniam Honduras	1 company, plus 3 companies Royal Foreign Artillery, distributed among these islands and mainland posts in small detachments.

At the beginning of 1805 Napoleon, in his encampment at Boulogne, set in motion his great naval plan which would enable him to obtain command of the English Channel for sufficient time

1793 to 1815 (continued)

to allow his invasion flotillas to cross safely and disgorge the "Army of England" on the southern shores of Britain. In its bare outline, Napoleon's plan instructed the French fleets at Toulon and Brest and the squadron at Rochfort to elude their British blockading forces, and, picking up the Spanish squadrons at Carthagena and Cadiz on their way, make for the West Indies where, joining together to form one mighty combined fleet, they would turn about, double back across the Atlantic, and, appearing in the Channel in overwhelming force, destroy the British Channel Fleet and any other naval forces Britain might have been able to concentrate there for her defence. The Rochefort squadron under Admiral Missiessy, was the first to escape. It slipped out of port on 11th January, and, evading the British blockading force under Admiral Graves, made for the West Indies. Missiessy's command consisted of 5 line-of-battleships (Flag *Majestueux* 120 guns) three frigates and two corvettes with embarked 3,500 troops under General Lagrange, and his orders were to create a diversion and as much trouble as possible in the West Indies, pending the arrival of Admiral Villeneuve with the Franco-Spanish fleet from Toulon and Cadiz. On 19th February Missiessy arrived safely at Fort Royal, Martinique, and, having landed some much needed munitions and supplies, sailed on the 21st for Dominica which, being nearest, he had decided to make his first objective.

Dominica, capital Roseau, was a typical West Indian island, volcanic in origin, with a pleasing fertile coastal strip, but with a mountainous, thickly wooded, roadless central massif, the whole measuring about 30 miles from north to south and about 15 from east to west. The colony's main defensive post was Fort Cabril—5—32 pdrs; 18—24 pdrs; 2—18 pdrs; 2—12 pdrs.—which stood guard over the fine anchorage of Prince Rupert's Bay at the northern end of the island. Roseau on the west coast, some 20 miles south of Fort Cabril, was protected by Fort Young—4—24 pdrs; 1—12 pdr—and Melville Battery—5—24 pdrs—while at the southern end of the island were two small forts, at Scott's Head—3—18 pdrs; 2—12 pdrs—and at Grand Bay—1—24 pdr; 1—18 pdr; 1—12 pdr. The garrison of the colony, under the command of General Prevost, consisted of:-

Captain Waller's Coy. R.A. stationed at			Fort Cabril
			Roseau
			Scott's Head
			Grand Bay
46th Foot	6 Coys	... ...	Roseau
	1 Coy.	... ...	Scott's Head
	1 Coy.	... ...	Grand Bay
1st West India Regt.	5 Coys...	...	Fort Cabril

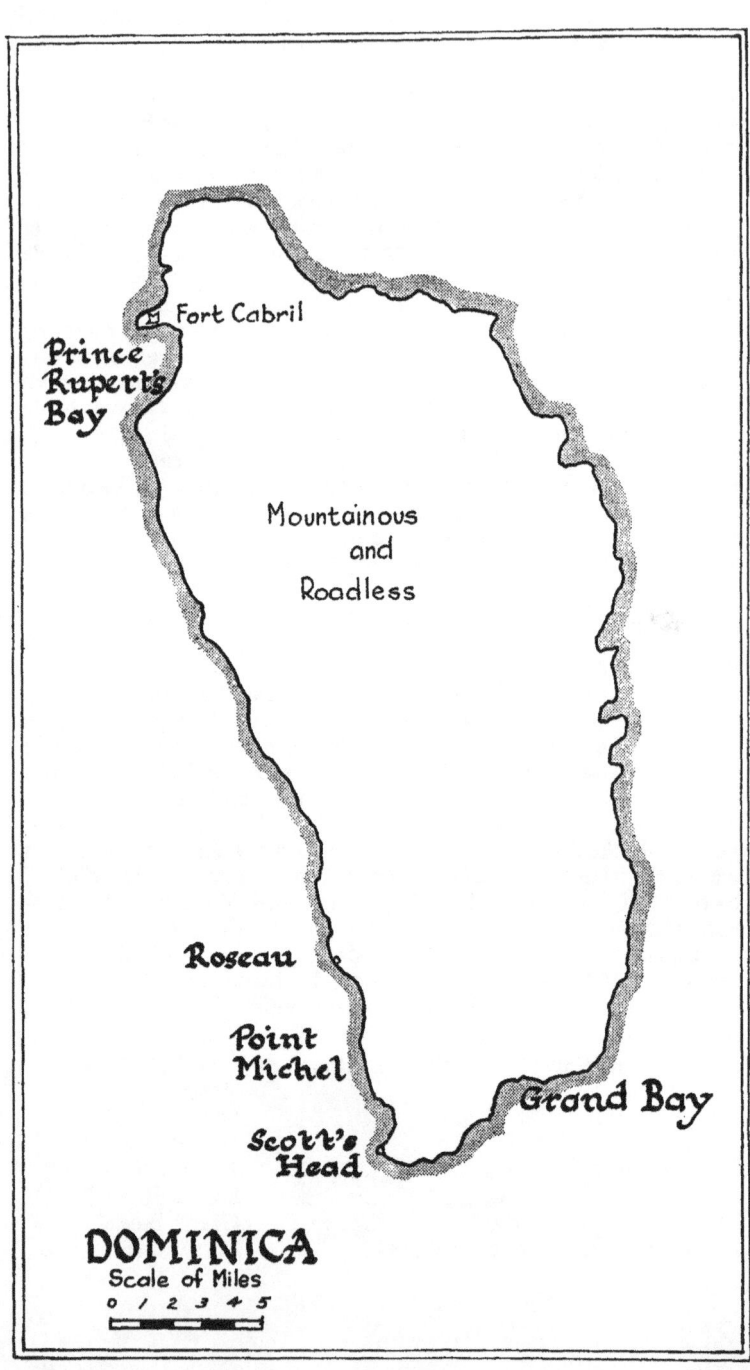

1793 to 1815 (continued)

(Negro Unit)	1 Coy.	... Roseau
York Light Infantry (Foreign Corps)	2 Coys...	... Roseau
St. George's Militia Regt.	...	... Roseau
St. Luke's Independent Militia Coy.		Roseau

in all about 1,200 men. The strength of the Royal Artillery on Dominica, commanded by Captain C. Waller, was one officer and 41 O.R.s who were scattered in small squads among the various forts, the "additional gunners" being provided at Fort Cabril by the West India Regiment, at Scott's Head and Grand Bay by 46th Foot, and at Roseau by the *militia*, one company of St. George's Regiment and St. Luke's Independent Company being specially trained in artillery duties: Melville Battery, on the sea-front at Roseau, was also assisted by merchant-seamen from the ships at anchor in the bay.

The French squadron was sighted approaching the island at dawn on 22nd February and the alarm was quickly given throughout the colony. Missiessy's plan was to raid Roseau and do as much damage as he could to the fortifications and shipping in the roadstead, so during the morning his squadron, anchoring in the bay, opened a heavy fire on Fort Young and Melville Battery. Under cover of this bombardment, Lagrange landed two bodies of troops, each about 900 strong, to the north and south of Roseau. The southern force, which was led by Lagrange himself, was opposed by two companies of the 46th, the Light Company of the West India Regiment, and two companies of the *Militia*, supported by two light field-guns—6 pdrs—brought out from Fort Young, and a severe and bitter struggle took place. Although the British failed to prevent the French from landing, they soon held up their advance and, thanks mainly to the fire of the two 6 pdrs., stopped them from penetrating beyond the approaches to the town. On the north side the French got ashore with little loss, in spite of a gallant resistance by a *militia* company, and quickly stormed the outlying defences, but here again they could not break through into Roseau, Prevost with the remainder of the garrison stubbornly defending the northern outskirts. By one o'clock afternoon the French seem to have shot their bolt, neither of their forces ashore apparently being able to advance, nor had their warships managed to silence the guns of Fort Young and Melville Battery which, although heavily outnumbered, had gallantly kept up the battle throughout the morning. However, about two o'clock, Lagrange threw in his last reserve, a battalion under General Claparède, and this, landing just below the town, rapidly overran Melville Battery thus threatening the rear of Prevost's defending forces, so that the British general had to issue orders at once for the evacuation of Roseau and the rapid withdrawal of his troops—by a cross-country march of some 25 miles—to Fort Cabril. Missiessy, however, had no intention of taking on such a hard nut as the fort, and, having destroyed the fortifications of Roseau, levied a tribute of £7,500, and sent 22 captured merchant men to Martinique, he sailed away on 28th February in search of his colleague Villeneuve and the combined Franco-Spanish fleet. In this little campaign the coast-artillery had three men wounded and lost 22 as prisoners of war.

Having waited in vain in West Indian waters for Villeneuve for a whole month Missiessy sailed for Europe again on 28th March, and it was not until 14th May that the Franco-Spanish fleet appeared at Martinique with 17 sail-of-the-line and nine frigates. The arrival of such a strong enemy force naturally caused great alarm among the British West Indian islands, and General Myers, commanding at Barbados, stated that none of the islands were safe for long with the enemy so strong. Napoleon, in his instructions to Villeneuve, had ordered him to wait 35 days in the West Indies for the Brest and Rochefort squadrons. The Rochefort squadron, as we know, had already arrived and left, Admiral Gauteaume at Brest had been unable to evade the British blockade, and *nemesis* in the shape of Nelson and his fleet was hot on Villeneuve's heels, so on 10th June the French admiral, having accomplished nothing in the West Indies, set sail for Europe much to the relief of the British. Without fear of interference the British were now able to set about mopping up the enemy held islands. The Danish islands of St. Thomas and St. John were taken in 1808, San Domingo and Martinique in 1809, Guadeloupe, St. Martin and St. Eustatius in 1810. Only once during this period did the French attempt any form of counter-stroke. During June 1806 a squadron of six line-of-battleships and two frigates under Admiral Willaumez, which had escaped from Brest, made a raid into the Caribbean and on 3rd July, arriving off St. Kitts, attempted to capture a convoy at anchor in the roadstead but failed to do so, being driven off by the coast-defence guns in position on Brimstone Hill. By 1811 there were no more lands to conquer, and coast-artillery detachments had to be found for the defences of both the British and occupied islands and of the mainland ports. In 1814 there were 13 companies of artillery in the West Indies, 10 of the Royal Artillery and three of the Royal Foreign Artillery, the headquarters of four R.A. and one Foreign being at Jamaica, six R.A. at Barbados, one Foreign at Trinidad, and one Foreign at St. Vincent, but the personnel were scattered in small detachments amongst 20 different garrisons:-

Jamaica	Martinique
Barbados	Guadeloupe
Trindad	St. Martin
St. Vincent	St. Eustatius
Antigua	St. Thomas

Dominica	St. John
Grenada	Honduras
St. Kitts	Demerara
St. Lucia	Berbice
Tobago	Surinam

and the yellow fever never ceased to take its deadly toll.

The Royal Foreign Artillery, which from 1798 until the end of the war, shared the perils of coast-artillery service in the West Indies with the Royal Artillery, had its origin during the last months of 1797 when the War Office decided to disband two foreign infantry regiments, Royal Etranger and Lowenstein's Fusiliers, in British employment in the West Indies. Most of the men of these two regiments were drafted into the 60th Foot, but from them also were formed two independent Companies of Foreign Artillery which continued in the West Indies, split into small detachments among the defences of the various islands, until the Peace of Amiens in March 1802 when they were disbanded. However, when war broke out again in May 1803, a new corps, the Royal Foreign Artillery, of three companies was raised for service in the West Indies. Two companies were formed from the remnants of the two former companies who were already in the Caribbean and of the Dutch Emigrant Artillery at Jamaica, and one new company, consisting mostly of Dutchmen, was sent out from England. The whole corps was disbanded after 1815.

All the West Indian islands maintained corps of *militia* from a very early date. We hear of the Antigua Militia assisting St. Kitts to repel the French in 1690 and the Jamaica Militia gallantly defeating some French detachments landed on the south-east coast of the island in 1693. These corps were raised from the white population only by ballot on the English *militia* system and consisted of artillery companies and infantry battalions—and even sometitmes in the larger islands of light horse squadrons. The artillery companies were specially trained to man the great guns of the coast-defences, the men performing as additional gunners to the small detachments of regular artillery. The West Indian *militia* corps saw considerable active service both in the War of American Independence and the French Revolutionary and Napoleonic War, but they had no very high reputation either for discipline or bravery. Nevertheless, we have seen the Dominica units acquitting themselves very well at Roseau in 1805. As has been stated the *militia* was raised entirely from the white population, although in some colonies during the long war of 1793-1815 mulattos were taken into the artillery companies, no pure-bred negroes, however—being almost entirely slaves—being ever enlisted. After 1815, Britain being paramount on the seas and the West Indian islands thus immune from the threat of attack, the majority of the *militia corps* were allowed to dissolve.

CHAPTER XIII

1793 to 1815. (concluded)

IT is now necessary to return to Europe. It will be remembered how in December 1793 the Allies abandoned Toulon to the French Republican armies and sailed out into the Mediterranean, their ships crowded with soldiers and refugees. The British fleet under Lord Hood, having assembled in Hyères Bay, then set sail for Corsica which island at that time was in insurrection against the French under the leadership of the great patriot Paoli. After operations which lasted from February to August 1794, the French were driven out of Corsica, and a British garrison was left to hold the island as a base for the *Mediterranean* fleet, coast-artillery detachments being stationed in the defences of Calvi, Bastia, and San Fiorenzo, these detachments being found from the same party of Royal Artillery, Captain Stephen's—now Captain Bentham's—Company plus details, which had served at Toulon. This company remained on coast defence duties in Corsica until the evacuation of the island by the British during October 1796 when it moved to Porto Ferrajo, Elba, which had just been seized by Nelson, where it continued until that place too was vacated in April 1797.

Right from the beginning of the war the Royal Navy attempted to keep a close blockade of all the coasts of France. In the early years this blockade was not always very efficient and was withdrawn during the winter months from the French Atlantic seaboard from Cherburg to Bayonne, but by 1805 it had become continuous, proficient, and vigorous and stretched from Hamburg to Leghorn. In 1795, however, it was still in its nonage, and in July of that year, Sir Sidney Smith—later of Acre fame—who was commanding a small squadron blockading the French coast from Havre to Cape Barfleur, came to the conclusion that his operations would be more easily carried out if he occupied the two islands of St. Marcouf which were situated about four miles from the enemy shore. Both the islands were very small, each about an acre in extent, bare and flat and quite uninhabited. The occupation of the two islands being successfully carried out by parties of seamen and Royal Marines from Sidney Smith's squadron, two gun vessels were then dismantled to provide the armament, simple fortifications were constructed, and huts built to house the garrisons.

The Navy soon realised that the defence of these two islands was rather outside their normal duties so it was arranged with the War Office and Board of Ordnance that an officer of the Royal Engineers with some Military Artificers should be sent to design and construct

1793 to 1815 (concluded)

more elaborate fortifications and a garrison of artillery and infantry should be provided. So in September Captain Hockings, R.E. arrived with a dozen artificers, and by January of next year two forts with blockhouses, ditches and stone revetments had been constructed on each island. When these were ready for occupation the military garrison appeared; in February two corporals and 12 gunners Royal Artillery from the Marching Companies stationed at Portsmouth, and in March a company of Invalid Infantry of one officer and 46 O.R.s. The gunners were divided equally between the two islands, but the Invalids remained together on one island as the naval officer in command, Lieut. Bourne R.N., still did not consider the military garrison strong enough to hold the islands and retained both seamen and marines to assist the soldiers.

The Marcouf islands proved of great value to the Navy. They were used as an advanced base by the frigates blockading Havre and the mouth of the Seine, they served as a jumping off place for small ships carrying out raids on the French coast, and were most useful as look-outs for reporting the movements of hostile ships, in fact they became an intolerable nuisance to the enemy, so much so that in the spring of 1798 the French decided to recapture them. The garrisons of the two islands in the spring of 1798 were as follows:-

	West Island	East Island
Royal Navy …	50	65
Royal Marines	125	44
Royal Artillery	10	10
Invalid Infantry	—	42
Totals …	185	161

all under the command of Lieut. Price R.N. Their armament was 2—32 pdrs: 6—24 pdrs: 12—6 pdrs. and 4 pdrs: 2—68 pdr. carronades: 3—24 pdr. carronades: 1—24 pdr. howitzer. During the afternoon of May 6th the look-outs on the islands reported a flotilla of enemy boats emerging from La Hogue so the alarm was immediately sounded, and the garrisons manned the defences. During the night, which was very dark and foggy, the French were able to approach the islands undetected, but at first light next morning the enemy flotilla was seen to be moving towards the south-western end of West Island, having in the centre a squadron of gun vessels, and on each wing a division of flat-boats carrying infantrymen. Under cover of the fire of the gun vessels, the two divisions of flat-boats tried to penetrate between the two islands and land their troops, but the defenders kept up such a hot cannonade that not a French Soldier was able to set foot on shore, several of the flat-boats being sunk and many of the attackers either killed or drowned, the fire of the carronades being particularly devastating. After about two hours the enemy gave up the attempt and pulled away for La Hogue, taking their wounded with them. The British casualties were small, only one killed and three wounded none among the Royal Artillery. Lieutenant Price, in his report on the action, gave high praise to the coast-gunners who, he said, had played a foremost part in repulsing the enemy. From this time onwards the two islands of St. Marcauf were left strictly alone by the French until the peace in March 1802 when they were evacuated by the British garrisons and handed back to their rightful owners.

On 19th May 1798 General Bonaparte, being determined to seize Egypt and destroy Britain's power in the east, set sail from Toulon with a great fleet of 33 ships-of-war and 230 transports crammed with some 36,000 troops. On 9th June this powerful flotilla arrived off Malta, then held by the Order of St. John of Jerusalem, and on 12th June the French occupied the island although the Order was strictly neutral and had taken no part in the war. On 18th June Bonaparte sailed off on his way to Egypt, leaving behind a garrison of about 3,500 troops under General Vaubois. The Maltese, neither liking their new masters nor the way they had seized their island, rose in rebellion against the French and during September set about investing General Vaubois and his garrison in Valetta. For two years the Maltese blockaded—and sometimes even vigorously besieged—the French in their fortress, being assisted from time to time by both the British Navy and army. At last on 4th September 1800, Vaubois, having no supplies left, was forced to surrender, so the French moved out, and the British moved in, and thus began the Birtish history of the famous naval-base and coast-artillery station in the Mediterranean. Six companies of Royal Artillery were promptly brought into to man the guns of the defences, five from home, and one from Gibraltar, and until 1815 there were never less than four companies in the garrison. From 1800 onwards Malta remained in possession of the British, an ownership which was not seriously challenged until the Second World War. In 1802 the "Malta Coast Artillery" was formed from 300 men of the Maltese Light Infantry (*I Cacciatori Maltesi*) an infantry unit which had been raised by the British commander during the blockade of Valetta. This "Malta Coast Artillery" was, through many vicissitudes, the direct ancestor of the Royal Malta Artillery.

Between the evacuation of Elba in April 1797 and the occupation of Malta in September 1800, Britain found herself in need of a safe fleet base in the western Mediterranean, so it was decided to capture and hold Minorca once again. This was carried out successfully in November 1798 by an expedition under General

1793 to 1815 (concluded)

Charles Stuart, two companies of Royal Artillery—one from Lisbon and one from Gibraltar—taking part in the expedition. These two companies stayed on in Minorca to man the coast defences, and their number was raised to three during 1800, this number remaining to form the coast-artillery garrison until the island was returned to the Spaniards at the Peace of Amiens. The British Government also resolved to occupy the island of Madeira—with or without the consent of Portugal—so on 23rd July 1801 a British squadron appeared in Funchal Bay and a landing force, which included one company of Royal Artillery from Portsmouth, took possession of the port and fortifications, the company R.A. continuing there as coast-artillery until the Peace in 1802. Madeira was occupied a second time by the British in December 1807.—Napoleon was now trying to force Portugal into war with Great Britain—and a company R.A. was again included in the garrison which remained in occupation until 1814.

The French invaded the Kingdom of Naples during the closing months of 1805, and a British expedition was hurriedly moved to Sicily in January 1806 to prevent the enemy from gaining possession of that invaluable island. Four companies of Royal Artillery eventually joined the army of occupation and moved into the defences of Milazzo, Messina and Syracuse, the three most important ports facing towards the toe of Italy which was now in the hands of the French. In May the island of Capri, off the Bay of Naples, was captured by a British naval squadron under Sir Sydney Smith, the defences being stromed by a combined party of seamen and Royal Marines, and was later garrisoned by a detachment of Royal Artillery—one Corporal and eight gunners—and five companies of the Royal Corsican Rangers. In June General Sir John Stuart invaded Calabria from Sicily with a British force but, except for the astounding victory Maida over the French general Reynier and the capture of the port of Scylla, effected very little and was back in Sicily with all his troops by the end of July.

Throughout 1807 the British were allowed to remain in possession of Scylla and the island of Capri undisturbed, but in January the following year Joseph Bonaparte, King of Naples, having been prodded into activity by his all-powerful brother, sent Reynier with some 3,000 men to Reggio with orders that the British must be driven out of Scylla back into Sicily. The coast-fort of Scylla was an old mediæval castle which had been adapted and strengthened to resist the bombardment of cannon. It stood upon a peninsular of rock, dominating the Calabrain shore of the Straits of Messina, and was garrisoned by a detachment of Royal Artillery of four N.C.O.s and 26 gunners under Lieut. W. Dunn R.A. and three companies (from 27th, 58th and 62nd Foot) of British infantry, all under the command of Major Robertson of the 35th Foot. Reynier took some time to get his heavy guns—4—24 pdrs: they had to be brought around by boat—into position within breaching distance of the castle's walls and so did not begin his bombardment until 6th February. By the 17th the old mediæval castle was collapsing in ruins about the garrison, most of the guns had been put out of action, and the British force in Messina was showing no signs of attempting relief, so Robertson withdrew his small garrison without the knowledge of the enemy through the postern gate down to the sea-shore where they were all collected by the Royal Navy and transported safely across to Messina.

Later in the year Joachim Murat, who had now replaced Joseph Bonaparte as King of Naples, being affronted by the sight of the British flag flying over Capri, determined to recapture that beautiful island, and therefore despatched General Lamargue from Naples on 3rd October with an expedition of two warships, 37 gunboats, and 3,000 soldiers, embarked in a large number of small craft, with orders to do so as quickly as possible. The British garrison at that date consisted of the small party of Royal Artillery (nine men) five companies of the Royal Corsican Rangers, and a battalion of the Royal Regiment of Malta, in all about 1,400 men under the command of Hudson Lowe, the future gaoler of Napoleon. The French successfully effected a landing below Anacapri during the night of the 4th/5th, and, taking the Maltese by surprise and in the flank, captured some 600 of them, thus reducing the British force by about one half. Lowe was therefore forced to concentrate the remainder in Capri itself where Lamargue proceeded to invest him both by land and sea, bringing round his gunboats and using them to bombard the British defences. Meanwhile, a relief force of two British frigates with transports carrying 600 men had set out from Messina, but, after driving the enemy gunboats back to Naples, was unable to give any further assistance, first on account of a storm which blew the ships away from the island and then of a calm which prevented them from returning. On 15th October, Lowe, having watched his one hope of succour lying becalmed on the horizon, got into communication with Lamargue and on the 16th agreed to capitulate on condition that he and his men were allowed to depart with their colours and arms. To this Lamargue assented, and by the 25th Murat was delighted to see his own royal Neapolitan flag flying over Capri. During the following years British troops, with coast-artillery and infantry garrisons in Milazzo Messina, and Syracuse, remained in Sicily until 1814.

To serve as a base for the ships engaged in blockading the north-western coast of Germany, it was decided early in 1808 to occupy the island of Heligoland with a garrison of artillery and infantry. The artillery were taken from the Invalid Companies R.A. at Woolwich—one officer and 30 O.R.s—and the infantry from one of the battalions of the Royal Veteran Regiment. The garrison was kept in Heligoland until January 1822, the Board of Ordnance only replacing the Invalid gunners when they either died or were unfit for any further duty. In March 1813, the Heligoland garrison was ordered to send a detachment of gunners and infantry to Cuxhaven whence it was thought the enemy had withdrawn, but unfortunately the enemy were still there and, after losing some prisoners, including seven Invalid gunners, the detachment hurriedly returned to Heligoland. The Invalid Companies R.A. were again called on to find an overseas garrison in September 1813. The Danish island of Anholt in the Kuttegat had been seized and occupied by the Royal Navy in May 1809 and garrisoned by Royal Marines, both artillery and infantry. These had gallantly survived a most determined attempt by the Danes to recapture the place in March 1811, and in the summer of 1813 it was decided to replace the Royal Marines by an Army garrison. So in September of that year a detachment of Royal Artillery Invalids—two N.C.O.s and 12 gunners—were sent from Woolwich and, together with three companies of 11th Royal Veteran Battalion, remained as garrison at Anholt until September 1814.

During the long years of war the French had occupied the Dalmation Islands, Corfu, and the Ionian Islands, and the British naval commander-in-chief in the Mediterranean in 1809, Admiral Lord Collingwood, considered it would be of great assistance to his blockade of those parts of Napoleon's empire which bordered on the Adriatic if Britain occupied at least some of those islands. So in September 1809 an expedition was sent from Sicily with the Ionian Islands as its objective, and the islands of Zante, Cephalonia, Ithica, and Cerigo were captured from their Italian garrisons with little difficulty, a company of Royal Artillery, headquarters Zante, being retained to man the coast-defences of the various islands. In January 1812 the British moved further up the Adriatic and seized the island of Lissa, leaving a garrison there of Royal Artillery (one Officer 28 O.R.s) and a battalion of the Sicilian Regiment. Lissa, modern Vis, was the headquarters of Marshal Tito, the Jugoslav partisan leader, for sometime during the Second World War, and the British officer, Brigadier Fitzroy Maclean, who was head of the Allied Military Mission to Tito, was much surprised to find the crown and royal cipher of King George III surmounting the gates of the two old forts commanding the entrance to the harbour. Later he discovered the forts were named *Fort St. George* and *Fort Wellington* and dated from the British occupation of 1812-14. Finally in July 1814 Corfu was occupied, and a company of Royal Artillery stationed there as part of the garrison.

Now to move to the western end of the Mediterranean. Although Britain during this long war with France had been twice at war with Spain, 1796-1802 and 1804-1808, Gibraltar had been left in peace, that is to say, compared with the hectic times of the previous wars against Spain during the eighteenth century. Always with a garrison of from three to five companies of Royal Artillery, the *Rock* had been continually called upon to provide artillery for expeditions such as those to Toulon, Minorca, Malta and Sicily but itself was neither attacked nor besieged. In 1808 Napoleon invaded the Iberian Peninsula, and Spain overnight became the ally of Britain instead of her enemy. The French, having overrun most of Spain, set about besieging Cadiz from the landside in February 1810, but as the Allies had command of the sea, the defenders were never in any great danger. However, the British, realising the vital importance of preventing the French from seizing the fortified bases around the Straits of Gibraltar, sent a force with two companies of Royal Artillery to Cadiz to assist the Spaniards in their defence, and despatched small garrisons from Gibraltar to Tarifa and Ceuta, that for Tarifa including one officer and 66 O.R.s R.A. for the coast-defences and that for Ceuta one Officer and 14 O.R.s.

Towards the end of 1811 the French commander of the seige operations against Cadiz, Marshal Victor, decided to vary the monotony of his activities by moving against Tarifa. For this he collected some 12,000 men—mostly for covering his attack on Tarifa from interference by the Spanish armies operating in southern Andalusia—and a siege train from the guns already in action against Cadiz, and, after some delay caused by terrible weather and the abominable state of Spanish roads, arrived before Tarifa on 22nd December. This coast-fortress was held by an Anglo-Spanish force of some 3,000 men, of which the British portion was the detachment R.A., five companies of infantry from various regiments, and a troop of 2nd Hussars, King's German Legion, while there was a small British naval squadron at anchor in the harbour. The commander of this force and governor of the fortress was the Spanish General Copons, but the moving spirit of the garrison was Major King of the 82nd Foot. The fortress itelf was an ancient stronghold consisting of a castle with outworks and towers connected to a fortified island in the harbour by a causeway some 500 yards long. The southern and western sides, being along the shore, and the island were well protected by the warships so the French had to concentrate their efforts against the northern and eastern fronts where there was a dry torrent bed which ran right up to the ramparts and then passed into the town. The French began their trenches opposite the eastern front on 23rd December and were able to open their bombardment with a battery of 10 heavy guns against the *Retiro*, the tower which barred the entrance made by the torrent bed, on the 8th, and by the evening of that day had made a practicable breach. During the next two days the breach was widened to 60 feet, and at 8 a.m. on the 31st the French launched their assault. It was carried out in the pouring rain, which somewhat damped the ardour of both sides, and mud and water at the bottom of the torrent bed made it difficult for the enemy to push forward. After a desperate struggle around the breach and portcullis of the tower, the French were forced to withdraw and for the time being give up the attack. The rain continued to descend in an absolute deluge throughout the next week, inundating the surrounding country, washing away the French earthworks, filling their trenches, and flooding their camp, until finally on 6th January they abandoned the whole enterprise, evacuated what was left of their siege works, and retired from in front of Tarifa, dragging their guns and their weary way through the mud back to Cadiz.

The French blockade of Cadiz lasted from February 1810 to August 1812 when Wellington's victory of Salamanca forced Marshal Soult to issue orders for the operations to cease and the troops engaged in them to retire northwards. Throughout those two and a half years the French were never able to set about a proper siege as the British command of the sea restricted their efforts to the landward side. There was always a chance however that the French might batter their way through the land defences, but the Spaniards, who formed the bulk of the defending force, were excellent soldiers behind cover and extremely difficult to dislodge from their fortifications. Moreover Cadiz was at the very extreme limit of the French communications, and throughout the blockade they found it almost impossible to sustain their force with drafts, munitions, and supplies. The first British assistance arrived as soon as the French, in February 1810, and consisted of two companies of Royal Artillery and three battalions of infantry: by the following February the British force had risen to 5,000 strong with five companies of Royal Artillery, all under General Sir Thomas Graham. It was this small army which, by a timely encircling movement, defeated Victor with his blockading force at the battle of Barrosa on 5th March 1810. However, after this set-back Victor did not abandon his exertions against Cadiz but, as soon as his troops had somewhat recovered, once more imposed a strict blockade on the landward side. The operations meandered on through 1811, Victor leaving for the Russian campaign and being replaced by General Vilatte, but the French were much too busily engaged with Wellington to be able to spare much effort for Cadiz. The great Marshal Soult himself, commander of all the French troops in southern Spain, came to have a look at the operations in March 1812 but, hearing of the Duke's approach to Badajoz, had to hurry away without being able to make any satisfactory arrangements for pressing on with the siege. This was the beginning of the end. The French before Cadiz were now on the defensive their business being rather to prevent the Anglo-Spanish force from breaking out into Andalusia than to try to batter their way into the city. In August Wellington's great victory at Salamanca forced Soult to withdraw from Andalusia altogether and give up the blockade of Cadiz so after two and a half years the Spanish sentries saw the French disappear away to the north for the last time. The five companies of Royal Artillery remained with the Spanish defending force right up to the end, their main duty being the manning of the defences of the Isla de Leon, the great island that covered the southern flank of the approaches to Cadiz.

By 1814 Napoleon's great empire had collapsed in ruins about him, and he was struggling desperately in France to save his capital and his throne. In Italy the French dominion was tottering, and Lord William Bentinck, who was commanding the British troops in Sicily, took advantage of the situation to land in March a small force at Leghorn with a view to advancing into Piedmont and cutting the lines of communication of the French corps in Lombardy. By 8th April Bentinck and his force was before Genoa, the great seaport and naval base of north-west Italy, and on the 21st the French commander surrendered the city to the British. Bentinck's force was then dispersed, but two companies of Royal Artillery were left to man the defences of Genoa, and there they remained until well into 1815.

During this long war of 22 years, Great Britain had been busy appropriating the oversea possessions of France, Spain and Holland. Among these were three which in after years were to become famous coast-artillery stations, the Cape of Good Hope, Ceylon, and Mauritius. The Cape of Good Hope was twice taken from the Dutch—being returned at the Peace of Amiens—in September 1795 and January 1806, and three companies of Royal Artillery were quartered there for the defence of Cape Town and Simonstown. Ceylon was captured—also from the Dutch—during 1795-96 but was occupied at first by troops of the East India Company. It was not until 1803 that it was decided that Ceylon should become a responsibility of the home government, and three companies of Royal Artillery were sent out to furnish the coast-artillery garrisons, one company being stationed each at Colombo, Galle, and Trincomalee. Mauritius was taken from the French by an expedition from India at the end of 1810, but a company of Royal Artillery was sent from Simonstown to take over the defences of Port Louis and remained there when the Indian troops returned to India. In 1815 a detachment of three Officers and 51 O.R.s Royal Artillery was sent to St. Helena from Portsmouth to reinforce the East India Company's St. Helena Artillery whilst Napoleon was a prisoner on the island. Finally, to demonstrate what can happen to

1793 to 1815 (concluded)

a busy trading port which foolishly was not protected by coast-defence, let us look at Sierra Leone in September 1794, when Britain without difficulty was maintaining everywhere her superiority at sea. A small French squadron under Captain Allemande was cruising with predatory intentions off the west coast of Africa, and on 28th September, entering Sierra Leone harbour which was entirely unprotected by a single gun, bombarded the town for nearly two hours, destroying the church, public buildings, go-downs and warehouses. The French then landed a party which plundered the stores and European residences and, before leaving, set fire to such structures as had survived the bombardment. Having done their worst for some eight hours, the enemy squadron then sailed away not having suffered one single casualty nor any damage to boat or ship.

The United States of America declared war on Great Britain on 17th July 1812. There were at that time four companies of Royal Artillery stationed in Canada with their headquarters at Quebec. Besides providing the garrison artillery for Quebec itself, these companies had also to find the men for mobile artillery and for the static guns of the many fortified posts which studded the frontier from the Richelieu River to Lake Superior. The frontier between Upper Canada—that is to say all Canada west of the River Ottawa—and the United States consisted almost entirely of the three great inland seas, Lakes Ontario, Erie, and Huron, and therefore the fortified posts, which were on the shores of these lakes, should be graded as coast-forts. The successful defence indeed of the frontier of Upper Canada depended almost entirely upon obtaining naval superiority on the *Great Lakes*, and the coast-forts on their shores served as protected bases for the war-vessels operating upon them. On Lake Ontario the British had a squadron of six small ships and on Lake Erie a similar squadron, all brigs and schooners with from two to 22 guns. To combat these the Americans could only produce the *Oneida* of 16 guns on Lake Ontario and the *Detroit*, 6 guns, on Lake Erie. There were no ships of war on Lake Huron. The British coast-forts were Kingston and Fort George at the east and west ends of Lake Ontario, Fort Erie and Amherstberg at the two ends of Lake Erie, and the blockhouse on St. Joseph Island at the north-western head of Lake Huron. All these fortified posts were garrisoned by artillery and infantry. To counter these the Americans had on Lake Ontario Sacketts Harbour, their main naval base, and Fort Niagara, and on Lake Erie Fort Buffalo, and Fort Detroit.

Operations began in August with the British capturing Fort Detroit and the brig of the same name which success seemed to set the seal on British naval superiority, but the American government in September sent Captain Isaac Chauncey, United States Navy, to Sacketts Harbour to command all their naval forces on the *Great*

Lakes with orders to destroy the British supremacy. Chauncey set to work with tremendous energy, seizing, requisitioning, building, and arming new ships. His first success was on Lake Erie where on 9th October he captured the British brig *Caledonia* and sunk the *Detroit* by a surprise attack right under the noses of the guns of Fort Erie. This gave the Americans the whip hand on Lake Erie. Later in the month the main body of American troops, based on Fort Niagara, tried to capture Fort George by crossing the Niagara river but were foiled at the battle of Queenstown. In April of the next year (1813) Chauncey led an amphibious expedition against Toronto which was the main British dockyard on Lake Ontario and where there were two ships under repair and two new ones under construction. Toronto had no coast-defences and only a small mixed garrison of British regulars, Canadian *fencibles*, and local *militia*, in all all about 600 strong. These had no chance against Chauncey's 1,800 seamen and soldiers and were forced to abandon the town after a short but sharp engagement. The Americans then set about pillaging the town, burning the ships on the stocks, and destroying the dockyard, towing away one armed brig which they found in the harbour.

After this success Chauncey decided to look for bigger game. He now had on Lake Ontario two brigs of 24 guns each, the *Oneida* of 16 guns, and several armed schooners, and the British as yet had nothing which could cope with him, their hopes for the spring of 1813 having gone up in smoke and flames at Toronto. Chauncey therefore determined to capture Fort George, the key to the passage between Lakes Ontario and Erie. Fort George at this time was garrisoned by a detachment R.A. from Captain Holcroft's Company (with men of the 41st Foot as "additional gunners") five companies of the 8th Foot, a company each from the Glengarry and Newfoundland Fencible Regiments, and about 300 Canadian *militia*, in all about 1,400 troops under command of General Vincent, and was armed with five heavy guns and several smaller ones. Covered by the fire of his warships and Fort Niagara, on 27th May Chauncey landed some 7,000 men on the low ground west of the mouth of the Niagara river. Vincent did not want to be attacked in his fort, so issuing out with almost all his force, attempted to drive the Americans back into the lake before they could get a firm foothold on land. Unfortunately he was hopelessly outnumbered and, after suffering heavy losses, was forced to retreat away westwards, sending off orders for both Forts George and Erie to be evacuated and abandoned at once. Chauncey now had control of both lakes and the vital passage between them.

However, Chauncey did not long continue to have matters all his own way. A brilliant young officer of the Royal Navy, Captain Sir James Yeo, had arrived to take command on Lake Ontario and

1793 to 1815 (concluded)

by the beginning of June had a flotilla of 13 war-vessels, including six gunboats, in all mounting 106 guns, upon the waters of the lake. During June Vincent and Yeo in co-operation drove back the American army which had attempted to advance westwards from the Niagara river, forcing them to abandon Fort Erie and concentrate their forces about Forts George and Niagara. Chauncey did not dare challenge Yeo, and by the end of the summer the British had regained command of Lake Ontario. Later in the year however they failed badly on Lake Erie, the British squadron on that lake being decisively beaten on 10th September by the Americans under Lieut. Perry U.S.N. In December, in the depth of winter, the British, now under General Gordon Drummond, made a surprise advance, recaptured Fort George, then, crossing the Niagara river, took in rapid succession Forts Niagara and Buffalo—which they utterly destroyed—burnt all the frontier villages, and, loaded with captured arms, ammunition, stores, and clothing, returned to winter quarters on the Niagara river before New Year's Day. The British were once again in control of the passage from Lake Ontario to Lake Erie. There is no need to continue further with this American War. During 1814, Napoleon having been defeated and exiled to Elba, the British were able to send considerable reinforcements to Canada and so take the offensive on a grand scale, the Americans everywhere being forced back on the defensive, with the result that the small coast-forts on the shores of the *Great Lakes* were left in peace, a peace in which they have continued to bask ever since.

With Napoleon's defeat at Waterloo on 18th June 1815 and his subsequent confinement on St. Helena these long wars at last came to an end. Coast-artillery had been kept busily employed during these 22 years. At the beginning the coast defences—few, neglected, and far between—of Great Britain and the Channel Islands had been manned by the Invalid Companies, in Ireland by small detachments from the Royal Irish Artillery, and abroad by 19 "Marching Companies" of the Royal Artillery. By the end of the war Great Britain and the Channel Islands were surrounded by a ring of coast forts and batteries, the small parties of Invalid gunners had been swamped by thousands of enthusiastic volunteers who, starting as "additional gunners" had by 1805 attained the status of true Gunner Volunteers, and no less than 10 "Marching Companies" were serving in the coast-defences at home. In Ireland the Royal Irish Artillery had disappeared, and four "Marching Companies" of the Royal Artillery were providing the personnel for the Irish coast-defences. Abroad coast-artillery was serving almost all over the world, from Anhalt to the Cape, and from Lake Erie to Trincomalee, and Malta, Capetown, Mauritius, and Ceylon had been added to the list of permanent coast-artillery stations. In 1813, no less than 42 Companies, Royal Artillery were serving as coast-artillery overseas.

The year 1815 did not only see the finish of the French Revolutionary and Napoleonic struggle but also the end of a century and a quarter of almost continuous war against France. During that period there had been little change in the methods and weapons of coast-artillery. This of course was because the main target itself had scarcely altered: for all those years it was always the wind-propelled great wooden-wall with its tiers of muzzle-loading, smooth-bore guns firing solid iron shot. However, it must be remembered that the British coast artillerymen, who during the eighteenth century were called upon to defend their coast forts and fortresses against enemy assault, were rarely faced by direct attack from the sea. The enemy well knew that their war-ships were no match for coast-batteries and invariably made their attempt to capture a coastal stronghold from the landward side. It was not warships and floating batteries with which the coast-gunners had to deal—except in the final phase of the great siege of Gibraltar, and it has been seen what a disaster for the French and Spaniards that was—but siege-batteries and infantry assault columns. Minorca, St. John's, Pensacola, Toulon, and many a Carribean coast fort were all captured from the landside. The coast-artillery had to be certain that the naval bases and ports which they had to protect were as well armed and strongly fortified on the land as on the seaside, for it was nearly always on the landside that they had to fight their battle. During the next century and a quarter this was forgotten. The stories of the great land assaults on Gibraltar and Minorca were allowed to sink into oblivion and their lessons to be lost and forgotten with inevitable disastrous results. However in 1815 there were no thoughts of disaster, coast-artillery, the fighting services, the country, the whole world, all were standing on the threshold of a new age. The "Industrial Revolution" was already well on its way.

CHAPTER XIV

1815 to 1856.

FOR a century and a quarter the great guns of the British coast-artillery had scarcely ceased to thunder. For the next 100 years, such was Britain's supremacy at sea, they did not have to fire one shot in anger. Nevertheless, while the coast-gunner of 1700 would have found his counterpart of 1815 firing the same guns from the same fortifications as he himself had, it is doubtful whether the defender of Britain's shores in the early years of the nineteenth century would even have recognised the weapons and defensive works of 1914 as belonging to his particular branch of the artillery service. This century from 1815 to 1914 saw a complete change in the weapons, materials, methods, and procedures of the fighting services, brought about by the development of the industrial revolution and the advance of scientific knowledge. The fighting services, however, were very conservative, and it took much longer for the new ideas and inventions to penetrate their minds and organisations than those of the commercial and industrial world. It took indeed the Crimean War (1854-56) to open the eyes of both sailors and soldiers to the fact that they were living in a new age and that the ships, weapons and methods of Trafalgar and Waterloo were no longer sufficient. It was during the period 1856 to 1914 that the great changes took place, and that coast-artillery was to alter out of almost all recognition.

Meanwhile, immediately after 1815, as after all wars, economy and reduction of the forces was the popular demand and government policy. Whereas there were 112 companies of Royal Artillery—including Invalids—in 1815, these had been reduced to 72 by 1820. Before proceeding any farther, a word must be said about the general organisation of the Royal Artillery at this period. In 1820 the 72 companies were organised in nine battalions, each of eight companies, and at long last in 1825 the companies were numbered in each battalion from 1 to 8. All the companies were really "garrison companies," there being no such units as permanently organised field-batteries. A company, if in garrison in a coast-fort or fortress, trained on both coast-defence guns and movable armament, if stationed inland, on movable armament only. Horses, harness, equipment, guns and vehicles for field-batteries were kept at Woolwich, Manchester, Weedon, Liverpool, Leith and certain stations in Ireland. To these stations each company would go in rotation for training as a field-battery, and then, having completed its training, return to its normal station and duties. In 1847-48 three more battalions were added to the Regiment, making a total of 96 companies. The following tables, taking four dates between 1815 and 1854, show how many companies were employed on coast-defence, both at home and abroad, during that period and where they were stationed.

Station	Number of Companies			
	1820	1830	1840	1850
Portsmouth and Gosport	1	1	1	3
Plymouth and Devonport	1	1	1	4
Dover	1	1	1	2
Landguard Fort				1
Hull				1
Liverpool				1
Leith Fort	2	2	1	2
Dublin Bay				1
Duncannon Fort (Waterford)			1	
Spike Island (Cork Harbour)				1
Jersey	1	1	1	1
Guernsey	1	1		1
Gibraltar	5	5	5	6
Malta	2	2	2	3
Corfu and Ionian Islands	3	2	3	3
Quebec	2	2	1	3
Kingston (Ontario)	1	2	2	1
Toronto			1	
Halifax (N.S.)	2	2	2	2
St. John (N.B.)	1	1	1	1
St. John's (Newfoundland)	1	1	1	1
Bermuda		1	1	2
Jamaica	3	3	3	2
Barbados	4	4	4	4
Bahamas				1
St. Helena	1		1	1
Cape of Good Hope	1	1	1	1
Mauritius	1	1	1	2
Colombo	1	1	1	1
Trincomalee	1	1	1	1
Hong Kong				1
TOTAL	36	36	37	54

The companies with headquarters at Barbados provided the detachments for the other West Indian islands.

In November 1816 a detachment of five Officers and 38 O.R.s was sent from the Cape to the remote island of Tristan da Cunha to prevent it being used as a base from which the escape of Napoleon from St. Helena might be organised, but the detachment was withdrawn in May of the following year. Up to 1833 detachments for the Irish coast-defences were found mostly from the companies stationed at Island Bridge, Dublin, but from that year onwards whole companies were increasingly moved into such stations as Duncannon Fort, Spike Island, Pigeon House Fort (Dublin Bay) etc. Aden was occupied in 1839 and Hong Kong in 1841, the coast-artillery for the defence of the former being provided by the Bombay Army of the East India Company, and for the latter by No. 1 Company 9th Battalion, Royal Artillery which was sent out from Woolwich in December 1841. In February 1845 the Board of Ordnance authorised the company at Hong Kong to raise a party of "Gun Lascars" from Asiatics, strength:-

1. *Jemadar*
2. *Havildars*
4 *Naiks*
81 *Lascars*

to be attached to the company and form part of its establishment. The authority was also extended to the companies in Ceylon. Thus we see the first beginnings of the famous Hong Kong-Singapore and Ceylon-Mauritius Battalions, Royal Artillery. In Malta three companies of the Royal Malta Fencibles were earmarked as "Artillery Companies" and helped to man the coast-defences. Their officers were all Maltese whose commissions were granted by the Queen.

The Invalid Battalion, Royal Artillery, was disbanded in 1819 but that was not the end of the Invalids. What was called the "Invalid Detachment" was retained with headquarters at Woolwich where resided its two officers, the Adjutant and the Quartermaster The Invalids themselves were as usual scattered among the coast-defences carrying out the duties of "district gunners", but as at first their strength was very low, only 107 in 1824, there were not enough of them to do this at all stations. As the century progressed, the strength of the Invalid Detachment increased until by 1859 it reached 450. There were then sufficient to provide "district establishments" for all the coast-defences in the British Isles, so that the system by which the Master Gunner with his squad of "district gunners" maintained, took care of, and accounted for the coast-guns, their ammunition, and stores was retained as it always had been. This system, which was as old as coast-artillery itself, lasted until the final disbandment in 1956, the Royal Artillery

Companies in garrison in the coast-defences never being responsible for the upkeep of the armament and its associated stores. Given below is the list of master-gunners serving in coast-defences of the United Kingdom in the year 1824, with beside each name the number of Invalid gunners the master-gunner had in his squad to assist him in his duties. Where no Invalids are shown, the master-gunner had either to borrow from the nearest R.A. Company in garrison or employ civilians.

Station	Master-Gunner	No. of Invalids.
Arundel Haven	A. Hoyland	
Berwick	Robert Dixon	
Blockhouse Fort (Gosport)	William Sharp	
Blatchington	Robert Kinniburgh	
Brighton	George Elliott	
Blackness Castle	John Hoyland	
Fort Cumberland (Ptsmth.)	John Keith	1
Cowes Castle	William Lockhart	
Calshot Castle	John Harcourt	1
Chatham	William Howie	5
Dartmouth Castle	George Aynge	
Dumbarton Castle	Romeo Drysdale	8
Eastney Fort (Portsmouth)	James Robertson	
Folkestone	No Master-Gunner	2
Castle Cornet (Guernsey)	Michael Knowles	2
Fort George (Guernsey)		2
Alderney	George Anderson	2
Sark		2
Gillingham Fort	No Master-Gunner	
Gosport	Alexander Matheson	
Gravesend	W. Scott	3
Greenock	No Master-Gunner	1
Holy Island	Charles Whyte	
Hurst Castle	Arthur Watson	
Hull	John Saunderson	3
Hastings	Thomas Ross	
Hythe	No Master-Gunner	
Elizabeth Castle (Jersey)	Alexander Fraser	2
Fort Regent (Jersey)	Kenneth Fowler	
Landguard Fort	James Barron	2
Lumps Fort (Portsmouth)	No Master-Gunner	
Fort Monkton (Gosport)	George Story	
Milford Haven	No Master-Gunner	
Newhaven	William Eddiston	
Pendennis Castle	James Russell	2

Portland Castle	John Stewart ...	
Portsmouth	John Webster ...	3
Plymouth Citadel ...	William Edmonds ...	5
Drake's Island (Plymouth)	George Mahon ...	
Devonport Defences ...	William Black ...	
Rye	Peter Bellow ...	
Scilly Islands	Jonathan Tewson ...	4
Sheerness	W. Buckingham ...	3
St. Mawes (Falmouth) ...	Charles Tate ...	
Southsea Castle	James Twales ...	
Sandown Fort	Robert Sutherland	
Seaford	M. Pinnock ...	
Scarborough Castle ...	Robert Purcell ...	3
Shetland	No Master-Gunner	2
Swanage	No Master-Gunner	2
Tilbury Fort	James Gordon ...	
Tynemouth	John Simple	1
Upnor Castle (Chatham)	John Murray ...	
North Yarmouth	Alexander James ...	1
Yarmouth (I.O.W.) ...	Thomas Lothian ...	
Dover	⎫	
Deal	⎬ Cinque Ports Det.	
Sandgate		
Walmer	⎭	
Duncannon Fort	Thomas Brennan ...	6
Cork	William Eldridge ...	
Bere Island	John Lillie	1
	John Earle	
Dunree	William McKenzie ...	
Charles Fort (Kinsale) ...	James Johnston ...	
Dublin Bay	Robert Leatham ...	8
	John Patterson ...	
Galway	James Moffat ...	4
Carrickfergus	Robert Garrick ...	2
Carrick Island	Samuel Robertson	
Tarbert Island	William Somerville	
Scattery Island	Andrew Daniel ...	
Whiddy Island	Ezekiel Wisdom ...	
Drogheda	No Master-Gunner	1
Lough Swilly	No Master-Gunner	3
Lough Foyle	No Master-Gunner	3
TOTALS ...	59	90

This list of master-gunners is taken from a return submitted to the Master General of the Ordnance by the Lieutenant-General (Lieut.-General Robert Douglas) in February 1824. At the bottom of the list Douglas added this note:- "The general duties of master-gunners are to see that the Ordnance, Carriages, Ammunition, and Stores are preserved in good order and report defects; to keep the batteries clean, to fire salutes where ordered, and hoist the flag; also to render quarterly and annual accounts of the Ordnance and Stores in their charge to the Principal Storekeeper's Office at the Tower. All the master-gunners detailed in this Return had served sixteen years or upwards in the Royal Artillery previous to being appointed master-gunner."

The office of master-gunner was a very ancient one. As soon as there were guns, there was a master-gunner in charge of them, responsible for their upkeep and efficiency, and in command of the gunners who served them. There was a master-gunner in charge of the guns at the siege of Calais in 1386, and four at the siege of Harfleur in 1415. Wherever there were guns there was a master-gunner, on board ship, in the field, or in a fort. The master-gunners on board ship gradually lost their pre-name and became "gunners" plain and simple. The master-gunners in the field disappeared with the arrival of commissioned "artillerists" with the "Trains of Artillery", but the master-gunner in the fort or fortress remained right down to 1956 when coast-artillery in the British Army came to an end. From the very earliest days in the coast-fort or fortress he was not only answerable for the care and maintenance of the ordnance and stores but also in executive command of the guns and gunners in action. It was only with the appearance of the commissioned officer amongst the ranks of coast-artillery that he gave this up and became solely the man who was responsible that the guns and ammunition, with all their associate stores, were ready for action in all respects when called upon and that all equipment, materials, and supplies were properly and accurately accounted for. As has been seen, the master-gunner with his squad of district-gunners was the central core of the coast-artillery from the days of Henry VIII onwards, in fact for most of the time they were the only coast-artillery at home, and it was entirely due to them that when the "additional gunners" were collected and mustered on the gun platforms, the guns were in proper order and the ammunition prepared. The master-gunner always had to struggle with the extreme parsimony of the Board of Ordnance, but he usually managed to produce the essentials when it came to action. Coast-Artillery owed a great debt of gratitude to its master-gunners.

Before the mechanical age—and indeed long after its inception —the business of moving the great guns, mounting and dismounting them from their carriages, shifting them from place to place, was a tremendous and complicated enterprise which was carried out by means of highly specialised gear. This gear consisted of:-

handspikes
levers
pulleys
jacks
rollers
sledges
sheers
capstans
derricks
gyns, etc. etc.

and were none too easy to handle. Expert knowledge was required before the master-gunner and his men could erect and use this apparatus properly and efficiently, and the place where the necessary instruction was imparted—the shrine of this art—was the Royal Military Repository at Woolwich, indeed this particular form of operation was known in coast-artillery popularly as "Scotch Up" and officially as "Repository Exercises" right down to the outbreak of the Second World War in 1939. The Royal Military Repository was founded at Woolwich in 1778 by Sir William Congreve and may rightly be considered the first School of Artillery. Here were taught siege operations, fortification, field-works, bridging, and above all the mounting, dismounting, and shifting of heavy ordnance, while Woolwich Common was used as an artillery range, all kinds of guns, howitzers, and mortars being fired across its open ground. As a company of Royal Artillery might be employed as either siege or coast artillery at any time, companies at home were sent in turn to the Repository to be put through a set course under the supervision of the Superintendent. "Repository Exercises" or "Scotch Up" always played a prominent part in both the training, and work of the coast-gunner. During the centuries of wooden-carriages, pieces were continually having to be dismounted so that the carriages might be changed and repaired. In fact in time of peace it was normal to keep a high proportion of the pieces dismounted so that the carriages could be stored under cover, and the wood protected from the weather. It was therefore obviously necessary for all coast-gunners to have a thorough knowledge of "Repository Exercises" so as to be able to remount the guns rapidly in the event of urgent need and to replace heavy equipment in case of casualties in action. The Repository continued to function as a school of instruction until 1900, but Woolwich Common ceased to be used as a range when the School of Gunnery opened at Shoeburyness in April 1859.

It was during the first half of the nineteenth century that the steam engine made its first appearance among the warships of the world. The British Admiralty was slow to adopt it, and it was not until June 1833 that the 10 gun brig *Falcon*, fitted with a high-pressure steam engine, underwent her first trials at Sheerness. Foreign powers were not so belated and, hoping to reduce Britain's great naval superiority, soon began turning out steam frigates and corvettes. The various navies started by fitting their new steam warships with paddles, but after 1845, it having been proved that the screw was definitely superior, paddles were abandoned. At first ships already in service, after being lengthened and otherwise altered, were fitted with steam engines and screws, so that by the time of the Crimean War (1854) the main battle fleets of the European powers still consisted of the great sailing wooden walls, a large number now being fitted with auxiliary engines and screws, supported by specially constructed steam frigates, corvettes and gunboats. All warships continued to be built of wood, but the terrible damage done to the allied fleets by the Russian coast-batteries at Sevastopol firing shell was forcing naval constructors to protect the wooden hulls with iron armour. The complete irocnlad lay only just in the future.

The appearance of the steam driven warship did not at first greatly affect coast-artillery gunnery. The speed of the mighty wooden-wall, even when fitted with an auxiliary steam-engine and manœuvering in calm waters, did not exceed 10 knots and that of the specially constructed steamer rarely 12. The smooth-bore, muzzle-loading gun was still the chief weapon, but the total destruction of the Turkish squadron by the Russians off Sinope, and, as has been stated above, the alarming damage caused to the British and French warships by the Russian coast-batteries at Sevastopol, proved that hollow shell filled with explosive was greatly superior to solid shot as anti-ship projectiles. This had already been suspected, and both warships and coast-batteries were being equipped with guns specially designed to fire shell. The most important of these were the 10 inch 84 cwt. and the 8 inch 65 cwt. It is to be noted that, as shell were light in comparison to their size, these new guns were designated by their calibre plus weight of piece and not by weight of shot fired by them. Both these guns were considered excellent for coast-defence purposes, and their projectiles had a most destructive effect upon any vessel. New and heavier types of normal smooth-bore guns were also constructed, 68 pdrs., and 56 pdrs., but these were apparently very inaccurate at long ranges. The extreme ranges of these weapons were:-

    10 inch 84 cwt.   3,170 yards
     8 inch 65 cwt.   2,900 yards
    68 pdr.   3,675 yards
    56 pdr.   3,550 yards.

As coast-defence batteries were expected to fire both rapidly and accurately, it was always necessary for their guns to be fitted with the best and quickest means available for giving line and elevation. During the first half of the nineteenth century these were the sights with which the Navy equipped their guns and consisted of dispart sights with brass tangent scales which without doubt constituted at that time the best means yet devised for laying quickly and correctly. Their use was gradually extended to all coast-defence guns which were thus able to deal without much difficulty with the early steam-driven warships. Carriages were still mostly made of wood although wrought iron was beginning to be used for the heavier pieces. The objections to wrought iron were its great weight, its tendency to splinter when struck by a shot, and its susceptibility to fracture and damage which would be difficult to repair. Coast-defence guns were normally mounted on sliding carriages with traversing platforms which enabled them to be traversed and run up with ease and speed. As has been pointed out above, shell packed with explosive was rapidly displacing solid shot as the anti-ship weapon; it was very much more effective. It was fitted with either a metallic or wooden fuze, the most famous of the latter being the "Boxer Fuze" which was produced by Captain Boxer, R.A. in 1849. The charge was still the bag of gunpowder, but it was now fired by the percussion tube, produced by Woolwich Arsenal in 1845, or the friction tube introduced in 1853.

The method of fortification was at last beginning to change, and after more than a century the Vauban system was slowly going out of fashion. This was mainly due to the great development of mortars and howitzers in siege warfare during the wars of 1792-1815 with the result that the enemy projectiles now arrived vertically rather than horizontally, and therefore guns mounted in fortifications had to be protected from above as well as from the front and flanks. This equally applied to coast-defence batteries as bomb-vessels, mounting both mortars and howitzers, were now the chief means employed by navies to deal with and silence their guns. The new method was known as the "Carnot System," after the great French war-minister and organiser, who introduced it into France and where engineers refashioned many of the French fortresses in accordance with it. As far as the design of coast-defence works was concerned, the main innovation was the placing of guns inside casemates instead of upon open ramparts, firing through embrasures and covered by parapets. Casemated batteries became the fashion and were considered especially suited to coast-defences. One, two, or three guns might be mounted in one casemate, but in coast-defence batteries it was normal to have only one. Almost all new coast-fortifications constructed between 1815 and 1870 were designed to have casemated batteries, sometimes in tiers one above the other. The casemate of course had its advantages, it gave protection overhead as well as on all sides, but apparently it also had its drawbacks, for it was discovered that a shot or shell, once entered within the casemate, being in a constricted space, did much more damage than in the open on a rampart.

The Crimean War lasted from March 1854 to March 1856. Coast-artillery in the British Army was not engaged in active operations during this war, the two Russian fleets—the Baltic and the Black Sea—being much too weak to threaten the naval supremacy of Great Britain, and too far away to attempt any raids on British naval bases or defended ports. The British base in the Crimea was the port of Balaclava which was defended by batteries and fortifications, but these were manned by the Royal Navy and Royal Marines. One casualty of the Crimean War was the Board of Ordnance with its Master General, Lieutenant-General, and civilian members. The terrible sufferings of the troops before Sevastopol during the winter of 1854-55, the complete breakdown of all supply and transport, and the utter collapse of military organisation and administration in the theatre of war caused a great outcry at home both in and outside Parliament with the result that a Parliamentary Committee was appointed to inquire into the general administration of the Army. Amongst the Committee's many recommendations was the abolition of the Board of Ordnance about which it had nothing good to say and a great deal to its discredit. So in May 1855 the "Letters Patent" were cancelled, the Department of the Board of Ordnance was dissolved, and the Royal Regiment of Artillery, the Corps of Royal Engineers, the guns, the coast-defences, the forts and fortresses, the ammunition, arms, equipment, and stores were at long last transferred to the War Department and found themselves under the joint rule of the Secretary of State for War and the Commander-in-Chief.

It is of interest to recall the contemporary views on coast-defence, and the following is an extract from a paper read to the Royal Artillery Institution in September 1855 by Captain J. B. Parkin R.A. (*Minutes of Proceedings of the Royal Artillery Institution* 1855):-

"In general Coast Batteries are intended for the defence of harbours, arsenals, and dockyards, sometimes of landing places only, when however they will rarely be very formidable. In either case much must depend on arranging, constructing, and arming them upon the following circumstances:-

(1) Upon the value and importance of the position geographically —as an extensive and valuable depot of war or of commerce, as a good and capacious harbour, or by a combination of some or all of these.

(2) Upon local peculiarities of situation—whether at the mouth of a river, in a bay or other inlet of the sea etc. more or less sheltered by land or rocks from the wind.

(3) Upon the number and nature of the approaches to that position, and the fewer (by water) the better. Whether the latter are open or confined, direct, or tortuous, free from obstruction, or difficult of navigation. How effectually commanded by land. The nearer to the main object of defence *i.e.* to the main defences themselves, the more important do these considerations become.

(4) Whether vessels can run near and past the batteries without being obliged to repass them (within range) to escape.

(5) And most importantly, upon the depth and description of the soundings, with reference to the nearest distance at which at high tide ships could either pass them (when the time they will be under fire must be considered) or at which they could take up positions for a regular battle. Whether in fact the object of the enemy is to pass the batteries only or is to destroy them. To ships propelled with steam, winds, and currents are comparatively of little consequence as they can always choose their own opportunity and are to a great extent independent of both.

With regard to the first point:- All very important positions should as nearly as possible be fortified so as to prevent any apparent possibility of their being assailed successfully, since, if there is any good reason to doubt this (as is too often the case) their capture or destruction may be worth a considerable risk and sacrifice on the part of the enemy. No expense therefore should be spared to render a maritime or other fortress which is exposed, or likely to be so, as nearly impregnable as it can be made. The means of conveying all reinforcements by a railway from an inland or secure depot will materially contribute to its strength, if the attack is to be a prolonged one.

The second and third points may be considered together. Bays, or other inlets of the sea, especially when the entrance is narrow, and the estuaries and mouths of rivers, provided the water is sufficiently deep, form frequently convenient harbours and eligible situations for dockyards, arsenals, and other warlike depots. They offer generally many facilities in fortifying. The approaches are often narrow channels, converging as they get nearer to the harbour. Mud or sand banks make them frequently tortuous and difficult of navigation to an enemy and many often subject him to the effects of a fire which he cannot conveniently or effectually return.

Such situations generally admit also of a more or less effective cross or flanking fire which should invariably be taken full advantage of and so as, if possible, to cross at the narrowest part of the channel or approach to the harbour; or otherwise at any other desirable point over which the enemy's ships must necessarily pass in order to make an attack.

Should there be any banks or islands above the level of high water, and which are within the distance of an extreme range from the main works, they should always be armed with powerful guns, the batteries built no higher than is absolutely necessary and retired as far as would be practicable from the front. If likely to fall into the hands of the enemy, the guns should be rendered unserviceable. These batteries should also if possible be assisted by a flanking fire from the works in the rear.

A most important consideration relative to the natural strength of a harbour consists in the nature and number of the approaches or channels conducting to it *i.e.* at how many distinct or independent points—by which I mean those points which are beyond the range of mutual defence the enemy might possibly attack—and also upon the front by which he may be able to advance to attack the batteries.

It is evident that the more numerous these independent points of attack are, and the more extended the front with which the enemy may be able to approach, the weaker will the position be and *vice versa*. As a general rule it would seem as if a position would be stronger when concave, that is to say, a concave series of convex batteries, and weaker when convex.

Now the more concentrated a fire is, the more destructive it may be expected to prove. Circumstances sometimes render this concentration of fire easy to effect on the part of the land defences, and on the contrary difficult on the part of attacking ships, and sometimes the reverse when however the position is a bad one or badly defended; as would be the case generally if the position were of a convex form which latter would cause rather a divergent than a convergent fire. When therefore the facilities for obtaining this concentration of fire are in favour of the enemy's fleet, as when very exposed to an open sea, the position is in this respect not so naturally strong, and its capabilities of defence must depend upon the command and nature of the ground.

With regard to the fourth particular:- Whenever it is the enemy's objective to run past the batteries only, his ships will generally do so as quickly as possible, and a few steady shots will do better than a hurried or unsteady fire.

As to the fifth particular:-A first class harbour, and the ways or channels leading to and from it, must be capable of floating and containing vessels of the greatest draught and dimensions, at least during high tides if not at all times. It must also be more or less surrounded by land so as to be sheltered from the wind and weather, and in short so as to render it tolerably secure; and it is on the relative depth of water in various localities and on the configuration and character of the ground that its strength in a military point of view depends. Of course the less water there is, the less near will the enemy's vessels be able to approach, or at any rate the smaller will they be, and the less formidable will therefore be his fire; consequently the less powerful need be the land defences; but when the depth of water permits of the enemy's ships of the line approaching within range, the question becomes a different one, and its solution must again depend very much upon the command and character of the ground.

In the fortification of a harbour or coast the first thing to be considered after a complete examination of the ground (taking all levels above high water mark) is at what distance from the batteries will the enemy's vessels of each class respectively be able to approach *i.e.* the nearest distance. Having decided this question, the next thing to be decided will be the best positions for the batteries with reference to the character of the ground. If the water is deep and the ground low, it may be necessary or desirable to retire the batteries a few hundred yards from the water's edge, the effect of which may perhaps be to render useless a proportion of the enemies' guns, those on the lower deck for instance, which are also the heaviest; then, by having heavy guns mounted on traversing platforms (or otherwise as may be best under the circumstances) a great advantage will be gained over him.

But it may be necessary or desirable that the batteries should not be so retired even though the water is deep and the ground low and flat; as for instance in the case of a harbour where cover is required both to intercept the enemy's view of the interior and for protection to the establishments, stores, ships etc. there from the effects of his fire. It is in this case that it appears to me casements must be resorted to, more especially for the more advanced works which defend the entrance to the harbour. Here we should have three tiers of guns: the stone used should be the hardest granite, and the works themselves should be rendered as fully capable as possible of their own defence, by being protected in flanks and rear and surrounded on the land side by a ditch.

All works which are somewhat advanced into the harbour and which are therefore exposed to the enemy's broadsides at short distances should be similarly built and protected. Those which are intended by their fire to enfilade the harbour, or to flank the former, may be constructed of earth; but the guns mounted in them should be heavy and pointing through embrasures unless at a considerable elevation above the water. Where casemates are employed it should be an object to obtain the heaviest possible fire with the least exposure of the interior; the power consequently of the batteries should consist more in the weight of metal than in the number of pieces so as to necessitate as few embrasures as possible". And so Captain Parkin continues for several more pages. His views, if somewhat prolix, would appear to have been sound.

CHAPTER XV

1856 to 1914.

WITH the year 1856 Coast Artillery in the British Army entered the last hundred years of its existence. It was a century during which Coast Artillery was to become a separate branch of the Royal Artillery—almost, one might say, of the Army—and as the result attain a level of organisation, skill, tactical doctrine, and above all *esprit de corps* which it had never reached before. It was also a century which was to see the old enemy, the warship, and the weapons with which it was armed—and therefore those of coast-artillery—alter out of all recognition, and the development of precision instruments with which to solve the problems set by the new ships and guns. It was finally a century which was to experience in its last quarter the shame, bitterness, confusion and disaster of Hong Kong and Singapore. It was indeed a century during which Coast Artillery in the British Army reached both the zenith and the nadir of its long career.

However, it is now necessary to turn back to the aftermath of the Crimean War. Up to the time of that war it had always been a basic principle of the Royal Artillery that a marching-company could at any time, with scarcely any further training, become a field-battery, a siege train, a fortress company, or a coast defence unit, and, if given the necessary equipment, carry out the duties on active service competently and efficiently. The Crimean War proved this principle to be false. It showed quite clearly that the short courses at Woolwich in the "Instruction Batteries" or at the Repository were quite insufficient to enable a company to step straight into the role of field battery or seige train with any hope of performing it effectively. At the beginning of the campaign neither the field nor the seige batteries knew their business properly. Once the war was over—and during the ensuing years—the problems of mountain, heavy and coast artillery also had to be taken into consideration so that the constant changes in organisation between 1856 and 1899 were a series of attempts to design a structure which would produce efficient field, mountain, seige, heavy, and coast units in time of war within the framework of one Regiment of Artillery. At the beginning, although it was realised that a unit must be organised and trained in peace time in the same role as it would undertake in war, the higher authorities of the Regiment found themselves unable to throw over the old principle all at once, and it was laid down that a unit of the Royal Artillery in time of peace, when trained in one role, could at any time be transferred to another, and that officers would continue to be on one list and liable to serve in any branch. This proved most unsatisfactory—chiefly due to the complete difference in function between the mounted and dismounted branches *i.e.* between field and coast artillery, and the resultant feeling that the latter was an inferior division of the Regiment—and, after several more unsuccessful attempts to solve the problem, in 1899 the Royal Artillery was split into two completely separate corps, but not, as might have been expected, into mounted and dismounted, movable and static, field-army and fortress, but instead into field only on the one side, and mountain, heavy, seige, and coast-defence on the other. This separation proved of enormous benefit to coast-artillery, and it was divided in this manner that the Regiment faced the outbreak of war in 1914. The problems of the organisation of artillery in groups higher than a battery, *i.e.* in lieutenant-colonels' or brigadier-generals' commands—turned out to be equally difficult to solve, indeed no really satisfactory solution was found until after the South African War.

The first re-organisation was carried out during 1859-60. It was designed with two main objects in view (1) To bring about decentralisation in command and control, and (2) To ensure that units were organised and trained in peacetime in the role which they would perform in war. The name "company" was abolished, and all units were to be known in future as batteries. These were divided into field and garrison brigades, each brigade consisting of from seven to 10 batteries each. There were nine garrison brigades with a total of 72 batteries, and their role was entirely coast, siege or fortress artillery. A brigade was commanded by a colonel, and the headquarters of the garrison-brigades were at:-

        Woolwich   (1st Brigade)
        Dover   (2nd Brigade)
        Plymouth   (3rd Brigade)
        Gibraltar   (5th Brigade)
        Quebec   (7th Brigade)
        Guernsey   (10th Brigade)
        Mauritius (12th Brigade)
        Portsmouth (15th Brigade)

The batteries were numbered from 1 upwards in each brigade, a battery being known for example as 4 Battery, 12 Brigade, or for short 4/12 Bty. It was laid down that batteries could be transferred from one brigade to another and from one role to another at any time, and that all should be prepared to act in any artillery capacity.

This organisation might have had a chance of being successful if all the brigades had had their batteries together in one station, like Fifth Garrison Brigade with all its eight batteries at Gibraltar, or even fairly close together like Third Garrison Brigade with six Batteries at Plymouth and Devonport and two at Portsmouth, but as there were many brigades whose batteries were scattered far and wide from headquarters like Twelfth Garrison Brigade with Brigade H.Q. at Mauritius and batteries at Sydney, Ceylon, Hong Kong, and the Cape, it was quite obvious it was not going to work. It made brigade control and administration almost impossible and regulation and supervision of training quite so. In fact most of the coast-artillery units were as independent as they always had been before the introduction of brigades.

The Indian Mutiny (1857-58) had brought about an influx of the Royal Artillery into India—by the end of 1858 there were 25 units on service in that country—and as one of the results of the great fundamental changes which took place in the constitution of the government and armies of India after the close of the Mutiny, it was decided to amalgamate the artilleries of the three presidential armies—Bombay, Madras and Bengal—with the Royal Artillery, this being carried out in February 1862. Until that year the defence of India's naval bases and commercial harbours had been in the hands of the artillery of the three native armies, the chief of these ports being the capitals of the presidencies, Bombay, Madras, and Calcutta, with innumerable lesser defended ports and forts scattered up and down the coasts, and overseas Singapore (1823) Aden (1839) and Rangoon (1853). From 1862 onwards these coast-defences became the responsibility of coast-artillery from the British Army, and we find "garrison-batteries" stationed at Bombay (Colaba) Madras (Fort St. George) Calcutta (Diamond Harbour) Rangoon, Singapore, Aden, and later Karachi. The absorbtion of the Indian artilleries brought about an increase of 10 brigades in the Royal Artillery, of which three were entirely garrison, and the total number of garrison batteries by 1864 was 88. It is to be noted that these now included mountain and heavy mobile batteries as well as coast, siege, and fortress.

The 1870s were the period of the Cardwell reforms which completely revolutionised the structure and organisation of the British Army, and the Royal Artillery did not escape. In 1874 there were 91 garrison batteries, 35 at home, 26 in India (including three mountain and four heavy) and 30 at other stations abroad, divided into 13 brigades with headquarters at Mauritius (2nd Brigade, nine batteries) Dover (3rd Brigade, eight batteries) Madras (5th Brigade, seven batteries) Bombay (6th Brigade, seven batteries) Halifax (7th Brigade, seven batteries) Plymouth (10th Brigade, seven batteries) Malta (12th Brigade, seven batteries) Gibraltar (15th Brigade, seven batteries) Woolwich (17th Brigade, seven batteries) Portsmouth (21st Brigade, seven batteries) and Jersey (22nd Brigade, seven batteries) and in the interior of India (two brigades). The new Cardwell system introduced the short-service period for the rank and file, and the scheme by which infantry battalions were linked together in pairs, one at home and one abroad, the former providing the necessary drafts for keeping the latter up to strength. In 1877 the Royal Artillery was re-organised in an attempt to bring it into line with the other fighting arms of the Army. The old brigades of 1859-60 were abolished, and new brigades were established with their headquarters at home only. Each brigade consisted of from 18 to 20 batteries of which some were at home and some abroad. The new brigades were organised entirely for the purpose of reliefs and drafts, the batteries of the brigade at home either replacing or feeding those abroad. They exercised no functions of command and control and very little of administration. There were five garrison brigades, numbered 7 to 11, each of 18 batteries and a depot, with headquarters at Portsmouth, Plymouth, Dover, Cork and Sheerness. An attempt was also made to organise some kind of tactical higher command, and it was laid down that coast-fortresses and forts should be allocated as and divided into "Artillery Districts," each district being under the command of a lieutenant-colonel.

This organisation did not last long, and in 1882 it was altered once again as the result of the recommendations of a committee under the chairmanship of Lord Morley. This fresh organisation was based neither on command and control, nor on administration, nor on reliefs and drafts, but this time on recruiting. The garrison artillery was divided into 11 brigades, each of eight or nine batteries and a depot, each brigade being allotted a territorial district—the name of which was added to its number—from which to draw its recruits. Its batteries were still divided between home and abroad and owed no allegiance to brigade headquarters except for the supply of drafts and recruits. These 1882 garrison-brigades were:-

1st Northern: headquarters and depot	Newcastle
2nd Lancashire:	Liverpool
3rd Eastern:	Yarmouth
4th Cinque Ports:	Dover
5th London:	Woolwich
6th Southern:	Portsmouth
7th Western:...	Plymouth
8th Scottish:	Leith
9th Welsh:	Pembroke Dock
10th North Irish:	Carrickfergus
11th South Irish:	Cork.

An advance however was made in tactical control, for each lieutenant-colonel in command of an Artillery District was now allowed an adjutant and small clerical staff. In 1884 there were 37 garrison batteries at home, 27 in India (including six mountain and six heavy) and 32 in other stations abroad.

The reorganisers were at it again in 1889, the 11 garrison brigades being abolished, and three garrison divisions established in their place. These, as with the brigades were territorial, being named Eastern (29 batteries, H.Q. Dover, depots, Dover, Woolwich and Great Yarmouth) Southern (42 batteries, H.Q. Portsmouth, depots Gosport, Liverpool, Dunbar, and Cork) and Western 25 batteries, (H.Q. Plymouth depots Devonport and Scarborough). There was yet a further advance in command and control, the colonels in charge of these divisions with their headquarters at Dover, Portsmouth, and Plymouth, being also appointed C.R.A. and made responsible for the command, training and administration of the garrison-batteries serving within the boundaries of their divisions, and for the supervision of the *militia* and volunteer artilleries which were now allotted—according to their counties and towns—to those divisions. Mountain artillery was at this time removed from the garrison-divisions and formed into a division of its own of 10 batteries.

In 1891 there was yet another major reorganisation of the garrison portion the of Royal Artillery, its object being, in the words of the Regimental Order "Improvement in the organisation and personnel of the Garrison Artillery, its greater efficiency in the duties of coast-defence, and increased means of instruction in the higher and more technical duties connected with the heavy armaments now in use." The system of three divisions, Eastern, Southern and Western, with their headquarters at Dover, Portsmouth, and Plymouth survived, but batteries were done away with, and companies substituted in their place. The establishments of the companies, both at home and abroad, were adjusted according to the armaments they had to man, the larger companies, known as "double companies" having eight officers and up to 240 men, the lesser or "single companies" five officers and as low as 100 men. The establishments included for the first time "specialists"— position-finders, range-finders, telephonists, layers and machine-gunners—who received extra pay on qualifying in their various grades. At the conclusion of this reorganisation the garrison artillery had 22 companies at home, 24 in India, and 21 in the colonies. In 1894 "double companies" were abolished, such units being divided to form two single companies, and in 1895 six companies were added to the establishment, and in 1898 11 more, bringing the total up to 99 companies.

Such was the situation when in June 1899 the Royal Artillery was divided into two distinct and separate corps, the Royal Horse and Royal Field Artillery to furnish the horse and field batteries, and the Royal Garrison Artillery to provide the mountain, heavy and siege batteries and to man the coast defences. The two corps combined still formed the Royal Regiment of Artillery, but otherwise were unconnected. It is now necessary to consider why this separation was judged by the War Office to be essential and what were the causes and steps which led up to it. The fundamental cause of course was the complete difference in function between field and coast artillery. Once permanent field batteries were established after the Crimean War, there was no affinity whatsoever in employment, methods, training, or outlook between the two. This cleavage was emphasised by the fact that the field artillery were mounted while the coast-artillery performed on their feet. In the hey-day of the horse the mounted-man considered himself greatly superior to anyone who had to carry out his duties without the aid of that remarkable animal. Moreover, with the introduction of high-powered warships and high-velocity rifled, breech-loading guns, scientific methods and precision instruments were required to hit and sink the steel-clad men-of-war, and the horseman was apt to regard with great suspicion anything that smacked of science. All this of course resulted in the officers of field batteries looking down somewhat upon their colleagues serving with coast-defence units and regarding a transfer to a garrison battery as a definite descent in their careers. This feeling became so marked that the higher authorities of the Regiment, well knowing of it, introduced the shameful custom of employing the transfer from field to garrison as a means of showing their displeasure with an officer who had committed some minor misdeed or been lazy and slack in carrying out his duties with a field battery.

There were other reasons for making service with coast-artillery batteries unpopular with both officers and other ranks. Peace time soldiering with the field-artillery, with their horses, pleasant stations, and social life, was much preferable to spending years in some isolated fort on a bleak and wind-swept island. Moreover there were many opportunities of seeing active service in the mounted batteries which never came the way of coast-gunners, and honours and promotion were apt to go to those who served on the field of battle. Furthermore, there being so many coast-artillery stations overseas, the officers and men of the garrison batteries spent too much of their service abroad. All this resulted in officers and men disliking their periods of service in coast-defence, spending most of their time making efforts to get away and back to field, and of course, as inevitable consequences, a standard of morale and efficiency among coast-artillery which was distressingly low, and a general shortage of recruits.

This unsatisfactory situation was not unknown to the higher authorities of the Royal Artillery, and a committee under the chairmanship of Major-General C. G. Arbuthnot D.A.G.R.A. was appointed in 1882 to consider it. This committee, having gone thoroughly into the whole matter, reported "that service in the garrison artillery, the most scientific branch of the arm and which should command the services of the best officers, instead of being sought after, is shunned, and if priority of choice were given to young officers on joining the Regiment according to their places in the batch, the best would select the field-artillery, and the garrison-artillery, which requires the most scientific officers, would only get those who, from idleness or want of ability, had failed to obtain a good place in their batch." This committee recommended, in order to make service in garrison-artillery more popular, the award of armament pay to the officers and an increase of pay for the men. However, these inducements proved insufficient, things went from bad to worse, and for the first time the idea of separation began to be entertained, with the result that in 1887 another committee, this time under the chairmanship of Lord Harris, was set up to review the whole organisation of the Royal Artillery, to recommend what could be done to increase the efficiency of the garrison artillery and to consider the advisability of dividing the mounted and dismounted branches into separate corps. This committee unfortunately could come to no unanimous opinion but stated, inter alia, that "On the subject of a division of the Royal Artillery into two corps, the evidence taken by the Committee shows a very general desire on the part of the officers of the Royal Artillery at regimental duty to have a separation between field and garrison artillery. The main cause of this desire arises from personal motives; and if it is admitted on the one hand that the majority of voices are in favour of separation, it must be allowed on the other hand that the weight of evidence is undoubtedly in favour of the existing system." This committee moreover strongly recommended a general increase in pay all round to the garrison branch to make it more attractive which the Treasury, with its usual procrastination, did not implement until July 1891.

At the same time as these pay concessions were granted, an order was issued which decreed that officers on first posting in the future would as heretofore be interchangeable between field and garrison artillery, but that after three years they would remain posted to whichever branch they were then serving with and that all further promotions would be made in that branch. For the next ten years officers already serving would continue to be interchangeable between the two branches, but, after the elapse of that period, would stay in the branch with which they were serving at that time. Promotion to lieutenant-colonel would henceforth be by selection, and officers would be selected from the two branches according to the nature of the duties they were performing. This order showed definitely which way the wind was blowing and that separation was rapidly becoming more than an idea. The promotion to lieutenant-colonel by selection did not last long—it was discontinued in January 1893—as it caused too many anomalies, senior majors being jumped by juniors because there were no vacancies to be filled in their own particular branch.

After two more committees, those of Lieut.-General Stirling and Major-General Marshall, had deliberated, the War Office at last determined to separate the Regiment into two corps, (1) Royal Horse and Royal Field Artillery, and (2) Royal Garrison Artillery, and this was duly carried out on 1st June 1899. It was, taking into consideration the conditions of the time, a wise decision. There had long since ceased to be any relationship between the operations of artillery in the field and operations of artillery in coast-defence, and the whole attitude of the mounted branch towards the dismounted gave the latter an inferiority complex which destroyed all morale and efficiency. Now that the two branches were separated into two distinct corps, all this was changed. The coast-artillery, which formed 80 per cent of the R.G.A., were now able to go their own way, officers, and men free from the depressing thought that they were serving in a branch of the Regiment which they should make every effort to leave at the earliest opportunity. They now had their own corps in which, for better or worse, they had to spend their whole military careers, and it was only natural that they should rapidly become proud of this corps and expend every effort to make it a good and effective one. The result was a swift and astonishing growth of enthusiasm, efficiency, skill, knowledge, morale, and *esprit de corps* so that, when war broke out in 1914, the coast-artillery in the British Army was second to none in the world.

At first there were no great changes in organisation—the South African War (1899-1902) kept everybody much too busy to have any time left for thinking about coast-defence problems—but by 1902 the Garrison-Divisions had been done away with and "Groups" substituted in their place, but these Groups:-

South Eastern	H.Q. and Depot Dover
Southern	H.Q. and Depot Gosport
Western	H.Q. and Depot Plymouth
North Western	H.Q. and Depot Liverpool
North Eastern	H.Q. and Depot Scarborough
Eastern Groups	H.Q. and Depot Great Yarmouth

were merely associations for the provision of recruits and drafts, and the allocation of *Militia* and Volunteer artilleries. They had no superior commanders or staffs. At the same time the command of coast-artillery at home and abroad was put on a sound basis, the stations being graded as coast-fortresses or defended-ports, and each garrison R.G.A. having its own commander and staff who were responsible for command and control, training, administration, and the artillery-defence of the fortress or defended port both in peace and war. This was a tremendous step forward, and at last gave coast-artillery a proper chain of command and responsibility, and senior officers R.G.A. the opportunity to exercise their time functions.

In 1907 the Groups in their turn disappeared, and the companies of the R.G.A. were now only connected together in the tactical commands at their various stations at home and abroad, and the *militia* and volunteer artilleries were definitely allotted to the coast-fortresses and defended ports at home. When war broke out in August 1914 there were 108 R.G.A. companies and batteries of which 66 companies were employed in the coast-defences in the United Kingdom, India and the Colonies. The table given below shows the number of coast-artillery batteries or companies at each station at six dates during the period 1856-1914:-

Station	1860	1870	1880	1890	1900	1910
Portsmouth, Gosport Isle of Wight	4	12	8	11	7	9
Plymouth and Devonport	6	5	4	6	4	4
Dover	6	6	4	4	3	2
Sheerness and Chatham		4	4	4	3	3
Shoeburyness	1	2	2	2	2	1
Pembroke Dock and Milford Haven		1	1	2	2	2
Weymouth and Portland			1	2	3	4
Languard Fort				1	1	
Tilbury Fort			1			
Falmouth						1
Tynemouth						2
Leith Fort					2	1
Cork Harbour			3	3	1	1
Dublin Bay		1				
Kinsale		1	1			
Shannon Mouth		1				
Lough Swilly					1	1
Jersey	1	1	1	1	1	1
Guernsey	3	1	1	1		1
Alderney	2	1	1			
Quebec	2	3				
Kingston (Ontario)		1				
Halifax (N.S.)	2	3	3	3	2	
Esquimault (B.C.)					1	
St. John (N.B.)	1					
Newfoundland	1	1				
Bermuda	1	3	2	2	2	2
Jamaica	1	1	1	1	1	1
Barbados	1	1	1	1		

History of Coast Artillery

St. Lucia					1	
Gibraltar	8	8	7	8	7	7
Malta	6	8	7	8	8	8
Corfu and Ionian Isls. ...	4					
Alexandria					1	1
Sierra Leone						1
St. Helena...	1	1	1		1	
Cape of Good Hope ...	1	2	1	2	2	2
Mauritius	1	1	1	1	2	1
Aden		2	3	2	3	3
Bombay		1	3	3	3	3
Madras		3	1	1	1	
Calcutta		1	1	2	1	1
Karachi				1	1	1
Rangoon		2	2	1	2	2
Ceylon	1	2	2	1	2	1
Singapore		2	1	2	2	2
HongKong	3	1	1	2	3	3
Sydney (N.S.W.)	1	1				
TOTALS	**58**	**84**	**70**	**78**	**76**	**72**

The Ionian Islands and Corfu were handed over to the new Kingdom of Greece in 1864 and the British garrisons withdrawn. The coast-artillery company at St. John New Brunswick, was reduced to a mere detachment during the Crimean War and finally taken away altogether in 1869. St. John's, Newfoundland—one of the first coast-artillery stations overseas—was evacuated, and Quebec and Kingston (Ontario) were handed over to Canadian forces during 1870-71. Sydney's defences were garrisoned from 1857 to 1870 by coast-artillery and then transferred to a New South Wales local corps. The coast-defences at Calcutta, Bombay, Madras, Aden, Rangoon and Singapore were taken over on the amalgamation of the Indian presidency artilleries with the Royal Artillery in 1862, Karachi becoming a defended port in 1881. The battery was withdrawn from St. Helena during 1889 as it was considered too unhealthy for a peacetime station, and in 1890 two special "Emergent Detachments" were formed at home and stationed at Granby Barracks, Devonport. These detachments were for St. Helena and Sierra Leone—which was also at that time very unhealthy—the scheme being that on the threat of war they should go out as quickly as possible to their proper stations. St. Helena recovered its healthy status in 1898, and a company R.G.A. was stationed there until 1906 when it was finally withdrawn and the island given up altogether as a coast-artillery station, and by 1901

Sierra Leone had all but stamped out yellow fever and malaria, with the result that a company R.G.A. was sent there as coast-artillery garrison from that year onwards. In 1906 the last imperial troops in Canada, the R.G.A. companies at Halifax and Esquimault, left the Dominion and the coast-defences at those two naval bases were handed over to the Royal Canadian Garrison Artillery. St. Lucia, the West Indian Island, was again a coast-artillery station from 1892 until 1905.

The reorganisation of the Royal Artillery in 1859 saw the end of the Invalid Detachment and of the Royal Artillery Invalids who had served coast-artillery in the British Army so long and so faithfully. At the time of its dissolution the Detachment's strength was 450 officers and men. In its place was established the "Coast Brigade", to consist of officers promoted from the ranks, master-gunners, and experienced N.C.O.s and men with not less than 12 years service. The duties of the personnel of the Coast-Brigade were the same as those of the Invalid Detachment, but they were expected to do them very much more efficiently. They were distributed in detachments throughout the coast-defences of the United Kingdom and, with their officers and master-gunners, were responsible for the care, maintenance, preservation, and upkeep of the guns, carriages, ammunition, instruments, and related stores, and for the accounting of them. During the period 1860-1880 there was a rapid increase in the number of coast fortifications and works so that by 1881 the Coast Brigade had risen to a strength of:-

    28 Officers
    131 Master-Gunners
    230 N.C.O.s
    680 Gunners

divided into 10 divisions with headquarters at:-

    Tynemouth
    Liverpool
    Portsmouth
    Plymouth
    Leith Fort
    Woolwich
    Folkestone
    Gravesend
    Dublin
    and Cork

with an over-all headquarters at Sheerness under a full-colonel and small staff. There was a Coast Brigade detachment in every coast fortress, fort, and defence-work in the United Kingdom, ranging from two officers and 83 men in the Portsmouth Defences (Portsmouth, Southsea Castle, Gosport, Spithead, and Isle of Wight) to two men at Hartlepool. In the reorganisation of 1891 the Coast Brigade was abolished as a separate establishment and entity, and instructions were issued that the personnel should be absorbed gradually into service companies. Its officers became District (Armament) Officers and its detachments "District Establishments". Their duties did not alter, except that their activities were extended to all stations abroad, and their numbers were now included in the establishments of the various Royal Artillery garrisons manning the coast-defences. It was again laid down that the N.C.O.s and men should be on long service engagements of not less than 12 years and experienced and skilled soldiers. In 1899, with the garrison-companies, the District Establishments became part of the Royal Garrison Artillery.

As a mark that the Royal Garrison Artillery had attained the status of a separate corps far advanced in efficiency, morale, and *esprit de corps* from the pre-separation garrison-companies, King Edward VII inspected the Royal Garrison Artillery, Portsmouth, in review order, at Clarence Barracks on 28th February 1905. On parade were six companies of coast-artillery (Nos. 16, 29, 32, 34, 37 and 42) plus No. 23 Siege Company and Nos. 26 and 108 Heavy Batteries, a total of 54 officers and 1333 other ranks, all under the command of Major-General R. A. Montgomery (late Royal Artillery) G.O.C. Southern District. The recently formed R.G.A. Band (Portsmouth), under the direction of Mr. George Miller, Bandmaster R.G.A., provided the martial music, the master-gunners in their cocked-hats were on duty at the entrance gates, and the ground was kept by recruits from No. 2 Depot R.G.A. Gosport. After having been welcomed by a "Royal Salute," the King inspected the companies drawn up in line, after which they marched past him first in column, then in close column, finally finishing with an advance in review order. After the parade, the King went to the Royal Artillery Mess where the officers were presented to him, and he drank in their company in champagne "to the health and prosperity of the very gallant Regiment of which I have the honour to be Colonel-in-Chief". In a Royal Artillery Order published after the parade the King expressed his approval of the parade at which he had inspected the Royal Garrison Artillery and his high appreciation of the fine appearance, turn out, and drill of the Companies that paraded before him and of all the arrangements made for the Review.

CHAPTER XVI

1856 to 1914. (*continued*)

IT has been continually stressed throughout this history how difficult Coast Artillery in the British Army found it to produce sufficient men to man their guns in time of war. At home there was never any intention to provide trained artillerymen to do so; these were merely required to carry out the skilled care and maintenance of the guns and ammunition under the direction of the master-gunner. When the guns had to be manned in earnest, the detachments were provided by "additional gunners" raised from the infantry garrison, the local county *militia*, the "town guard", certain civilians, such as dockyard workers or tin miners, hastily enrolled, or, as during the French Revolutionary and Napoleonic War, from properly organised corps of Artillery Volunteers. Abroad the coast-artillery garrisons had been furnished by the "marching companies", but these were never sufficiently strong to man even half their guns, and "additional gunners" had always to be borrowed either from the fleet, the infantry, or local *militia* corps. This system was evidently unsatisfactory and inefficient. At home the Artillery Volunteers of 1794-1815 had produced a fairly serviceable answer, but these had all been disbanded once Napoleon had been safely tucked away in St. Helena. Abroad we have seen how attempts were made to solve the problem by training special companies of the Royal Malta Fencibles in artillery work, and the raising of half companies of "Asiatic Gun Lascars," to be attached to the regular coast-artillery companies in Hong Kong and Ceylon.

With the various re-organisations between 1856 and 1899, and the formation of the Royal Garrison Artillery in the latter year, the strength of the coast-artillrey both at home and abroad was greatly increased, but the coast-defences of the Empire expanded at the same time, and the strengths never caught up with them. Moreover, as coast-defence guns had to be manned practically continuously in time of war, one meagre detachment per gun was not sufficient, and enlarged detachments had to be constituted and arranged in watches and reliefs. This organisation demanded large numbers and therefore high costs which the Treasury would never authorize in time of peace, and yet the coast-defences had to be in all respects ready for action even before the actual outbreak of war. In addition to these difficulties, the guns of the period under review were no longer the simple smooth-bore, muzzle loaders of the preceding centuries, but developed into high-velocity, rifled, breech-loading weapons which required complicated instruments to control them and highly trained gunners to serve them. Hastily raised or borrowed "additional gunners" were obviously no longer sufficient or efficient.

Abroad the solution to these problems was thought to lie in the establishment of regular coast-artillery corps formed from local natives who would not require such high rates of pay as soldiers sent out from Britain. These local corps were not to form the sole coast-artillery of any naval base or defended port but were to be additional to the garrison companies already stationed there. In pursuance of this policy the Royal Malta Fencibles in 1861 ceased to be an "infantry with artillery companies" unit and became the Royal Malta Fencible Artillery, altered to Royal Malta Artillery in 1889, a full-time regular corps of the British Army, and part of the coast-artillery branch of the Royal Artillery with their own officers and the primary role of manning a considerable portion of the coast-defences of Malta.

Further east the system of "Asiatic Gun Lascars" had been extended from Hong Kong and Ceylon to Singapore and Mauritius, the "lascars" being enlisted from Indians who had settled locally, but these were rather auxiliaries of the British regular companies and not very satisfactory ones at that. So in 1891 it was decided to re-organise them, form them into distinct and separate units, and to recruit them direct from India from among the brave and warlike Punjabi Mussalmuns and Sikhs. These "local coast-artillery" units went through several changes of name and organisation between 1892 and 1914:-

- 1892 Organised as separate "Asiatic Artillery Companies" at Hong Kong, Singapore, Ceylon and Mauritius.
- 1893 Renamed "Local Companies R.A."
- 1898 Organised in two battalions:-
  Hong Kong—Singapore Bn., R.A. H.Q. Hong Kong
  Ceylon—Mauritius Bn., R.A. H.Q. Ceylon.
- 1899 Renamed Hong Kong—Singapore Bn., R.G.A.
  Ceylon—Mauritius Bn., R.G.A.
- 1907 Ceylon—Mauritius Bn. abolished.
  Ceylon companies disbanded.
  Mauritius company absorbed into Hong Kong—Singapore Bn.
  Thus Hong Kong—Singapore Bn., R.G.A. with companies in Hong Kong, Singapore and Mauritius and H.Q. at Hong Kong.

their primary role throughout being to man an integral part of the coast-defences of Hong Kong, Singapore, Ceylon (until 1907) and Mauritius. The same system was pursued in the West Indies and West Africa. In 1892 regular local companies were raised at Jamaica,

1856 to 1914 (continued)

St. Lucia and Sierra Leone from among the negroes and became part of the coast-artillery garrisons of those places. In 1903 they were combined together as the "West Indian Battalion R.G.A." with headquarters in Jamaica, but in 1907 the Jamaica and St. Lucia companies were disbanded, the surviving company becoming the "Sierra Leone Company R.G.A.," the only regular negro Royal Artillery unit. All these local artilleries—with the exception of the Royal Malta Artillery—were officered by British officers from the Royal Garrison Artillery, normally for a term of three years. In India the coast-artillery companies R.G.A. were expanded by having an establishment of Indian ranks who, fully trained as coast-gunners, provided the extra numbers necessary for manning the defences.

Before finishing with overseas some of the more famous coast-artillery *militia* and volunteer corps, who helped to man the coast-defences during the period under review, must be mentioned. These corps played their full part in manning the coast-defences—in some cases they replaced regular units disbanded for reasons of economy such as at Jamaica and Ceylon—but were part-time soldiers and could not give many hours to training. In the West Indies there were the Jamaica Militia Artillery (raised 1891) and the Bermuda Militia Artillery (raised 1895), this latter corps being paid for out of War Department funds and therefore forming part of the British Army. In Cape Colony there was the Cape Garrison Artillery (raised 1891) who assisted in the coast-defence of Capetown and Simonstown, and in Ceylon the Ceylon Garrison Artillery (raised 1888). All these corps had regular R.G.A. officer and N.C.O. instructors whose business it was to train them and turn them into efficient coast-gunners.

In the United Kingdom it was decided that the reinforcements required on the outbreak of war to complete the manning of the coast-defences should be found from the various corps of *militia* and Volunteer Artillery. These belonged to two very different part-time military forces of the Crown. The *Militia* were the direct descendants of the "Old Constitutional Force" which had consisted entirely of infantry battalions. Allowed to lapse during the first half of the nineteenth century it was revived and reorganised during the Crimean War on an entirely voluntary basis, the men doing 76 continuous days recruit training on enlistment and then coming up for a month's embodied service with their unit each year. It was in this newly re-organised *Militia* of the Crimea War period that coast-artillery units first made their appearance, there being already 22 such corps in existence by 1856 most of which had been embodied during the war to man the coast-defences at home. The number had risen to 28 by 1850 and to 35 by 1882. With the reorganisation of the garrison-artillery in the latter year into territorial divisions, the various *militia* artillery corps were allotted to the divisions within which they were raised thus:-

Division	Militia Artillery Corps
Northern	Durham
	Northumberland
	Yorkshire.
Lancashire	Lancashire
Eastern	Norfolk
	Suffolk
Cinque Ports	Kent
	Sussex
Southern	Hampshire
	Isle of Wight
Western	Cornwall and Devon Miners
	Devon
Scottish	Berwick and Haddington
	Edinburgh
	Fife
	Forfar and Kincardine
	Argyll and Bute
Welsh	Glamorgan
	Carmarthen
	Pembroke
	Cardigan
North Irish	Antrim
	Donegal
	Dublin
	Galway
	Mid-Ulster
	Wicklow
	Sligo
	Londonderry
South Irish	West Cork
	Cork City
	Limerick
	Tipperary
	Waterford
	Clare

These *militia* corps each had a regular R.A. adjutant and gunnery-instructor.

The amount of *militia* artillery in any one division had no relation whatsoever to the coast-defences in it which had to be manned, but was solely due to the number of corps which could be raised within its boundaries. Thus, as can be seen, there were far too many corps in Ireland, where the attraction of *militia* pay and bounties proved irresistible, and not enough in the South of England where lay the principal coast fortresses and forts. When in 1889 the garrison-divisions were reduced to three, Eastern, Southern and Western, the *militia-corps* were re-allotted, four to Eastern, 21 to

1856 to 1914 (continued)

Southern and nine to Western (The Galway having been disbanded) and were detailed to the defences which needed them to complete their manning requirements, and it was to these defences that the various corps had to come for their annual month's embodiment and training. This in many cases entailed long journeys and meant that the militiamen only saw the defences, which they were to man on mobilization, once a year. Now two of the essential conditions for the efficient manning of coast defences by part-time artillerymen were (a) That they normally lived close to their allotted defences so that in case of sudden emergency they could be called out and mobilized rapidly, and (b) That they should be thoroughly familiar with and able to train regularly on the guns and defence works which they would have to man in case of war. The majority of *militia* artillery units fulfilled neither of these conditions, and when, during the South African War, they were mobilized and manned their allotted coast-defences, all these weaknesses—besides many others—were discovered, so that, when in April 1908 the Territorial Force was formed from the Volunteers, *militia* coast-artillery corps were abolished altogether except in Ireland and the Channel Islands where there were to be no Territorials. "The Antrim Artillery" and "The Cork Artillery" were retained to assist in the manning of the North and South Irish Coast Defences, while in the Channel Isles the ancient and famous *Militia* of those islands was as always responsible for their defence, this force never having been allowed to lapse or been abolished. In 1881 the three artillery corps:-

Royal Jersey Artillery
Royal Guernsey Artillery } Channel Islands Militia
Royal Alderney Artillery

were re-organised as coast-artillery, formed into garrison-companies and allocated to the coast-defences of their own islands which they were to man on mobilisation.

As has already been pointed out, the vast host of Volunteers, which had come into existence during the French Revolutionary and Napoleonic War, was disbanded as soon as Napoleon had been finally defeated and exiled to St. Helena, and Britain, supreme at sea, remained without Volunteers until after the Crimean War. However in the year 1859 the victories of Napoleon III in northern Italy, the general aggressive attitude of France towards this country, and the bellicose outpourings of the Parisian press caused uneasiness in Great Britain and brought to the notice of both government and people the grave lack of troops for home defence. The dread of French invasion spread alarm throughout the land with the result that there was a tremendous urge to form volunteer units and a general desire to join them, and in July 1859 the government authorised the formation of volunteer corps throughout the country. These volunteers received no pay nor uniform—except when called out on actual service—but their units received an annual capitation grant of 35 shillings per each efficient volunteer which went towards furnishing the clothing. The drill and training was carried out in the evenings, and units went into camp for a week—later two weeks —each year.

Amongst these "Volunteers of 1859" were several artillery corps, the forerunners of many coast-artillery units of the future Territorial Army, these being (July 1862) :-

Northumberland	H.Q. Tynemouth
Hampshire	H.Q. Portsmouth
Devon	H.Q. Devonport
Sussex	H.Q. Brighton
Edinburgh	H.Q. Edinburgh
Cornwall	H.Q. Falmouth
Midlothian	H.Q. Leith
Kent	H.Q. Gravesend
Forfar	H.Q. Dundee
Lancashire	H.Q. Liverpool
Kincardine	H.Q. Montrose
Cinque Ports	H.Q. Dover
Renfrew	H.Q. Greenock
Dorset	H.Q. Weymouth
Fife	H.Q. Kirkcaldy
Glamorgan	H.Q. Cardiff
Haddington	H.Q. Dunbar
Lanarkshire	H.Q. Glasgow
East Riding	H.Q. Scarborough
Pembrokeshire	H.Q. Tenby
Cheshire	H.Q. Chester
Aberdeen	H.Q. Aberdeen
Dumbarton	H.Q. Dumbarton
Durham	H.Q. Sunderland
Orkney	H.Q. Kirkwall
Cromarty	H.Q. Cromarty
Suffolk	H.Q. Ipswich

The volunteer artillery-corps were never properly organised as either part of the field-army or coast-artillery. They were mostly classified as "position artillery," given such armament as was out of date and could be spared, and carried out their drill and training on these weapons firmly fixed to platforms let into the ground. What their role was to be in case of enemy invasion was never divulged. Some of the corps, whose headquarters were on the coast and near coast-defence works, trained as coast-artillery and were presumed to have their places on general mobilisation in the coast-defences. In 1889 the various volunteer artillery-corps were allotted to the garrison-divisions, and in 1899 all became Royal

Garrison Artillery Volunteers, but their roles were still vague and undefined, some being "position artillery", some semi-mobile heavy batteries, and some coast-artillery. It was not until 1908, on the formation of the Territorial Force, that the situation was properly clarified, and the artillery of that Force allocated to horse, field, and garrison, the last being divided into mountain, heavy and coast-defence.

It was on 1st April 1908 that the Territorial Force came into being, and the old Volunteers were abolished. The object of this new volunteer force was to provide a Home Defence Army properly organised into field divisions and coast-artillery units, which being located close to and around naval-bases and defended ports, would be available to man their coast-defences rapidly and without delay in case of general mobilisation, and in peace time be able to train on and become familiar with the guns and works which they were to man in war. These Territorial coast-artillery units were expected to be capable of taking their places beside the regular R.G.A. companies, and each had a regular R.G.A. adjutant and N.C.O. instructors. The Territorial gunner enlisted for a period of four years after which he could re-engage for one, two, three or four more. As a recruit he had to do 45 drills; then he had to do 20 drills per year and attend the annual 15 days training camp. He only received pay when at camp or when called out for actual service, and his uniform was issued to him free by the County Territorial Association. He was liable for home service only, but could volunteer for foreign service in time of war. The Territorial Force coast-artillery units attended their annual training camps at permanent coast-defences—preferably those in which they were going to serve in war—and carried out their firing practice seawards during the 15 days they were there. The National Artillery Association presented a trophy, known as the Kings Cup, for the company obtaining the highest figure of merit in any one year, provided the firing was carried out from an authorised coast-defence work. When war broke out in August 1914, the R.G.A. Coast-artillery units of the Territorial Force were:-

Unit.	Peacetime Locations.
Tynemouth	North Shields
(4 companies)	Seaton Delaval
	Blyth
Hampshire	Southampton
(7 Companies)	Eastleigh
	Portsmouth
	Woolston
Devon	Plymouth
(4 Companies)	Devonport
Sussex	Brighton
(2 Companies)	Lewes

Forth (6 Companies)	Edinburgh Kirkcaldy Bruntisland
Cornwall (5 Companies)	Falmouth Looe Marazion St. Ives Truro
Kent (3 Companies)	Sheerness Rochester Gravesend Dover
Clyde (3 Companies)	Port Glasgow Helensburgh Dumbarton
North Scottish (4 Companies)	Broughty Ferry Aberdeen Montrose Cromarty
Essex and Suffolk (4 Companies)	Harwich Stratford Southend Ipswich
Lancashire & Cheshire (8 Companies)	Liverpool Seacombe Barrow
Dorset (8 Companies)	Weymouth Swanage Poole Portland
Glamorgan (5 Companies)	Cardiff Penarth Barry
East Riding (4 Companies)	Hull
Pembroke (3 Companies)	Milford Haven Saundersfoot Fishguard
Durham (4 Companies)	West Hartlepool Sunderland
Orkney (7 Companies)	Kirkwall Sanday Shapensay Stromness Evie Holm

CHAPTER XVII

1856 to 1914. (concluded)

AT the close of the Crimean War the battleship was a vessel of some 3,700 tons, built of wood with yards, masts and sails, with a screw driven by a 700 horsepower engine producing a speed of 10 knots. Its armament was about 130 smooth-bore, muzzle-loading guns. At the outbreak of war in 1914, Britain's largest battleship was the *Iron Duke* of 25,000 tons, constructed of steel, with engines of 30,000 horse power developing a speed of 22 knots. Her armament was 10—13.5 inch and 12—6 inch guns, all rifled breech loaders. The period under review also saw the introduction of torpedoes, torpedo-boats, torpedo-boat destroyers, and submarines. It can easily be realised that these tremendous changes altered almost completely the tactical problems to be solved by the coast-gunner. The Crimean War had proved that the wooden-wall was helpless against shell-fire, even spherical shell with an uncertain fuze fired from a smooth-bore gun. After the war the naval powers began to seek an antidote to the shell, and, with France in the lead, several experiments were carried out with armour plate and iron hulls. As the result of these the ironclad warship was born in 1858 (the French frigate *La Gloire*), and in June the next year Britain laid down the *Warrior*, the first British ironclad, and the scientific era of naval construction had begun. The *Warrior* carried a belt of armour 4½ inches thick, which was expected to resist the 68 pdr. shell, given machinery to produce 14 knots, and armed with 48 smooth-bore muzzle loaders.

The ironclad developed rapidly, and it was soon found that sperical shell had very little effect against the armoured warship. The *Bellerophon* (1863), constructed largely of steel instead of iron, was considered to be proof against the spherical shell's most violent efforts, and the armament experts started to search for a projectile that would penetrate steel, and a gun which could fire it. The long struggle between armour and artillery had now started, and the type of gun produced by Sir William Armstrong in 1859 was thought to be the contemporary answer. The Armstrong gun was a rifled breech loader, the first of its kind, which fired a cylindrical shaped shell with a pointed nose. It was accepted with joy and relief by the British Government who ordered 6 pdrs, 20 pdrs., 40 pdrs. and 7 inch guns both for their warships and coast defences. However, the Armstrong R.B.L. gun did not prove a success, there were continual difficulties with its breech mechanism, which is not surprising considering it was the first of its sort to be put into service, and in 1863 a special committee was appointed to examine and report upon the comparative efficiency of rifled breech loaders and rifled muzzle loaders, it now being beyond question that the rifled gun firing a cylindrical, pointed projectile was the only type of weapon which could deal effectively with the armoured warship.

The committee, after long deliberation and many experiments, reported in August that it considered the B.L. gun far inferior to the M.L. gun, that the former could not be compared with the latter in efficiency for active service, and that the muzzle-loading system, including gun and ammunition, was on the whole far superior to the breech-loading system for the serivce of artillery in the field. The committee finally stated that R.M.L. guns could be loaded and worked with perfect ease and abundant rapidity. So the government decided for R.M.L. guns, and R.M.L. guns it became both for ships and coast-defences until the "nineties". To deal successfully with the armour of a warship these guns had to be of tremendous weight and power, so during the "seventies and eighties" we find coast-defences being issued with 12.5 inch 38 ton, 10 inch 25 ton and 9 inch 12 ton R.M.L.s. These great guns fired either solid shot or hollow shell packed with explosive, both with specially hardened noses for piercing armour, these projectiles being fitted with studs or driving bands for gripping the rifling and being propelled on their way by slow-burning powder. *"Cascable,"* writing in the *Gunner* Magazine for January 1957, gives a vivid description of serving these great guns:-

"Loading from the muzzle was in this order: charge: next the gas-check, driving band—a soft, copper, flat-bottomed dish whose circumferential side carried the requisite number of projections to fit between the gun-lands exactly, and therefore rotated as it was rammed to the head of the cartridge—finally the projectile. Its base was serrated in order to allow the gas-check to grip on by the heat of discharge and so give the necessary rotation to the shell which of course had to be centred on the base at the time of loading by a comic little gadget known as a wedge-wad. This consisted of two minute scotchs, each about the length of a thumb, an inch wide and about half-an-inch high, with a very gentle gradient; these soft wood scotchs were connected by a piece of cane bent to the inner circumference of the bore. Just before the shell was rammed home, the wedge was placed under the shoulder, and thus the projectile arrived duly centred on its gas check for it took the wedge with it. Means of ignition were by friction or electric tubes. With the 12.5 inch gun the cascable loop was dispensed with, and its place taken by a kind of miniature breech mechanism that housed the tube in the vent which, in these cases, was axial. The 10 inch and under retained the cascable loop, and the vent was radial, situated two-thirds up the side of the gun in order to facilitate insertion and serving for, when the gun was in the loading position at the top of the inclined steel slide, the "top-side" vent hole would have been quite out of reach. The electric tube for the 12.5 inch was a rather longer metal casing than used in more modern times, and had looped terminals to which was connected the firing flex, so had the 10 inch type, but their "tubes electric" had a very long quill body! . . . . . The steel gun-port covers swung open, mantlets closed yet allowing the muzzle to pass between, and the gun loaded: gunners then rushed in with steel-shod handspikes either side to ease the retaining friction plates; the gun on its steel upper carriage then travelled majestically down the inclined steel slide of the under-carriage like a ship on the slips, to be eased by two spiral-spring stops on the lower transom. The piece was then laid for line and elevation. Rate of fire about three minutes per round, depending of course on case of laying".

In 1856 at the close of the Crimean War, when all this contention about R.B.L. and R.M.L. guns may be said to have begun, the coast-defences both at home and abroad were armed with very large numbers of old type, smooth-bore, muzzle-loading guns, there being 600 of these at Gibraltar alone. It was obviously impossible to replace them all at once by rifled weapons, both from the point of view of expense and production. Fortunately in 1863 Major Sir William Palliser invented a method of lining smooth-bore, cast-iron guns with wrought-iron, rifled tubes, and during the latter part of the 'sixties large numbers of smooth bores at naval-bases and defended ports throughout the Empire were so converted. Smooth-bore 32 pdrs. were altered into rifled 64 pdrs., smooth-bore 68 pdrs. into rifled 80 pdrs. and so on. Coast-defence always being begrudged money, this change from smooth-bore to rifled ordnance was spread over many years, took a long time to carry out and had not yet been altogether completed when in the 'nineties the breach-loader made its come-back and swept all muzzle-loaders into limbo for ever. The table on following page gives the armaments at some of the defended bases and ports, both at home and abroad, at two dates during this period of change.

During the late 'seventies, the Elswick Ordnance Factory turned out a number of 6 inch breech-loading guns with a novel type of breech mechanism. These, after many experiments and tests, proved a great success with the result that throughout the early 'eighties the contest between the B.L. and M.L. raged fiercely and without intermission. By 1885 the battle had been won and lost, all the experts were agreed that the breech-loader was by far and away the superior weapon, and at one stroke every muzzle-loading gun became obsolete. Ships and coast-defences had all to be re-armed. Meanwhile warships continued to develop in size and power. By 1891 Britain's most modern battleship was the *Barfleur*, 10,500 tons, 13,000 horse-power developing 18 knots, with 4—10 inch

History of Coast Artillery

	1867			1881		
Station	R.M.L.	Smooth Bores	Carronades	R.M.L.	Smooth Bores convert. to R.M.L.	Smooth Bores
Portsmouth, Gosport, I.o.W.	46	540	64	85	233	119
Plymouth	4	128	10	81	103	90
Pembroke and Milford Haven	20	121	7	83	16	44
Dover	15	58	53	20	45	108
Languard Fort ...		12		12	3	3
Sheerness	15	86	2	37	36	80
Leith Fort		14				13
Dumbarton		14				14
Cork		59	3	32	10	16
Duncannon Fort ...		23	2		4	15
Jersey		122	28		21	120
Guernsey	2	135			14	122
Gibraltar	32	566	51	41	155	276
Malta	27	599	244	48	229	342
Bermuda	20	116	31	34	7	93
Jamaica		68	12			47
Sierra Leone ...		20				57
Mauritius		68	14	2	4	64
Capetown		92		12	2	66
Ceylon		114		12	11	43
Hong Kong		14	5		6	11
Halifax (N.S.) ...	29	119	6	69	19	39

B.L. guns mounted in barbettes and 10—4.7 Q.F. guns; and her most modern armoured-cruiser the *Immortalite*, 5,600 tons, 8,500 horsepower developing 18.5 knots, with 2—9.2 inch and 10—6 inch B.L. guns. The armour of the *Barfleur* was 12 inches thick and of the *Immortalite* 10 inches. Masts, yards and sails had disappeared altogether, and warships were able to manœuvre at speed under most weather conditions. In 1870 Whitehead successfully carried out experiments under the eyes of the Board of Admiralty with his self-propelled torpedo and sank an old hulk anchored in the Medway at 130 yards range, with the result that in 1877 the Admiralty ordered the first torpedo-boat to be laid down. By 1891 a first-class torpedo-boat displaced 150 tons, had a 1,600 horsepower engine

1856 to 1914 (concluded)

producing 21 knots, and carried three torpedo tubes. Warships could no longer ride safely at anchor in their defended harbours.

All this presented new problems to coast-artillery which could not be solved by muzzle-loading guns, but nearly all the money available was being expended on the Navy, and there was very little left for the re-arming of the coast-defences. The new battleships, cruisers and torpedo-boats had to be stopped and sunk, but the coast-artillery had not the means with which to do it. No move to replace the heavy R.M.L. guns was made before 1899, and then only very slowly, some 10 inch, 9 inch and 6 inch B.L. starting to appear in coast-batteries. The threat of attack on ships riding at anchor in harbour by fast torpedo-boats was met by the issue of the first quick-firing guns ever received by coast-artillery, the 3 pdr. and 6 pdr. Q.F. Nordenfelt and Hotchkiss guns, firing from fixed pivots. By 1890 these were being delivered and were mounted at the entrances of defended harbours, but torpedo-boats were already growing too fast and too large for such light guns.

The situation at the turn of the century as regards coast-defence guns was chaotic. Both at home and abroad, the ports and harbours were defended by a medley of breech and muzzle-loaders, some very modern, some very old—but mostly very old—and it was only gradually that order was created out of the confusion. It was a sound appreciation of the forms of attack to which bases and ports protected by coast-defences might be submitted that clarified the problem, and enabled coast-artillery to find the solution. These probable forms of attack were declared to be:-

(a) Bombardment at long range by battleships or heavy cruisers.
(b) Bombardment at medium range by light cruisers.
(c) Attempt to break down naval obstructions or block the entrance to port or harbour.
(d) Attack by torpedo-craft at night.

To deal with these threats it was decided to rely upon four guns:-

9.2 inch B.L.
6 inch B.L.
4.7 inch Q.F.
12 pdr., Q.F.

and re-armament of the coast-defences, both at home and abroad, with these weapons was started during the first five years of the century. It is to be remembered that at this period battleships were already armed with 12 inch guns, heavy cruisers with 9.2 inch and light cruisers with 6 inch, and that torpedo-boat destroyers had reached 550 tons displacement and could steam at 27 knots.

The Japanese attack upon and their capture of the great Russian coast-fortress of Port Arthur in the Far East in 1904-05 confirmed these probable forms of attack and taught many other lessons besides, (The most important and vital one that a coast-fortress is most vulnerable upon its landward side was apparently neither appreciated nor absorbed) but also demonstrated the extreme caution with which warships approached within range of coast-batteries, no admiral apparently being prepared to risk his larger ships in a contest with fixed armament. However, the appearance of the all big-gun battleship, the *Dreadnought* and *super-Dreadnoughts*, during the years following 1905 showed that coast-defences might have to deal with even more powerful opponents than had been reckoned with before that date, and in 1906 a committee was appointed under the presidency of General Sir John Owen to consider defences both at home and abroad. The Committee in its report condemned the large number of obsolete R.M.L. and R.B.L. guns still to be found in the defences and recommended they should be cleared away as soon as possible as they were useless against modern warships and absorbed large numbers of men to man them. They confirmed that the only guns really suitable for employment in coast-defences were the 9.2 inch B.L., the 6 inch B.L., the 4.7. inch Q.F., and the 12 pdr. Q.F., but they thought the 4.7 no longer sufficiently powerful enough to deal with the light cruiser although it might be suitable for stopping unarmed blockships. They urged that production should be concentrated on these weapons and that rearmament with them of the coast-defences throughout the Empire should be hurried forward as soon as possible. It was appreciated that the 9.2 inch gun was probably no longer heavy enough to deal with the new great battleship, but it realised that it was unlikely that anything larger would be available for coast-artillery for some years ahead. The Committee finally submitted detailed lists of the alterations and additions recommended by them at each naval base and defended port at home and abroad.

As it was with these four types of guns that the coast-artillery were armed when war broke out in 1914, it is necessary to take a closer look at them:-

(a) 9.2" *B.L. Gun Mark X*
   Carriage Garrison Barbette Mark V allowing 15° Elevation.
   Range 17,400 yards
   Projectiles— Common shell   Weight 380 lbs.
                 Armour piercing
                 Lyddite
                 Shrapnel
   Charge—Cordite   M.D.

(b) 6" *B.L. Gun Mark VII*
   Carriage Garrison Mark II allowing 16° Elevation
   Range 12,600 yards
   Projectiles— Armour Piercing   Weight 100 lbs.

            Common Lyddite
            Shrapnel
      Charge—Cordite M.D.
(c) 4.7" *Q.F. Mark III*
    Carriage Garrison Mark IV allowing 20° Elevation
    Range—11,800 yards
    Projectiles— Common Shell   Weight 45 lbs.
            Common Lyddite
            Armour Piercing
      Charge—Cordite   Mark I.
(d) 12 *pdr*. 12 *cwt. Q.F.*
    Carriage Garrison Mark II allowing 20° Elevation
    Range 8,000 yards
    Projectiles— Common Shell   Weight 12 lbs.
            Lyddite
      Charge—Cordite M.D.

The fuses of the projectiles of these four guns were in the base, and the charges were fired by means of an electric tube.

The tremendous progress in warships and guns during the period under review presented the coast-gunner with a set of problems he had never previously had to solve. The increase in the speed of warships and in the range of guns made it probable that his target would be moving at a fast rate at a far distance from his battery. Engaging the enemy at point blank range or laying with the old tangent and foresight as had been done for so many years, was no longer of any use. Some instrument which would discover the varying range of the target as it steamed through the water was absolutely necessary. The solution of this problem was entirely due to the inventive genius of Capt. H. S. Watkin R.A. who, whilst stationed at Gibraltar during the 'seventies, set his mind to work on these difficulties and as the result produced the Depression Range Finder and the Position Finder. Both of these instruments made their appearance during the 'eighties. The D.R.F., when laid on the bow-waterline of the target, recorded the range of the target as it moved through the sea, the P.F. both the range and bearing. The discovery and development of electricity enabled this vital information to be transmitted direct to the guns by the means of dials so that, in the case of the D.R.F. with the line-layer laying direct on his target, the gun was given the correct elevation, and, in the case of the P.F. the correct line and elevation without the layer seeing the target at all. These instruments were gradually installed in the coast-defences at home and abroad during the 'nineties and the early years of the twentieth century, the P.F.s to serve the heavy guns (9.2") and the D.R.F.s the medium ones (6").

However, although the P.F. and D.R.F. solved the problem of finding the direction and range of the target when it was steaming at speed at some distance from the guns, they did not enable the coast-gunner to deal effectively with the torpedo-boat destroyer or blockship making for the entrance of a harbour under cover of darkness. Throughout his existence the coast-gunner had found great difficulty in illuminating his target at night so that he could see to shoot at it. He had to depend mainly on the burning carcass fired from his own guns. The discovery of electricity had fortunately altered all this by making possible the searchlight, or "defence electric light" as it was called in coast-defence. The D.E.L. was employed either to produce a concentrated, movable beam which could search for, pick up, and follow a hostile target, or, used in a cluster, to give a mass of fixed, dispersed beams which would illuminate a stretch of narrow water through which T.B.D. or blockships had to pass to penetrate the entrance to a harbour. The D.E.L. were manned and operated by the Fortress Companies, Royal Engineers, who thus became an integral and essential part of coast-artillery without whom the coast-gunner could not function at night.

Although the D.E.L. enabled the target to be seen at night, it did not ensure that it would be hit and sunk by the guns. Destroyers, attempting to break into a harbour to torpedo the warships there, would be steaming at speed and would not remain long within the "illuminated area" of the fixed searchlight beams. There would be no time for ranging, the targets would have to be hit from the first shot. The invention of the automatic sight solved this problem. This sight, when layed by the layer on the bow water-line of the target, at short ranges automatically gave the gun the necessary elevation to cover the range. With the comparatively limited illuminating capacity of D.E.L., and the narrow waters at the approaches to most harbours, it was not expected to have to use the auto-sight at anything but short ranges. The sight, being mechanical, was not of course perfect, but, if properly adjusted and used, was calculated to produce the first rounds either on or very close to the target, and was therefore fitted to all guns which might have to deal with enemy warships at close ranges at night.

In order to understand the principles of all these new instruments and methods and be able to handle and employ them so that they could produce effective and quick "target sinking" results, the coast-gunner had to study electricity and mathematics, and to consider the effects of wind, weather, and temperature upon the ballistics of his projectiles. Thus slowly but surely the coast-artillery officer became the pioneer in exact and accurate shooting, and the leader in applying scientific methods to solving the problems of gunnery. The place where all these new ideas, methods, instruments, and weapons were tested, tried out, expounded, and taught was the School of Gunnery at Shoeburyness. When Woolwich Common

1856 to 1914 (concluded)

and Plumstead Marshes in the early 'fifties proved too small and too surrounded by buildings to be safely used with the ever increasing range of guns for artillery practice, the Board of Ordnance bought some land at Shoeburyness to be employed for that purpose. In the early days units went down only in the summer to carry out their practice there, and it was not until 1856 that a permanent establishment was installed. In that year the Secretary of State for War Announced:-

> "I am quite prepared to take the necessary steps to establish Shoeburyness as a place for experimental practice under the control of the Scientific Committee as soon as the new range in the marshes is complete. I understand that the range gives 1,500 yards to the butts and 1,500 yards beyond them, quite sufficient for all purposes of artillery instruction and practice. Under these circumstances there is no reason why Shoeburyness should not be placed entirely under control of the Director-General of Artillery and Select Committee. There will be many things there for officers, N.C.O.s and men to see and profit by, and arrangements can be made for this end without interfering with experiments".

The experimental and instructional work were divided in 1859 when on 1st April of that year the School of Gunnery was established with Colonel J. W. Mitchell, R.A. as Commandant and Superintendent with a staff of three Instructors in Gunnery, an Assistant Superintendent of Experiments, one Instructor of Musketry, a brigade-major, and adjutant, and a permanent party of 14 N.C.O.s and men. At first the School dealt with the instruction of all branches of artillery, but in 1899, when the Regiment was divided, the School was split into two separate establishments, one for R.H. and R.F.A. and one for R.G.A., the former carrying out most of its instructional work during the summer at Okehampton ranges, and the latter dealing almost solely with coast-artillery. The chief duties of the R.G.A. School was the instruction of officers and other ranks in the theory and practice of coast-defence, the courses undertaken during the period immediately prior to 1914 being:-

(a) Gunnery Staff Course, Officers.
(b) Gunnery Staff Course, N.C.O.s.
(c) Junior Officers' Courses.
(d) Short courses.
(e) Quick-Firing courses.

The object of these first two courses was to turn out Instructors of Gunnery (Officers) and Assistant Instructors of Gunnery (Warrant Officers). The course lasted a year and included study for three months at the Ordnance College Woolwich under the Director of

Artillery Studies, and a month at Portsmouth for attendance at practice-camp and a visit to the great naval establishments there. The Junior Officers' Course was for young officers who had just been commissioned and lasted eight weeks, and short courses were refreshers for both officers and men. However, the glory of the School was the Q.F. Course. The object of this course was to train officers, N.C.O.s and men as instructors in anti-torpedo craft defence; to turn out as large a number of trained layers as possible; and to teach shooting by night with anti-torpedo craft guns. The whole course was carried out with tremendous energy, enthusiasm, and speed, and no-one who ever went through it and carried out 12 pdr. Q.F. practice at night from "The Gantry" will ever forget the experience.

In 1888 the School instituted a practice-camp among the permanent defences of Portsmouth at *Golden Hill*, Isle of Wight. Here it was possible to carry out "practice seawards" with all the instruments in their service positions and under circumstances similar to a coast-fortress in time of war, and so to *Golden Hill* came the courses and units to carry out their practice. There was soon a demand for similar practice-camps to be held at other stations, and by 1895 gunnery-instructors had been sent from Shoeburyness to Plymouth, Dover, Cork, Gibraltar, Malta and Bermuda to institute "branch" practice-camps which were all considered to be parts of the School of Gunnery under the supervision of the Commandant. This lasted until 1904 when the branch practice-camps were abolished, and, command and training being now considered indivisible, all coast-artillery units carried out their annual practice seawards under the command and supervision of their own commanding officers, assisted by instructors-of-gunnery. As early as 1886 targets—known as Hong Kong targets— which could be towed by a launch with a long rope let out astern, had been constructed, and later winches were installed in the launches which enabled the targets to be hauled in at fast speed simulating a torpedo-boat or destroyer. R.G.A. companies carried out their annual practice-seawards against these targets, both by day and night, as far as possible from the batteries and defences which they would man in war. The results were then entered up in large and complicated practice-reports from which, after a lot of calculation, could be deduced the "figure of merit" of the practising unit. Company-practice was followed by fortress-practice during which the organisation, command-system, communications, batteries, and guns of the fortress were tested out together as a whole.

The Royal Military Repository at Woolwich was closed down in 1900, and instruction in the art of "Scotch Up" was transferred to Shoeburyness where it formed a most important and essential part of the officers and N.C.O.s Gunnery Staff Courses. During these the drill for moving heavy ordnance, parbuckling, the use of sleighs, mounting and dismounting of guns, and employment of sheers and derricks were taught. *Garrison Artillery Training* (1911) Vol. III in its instruction to "Moving Ordnance," stated:-

"The importance of moving ordnance expeditiously has increased with the introduction of complicated mechanisms which are liable to injury when struck by fragments of shell or flying debris. In order to repair such damage it may be necessary to move guns and mountings with the utmost speed. No shift can be considered to have been properly carried out unless it has been effected without accident, or risk of accident, to men, damage to stores, or unnecessary delay".

Every officer, N.C.O. and man of the coast-artillery was expected to be able to play his part efficiently when called upon to carry out a gun-shift, and "Scotch Up" formed a very important part of their training.

As far back as 1859 a Royal Commission had been appointed by Sidney Herbert, the then Secretary of State for War, to consider the defences of the United Kingdom with special reference to:-

Portsmouth (including the Isle of Wight and Spithead)
Plymouth
Portland
Pembroke
Dover
Chatham and Medway.

The Commission, under the Chairmanship of Major-General Harry Jones, late Royal Engineers, Governor of the Royal Military College, consisted of two other major-generals, two rear-admirals, Colonel J. H. Lefroy, R.A., and a civil servant from the Treasury. The reasons for appointing such a commission were stated to be:-
(1) The introduction of rifled ordnance which had created a revolution in practice of artillery.
(2) The introduction of steam which had revolutionised the state of naval warfare.

The Commission first laid down what they considered to be the objects of coast-artillery and coast-defence works:-
(1) For the immediate defence of a harbour, to prevent an enemy running in his fleet and destroying dockyards, store-houses and shipping.
(2) To prevent an enemy obtaining a footing upon any part of the shore to landward and effecting the destruction of naval establishments by a force landed for that object.
(3) The protection of ports, anchorages and dockyards against bombardment by sea.

The Commission, having inspected with great detail the naval bases and defended ports and taken a vast amount of evidence, rendered its report in February 1860. This report, together with maps, minutes of evidence, and appendices, ran to 160 pages, and is of great interest because it was largely on its recommendations that the coast-defences of the United Kingdom were still based when war broke out in 1914. It is intended to record here only some of its main recommendations regarding coast-defences, but the Commission, in its wisdom, fully appreciated that a coast-fortress was most vulnerable on its landward side, and it was as the result of its counsel that the rings of forts protecting the great coast-fortresses of Portsmouth, Plymouth and Chatham on the landward side were built in the 'sixties. These forts have long been out of date—although they provided invaluable bomb-proof headquarters during the Second World War—but they are a lasting memorial to the sound understanding and knowledge of the Commission of the problem of the defence of the coast-fortress, an understanding and knowledge which their successors unfortunately did not appear to possess.

The Commission dealt carefully with each place in its turn, and its main recommendations were:-

Portsmouth.
   (1) To safeguard *Spithead Anchorage*, the entrance to the main harbour, and guard against bombardment:-
      (*a*) Forts to be constructed in *Spithead Anchorage* on the following shoals:-
         No Mans Land
         Horse Sand   } (The genesis of the famous
         Spit Sand   } "*Spithead*" forts)
      (*b*) Additional batteries to be constructed at:-
         Southsea Castle
         Gilkicker Point
         Nettlestone Point
         Appley House.
   (2) To cover the *Needles Passage*, additional batteries to be constructed at:-
         Cliff End
         Totland Point
         Hatherwood Point
         Needles Point
         Hill Farm
         Hurst Castle
   (3) To cover the Isle of Wight, additional batteries to be constructed at:-
         Sandown Bay
         Yaverland
         Bembridge Down

1856 to 1914 (concluded)

    St. Helen's Point
    Atherfield Point
    Brixton
    Brook.

Plymouth.
  (1) To defend the entrance to the *Hamoaze*, additional guns to be added to the batteries already at Eastern and Western King and an additional battery to be erected on Mount Edgcumbe.
  (2) To ensure the security of the *Sound* as an anchorage for our own ships and against its occupation by an enemy, and to prevent the bombardment of the dockyard at a long range, additional batteries to be constructed at:-
       Picklecombe
       Staddon Point
       Breakwater
       Hooe Lake Point
       Drake's Island
       Whitesand Bay
       Knattenbury Hill.

Pembroke.
To prevent an enemy fleet running up Milford Haven and to protect the dockyard against bombardment, additional batteries to be constructed at:-
    Stack Rock
    South Hook Point
    Chapel Bay
    Signal Staff Point
    Popton Point.

Portland.
To prevent a squadron, superior in force at the time, running into the harbour and capturing or destroying any ships lying at anchor, additional batteries to be constructed at:-
    Verne Hill
    Inner Pier, Southern Entrance
    Disdale Point
    Blacknor Point
    The Nothe.

Chatham and Medway.
  (1) To ensure the security of Sheerness Dockyard against bombardment, to guard against the occupation of the anchorage in the entrance to the Medway, and to deny the navigation of the river to an enemy, a powerful work should be constructed at Garrison Point, and additional batteries erected at:-

Grain Spit
Okeham Ness.

(2) To ensure the security of Chatham Dockyard against bombardment, the existing and proposed defences of the Medway are considered sufficient.

Dover.

To secure the harbour from bombardment, the Castle should be remodelled and strengthened to be suitable for the mounting of modern weapons, and additional batteries should be constructed at Western Heights

Cork.

To secure the harbour for the service of the fleet and to protect the naval establishment at Haulbowline, Camden and Carlisle Forts should be remodelled, additional guns should be mounted on Fort Westmoreland, Spike Island, and additional batteries should be constructed at:-

Cork Beg
Queenstown
White Point

and the tidal harbour of Youghal should be protected by a battery of four or five guns.

So recommended the Royal Commission in 1860. At that date the casemate was still the approved form of fortification and protection for the coast-defence gun, but, as has already been pointed out, it had many disadvantages, the main ones being that it was very conspicuous and furnished a well-defined mark for a bombarding warship, it prevented smoke from escaping which, after a few rounds, seriously interfered with the service of the gun, and it greatly restricted the field of fire, thus compelling the employment of a large number of weapons to cover a comparatively small area of water. As guns became more powerful, it was evident that the casemate was wasteful and obstructive, and moreover could not keep out shells fired from rifled guns. So from 1880 onwards, when new works were built, they were constructed *"en barbette,"* that is to say an earthen terrace behind rampart and parapet, the terrace serving as platform for the guns. This, up to 1914, was the normal form of fortification for coast-defence batteries of all sizes and calibres.

CHAPTER XVIII

1914 to 1918.

COAST Artillery in the British Army had never been more ready for war than it was in 1914. The R.G.A. companies in the coast-defences at home and abroad were well led, well trained, enthusiastic and of high morale. Their colleagues, the Hong Kong—Singapore R.G.A., the Sierra Leone R.G.A., the *Militia* and Volunteer Artilleries abroad, and the Territorial Force R.G.A. at home were prepared and competent to take their places alongside them. The coast-defences throughout the Empire had been re-armed with modern guns during the previous 15 years, ammunition was plentiful, and the latest instruments and electrical appliances had been installed. Much thought and study had been given to the organisation of coast-defences in war and the best methods of fighting the guns. In short coast-artillery was in all respects ready for the enemy whenever he might show himself within range of the guns.

That enemy, Germany, since 1900 had set about building herself a great fleet which would be able to challenge Britain's naval supremacy in the narrow seas and threaten her trade-routes across the oceans of the world. By the summer of 1914, she had afloat in ports either in Germany or abroad:-

- 13 Super-Dreadnought Battleships
- 20 Other Battleships
- 4 Battle-Cruisers
- 9 Heavy Cruisers
- 39 Light Cruisers
- 142 Destroyers

Her newest and most powerful ship in each class was:-

Class	Name	Main Armament	Speed	Displacement
(a) Super-Dreadnought	Frederick der Grosse	10-12 inch	22.4 knots	24,300 tons
(b) Battle-Cruiser	Seydlitz ...	10-11 inch	29.2 knots	24,600 tons
(c) Heavy Cruiser	Blucher ...	12-8 inch	25.3 knots	15,500 tons
(d) Light Cruiser	Karlsruhe	12-4 inch	27 knots	4,800 tons
(e) Destroyer ...	S.24	2-23 pdrs. 4 torp. tbs.	32.5 knots	550 tons

worthy and formidable opponents for the British coast-artillery.

As has already been said, much thought and study had been given before 1914 to the organisation for war of coast-defences and the best methods of fighting the heavy, medium, and light guns with which they were armed. Each naval-base or defended port, together with its coast-defences, landward defences, establishments etc. and garrison of artillery, engineers, infantry and other arms was organised as a self-contained coast-fortress with a general officer as fortress-commander. Within the fortress the Royal Navy, which was not under the fortress-commander but worked in close co-operation with him, was responsible for obstructions by sea, such as mines and booms protecting the entrance to the harbour. The Royal Artillery was responsible for artillery defence, both seawards and landwards, with the fixed and moveable armament, and for the tactical control of D.E.L. The Royal Engineers were responsible for the maintenance of fortifications, communications and D.E.L. and for carrying out the normal engineer services. The Infantry were responsible for manning the ramparts and entrenchments, for local protection of guns, D.E.L., and other important installations, and for guarding possible landing-places.

The senior Royal Artillery officer was C.R.A. of the fortress, and, besides commanding the whole of the artillery, was expected to be the technical adviser of the fortress-commander in all artillery matters. Next below the C.R.A. came the fire-commanders, each commanding a group of coast-defence batteries located in the same area, the group being known as a fire-command. These batteries grouped together in a fire-command could be of various types, heavy, medium, and/or anti-torpedo craft. Then came the battery-commanders with their batteries. A battery might consist of one, two three or four guns, but all its guns were of the same type. The D.E.L., which were worked and maintained by the Fortress Company R.E., were under the direction of fire-commanders, except for fighting-lights, *i.e.* moving beams, allotted to serve specific batteries, which were under the control of the battery-commander concerned.

The possible forms of attack on a coast-fortress were thought to be:-

    (1) Attack by one or more battleships
    (2) Attack by one or more cruisers
    (3) Attack by blockers
    (4) Attack by boomsmashers
    (5) Attack by torpedo-craft
    (6) Land Raid.

The 9.2 inch (heavy) guns were expected to deal with (1); the 9.2 inch and 6 inch (medium) with (2); the 6 inch and 4.7 inch with (3) and (4); the 4.7 inch and 12 pdrs. (anti-torpedo craft) with (5); and the movable armament with (6) if the landing was made beyond the range of the coast-defence guns: otherwise all guns were expected to open fire on ships and boats carrying a landing force. The fire-commander was responsible for the control of all the batteries in his command. It was his duty to identify attacking enemy men-of-war and to select the targets for his batteries and indicate them to his battery commanders. By night he also regulated the observation lights and the fixed lights of any illuminated area.

The foremost duty of the battery-commander was the direction of the fire of his battery, and it was his business to engage enemy ships effectively as soon as he received information from the F.C. and they came within range of his guns. If the B.C. used a P.F. the instrument could provide both ranges and bearings in which case there were no layers on the sights, elevations and directions being transmitted by dials to the guns. (Case III: for heavy guns and sometimes for medium guns.) If the B.C. used a P.F. for ranges only or a D.R.F. the guns received elevations alone from the instrument, and the line-layers had to be on their sights, laying direct on the bow-water line of the target, and so giving the gun the necessary direction (CaseII: for both heavy and medium guns.) If the range was short—especially at night—no independent range-finder was used, both layers were on their sights laying on the bow-waterline of the target, the auto-sight giving elevation and the other sight direction. (Autosights for medium guns and always for A.T.C. Q.F. guns.) The B.C. could engage his target either by battery-fire or gun-fire, if by the former the guns were fired in succession, if by the latter they were fired independently and as rapidly as possible. At night the B.C. had to direct his fighting lights as well.

To deal with and stop an attack by destroyers at night was the special duty of the anti-torpedo craft guns. The "illuminated area," produced by the fixed D.E.L., was normally so designed that ships, attempting to enter the protected harbour, would be forced to pass through it. Nevertheless, warning of an attack by hostile destroyers could only be expected a very few minutes before they entered the illuminated area and the A.T.C. guns had to open fire, and targets would probably be numerous, speedy, and only under fire for a very short time. The arrangements made to meet such a night attack had therefore to admit of rapid and accurate fire being opened up at a moment's notice, and to have an efficient scheme for the distribution of fire which had to be thoroughly known by all concerned. To enable fire to be opened without delay, the detachments were organised in watches, the watch on duty being awake and alert close to the guns with a look-out peering continuously into the illuminated area: the distribution of fire scheme was thought out and drawn up most carefully by the F.C. and taught to all officers, gun-captains, and layers manning the A.T.C. guns. The secret of success with A.T.C. guns was alert look-outs, rapid opening of fire, accurate laying, a fast rate of fire, and a good and well learnt scheme of distribution of fire. Each coast-fortress was also provided with a selection of movable-armament—field-guns and howitzers—which were to be manned by the coast-artillery in case of emergency and used to assist in the defence of the landward front.

Each naval base or defended port had a Port War Signal Station, manned by the Royal Navy, the main function of which was the identification of men-of-war approaching or within sight of the port, and the passing on of that identification without delay to naval headquarters and the coast-artillery, there being an officer of the latter known as "the Selected Military Officer," specially appointed to keep liaison with the P.W.S.S. and prepared to give orders for opening fire should an enemy warship unexpectedly be disclosed or an unidentified one unable to respond with the proper signals. At defended commercial-ports there was also the "Examination Service," the object of which was to prevent hostile vessels—other than warships—from entering the port and to facilitate the safe entry of friendly vessels. Each port had an area detailed as an "Examination Anchorage" within which all vessels entering the port had to anchor and await examination by the Examination Officer, an officer of the port-authority who carried out his duties from the Examination Vessel. The work of the Examination Officer and the Examination Anchorage itself were covered by a specially detailed battery of the coast-defences known as the Examination Battery, whose duty it was to "bring to" with a round across the bows any merchant vessel which failed to stop for examination or to sink any ship which resisted the Examination Officer or was obviously hostile. The examination Battery had therefore always to be manned and on the alert throughout the hours of daylight.

It is obvious from all the foregoing that the organisation for war of a coast-fortress, its efficient operation, and the competent fighting of the guns of the coast-defences required very careful preparation in time of peace. As far as coast-artillery were concerned, this entailed, besides the most essential training of the men, the careful preparation of mobilisation-schemes, the upkeep for each fort or work of the "Fort Record Book" which provided all the information that could possibly be required concerning that fort or work, the compilation of a "Fighting Book" for each battery, and the drawing-up of "Manning Tables" which showed the number of men required to man the guns of the coast-defences and their distribution: these last were made-out in accordance with the "Rules for Calculating R.G.A. Garrisons". So that when, on the evening of the 29th July 1914, the "Precautionary Period" was put into operation, the coast-defences throughout the British Empire were quietly and efficiently manned without fuss or publicity. Given below is a list of the naval-bases and defended ports at home and abroad which were guarded by the Coast Artillery of the British Army, together with their coast-artillery garrisons and their armaments. It is regretted that the armament figures may not be absolutely accurate as the official records for the period 1907-1914 have been lost or destroyed, but they are as nearly correct as it has been possible to get them. Only "modern" guns have been included:-

UNITED KINGDOM

(1) *Eastern Coast Defences*
Defended Ports of Medway and Thames
R.G.A. Garrisons:-
Nos. 2, 14, 18, 19 and 22 Coys. R.G.A.
Two Coys. Essex and Suffolk R.G.A. (T.F.)
Two Coys., Kent R.G.A. (T.F.)
Armament:-
Sheerness   4—9.2 inch
                 6—6 inch
                 4—4.7 inch
                 6—12 pdrs.
Slough Fort   2—9.2 inch
                 2—6 inch
Coalhouse Fort   4—6 inch

(2) *Harwich Coast Defences*
Defended Port of Harwich
R.G.A. Garrison:-
No. 13 Coy. R.G.A.
Two Coys., Essex & Suffolk R.G.A. (T.F.)
Armament:-
Landguard Fort 2—6 inch
                2—4.7 inch
Harwich   2—6 inch
                2—4.7 inch.

(3) *South Eastern Coast Defences*
Defended Ports of Dover and Newhaven
R.G.A. Garrisons:-
Dover:
Nos. 40 and 46 Coys. R.G.A.
One Coy., Kent R.G.A. (T.F.)
Newhaven:
Two Coys., Sussex R.G.A. (T.F.)
Armament:-
Dover   5—9.2 inch
            6—6 inch
            5—12 pdr.
Newhaven   2—6 inch.

(4) *North Eastern Coast Defences.*
Defended Ports of Tyne, Tees, Hartlepool and Humber
R.G.A. Garrisons:-
Tyne:
Nos. 12 and 47 Coys. R.G.A.
Tynemouth R.G.A. (T.F.)
Tees and Hartlepool:
Durham R.G.A. (T.F.)
Humber:
East Riding R.G.A. (T.F.)
Armament:-

Tyne	2—9.2 inch
	6—6 inch
Tees	4—6 inch
	2—4.7 inch
Hartlepool	3—6 inch
Humber	4—6 inch
	4—4.7 inch

(5) *Southern Coast Defences.*
Defended Ports of Portsmouth and Portland
R.G.A. Garrisons:-
Portsmouth
Nos. 11, 29, 32, 33, 34, 37, 42 and 67 Coys. R.G.A.
Hampshire R.G.A. (T.F.)
Portland
Nos. 16, 28, and 30 Coys. R.G.A.
Dorset R.G.A. (T.F.)
Armament:-

Portsmouth and Gosport	2—9.2 inch
	2—6 inch
	20—12 pdrs.
Isle of Wight	12—9.2 inch
	9—6 inch
	3—12 pdr.
Spithead Forts	8—6 inch
	2—4.7 inch
Hurst and Calshot Castles	2—4.7 inch
	8—12 pdrs.
Portland	6—9.2 inch
	10—6 inch
	8—12 pdrs.

(6) *South-Western Coast Defences.*
Defended Ports of Plymouth and Falmouth.
R.G.A. Garrisons:-

1914 to 1918

Plymouth:
    Nos. 36, 38, 45 and 41 Coys., R.G.A.
    Devon R.G.A. (T.F.)
Falmouth:
    Cornwall R.G.A. (T.F.)
Armament:-

Plymouth	8—9.2 inch
	13—6 inch
	3—4.7 inch
	15—12 pdrs.
Falmouth	4—6 inch

(7) *North-Western Coast Defences.*

Defended Ports of Mersey and Barrow
    R.G.A. Garrison:-
        Lancashire and Cheshire R.G.A. (T.F.)
    Armament:-

Mersey	6—6 inch
	2—4.7 inch
Barrow	2—6 inch

(8) Defended Ports of Milford Haven, Cardiff, Barry and Swansea.
    R.G.A. Garrisons:-
    Milford Haven:
        Nos. 44 and 57 Coys. R.G.A.
        Pembroke R.G.A. (T.F.)
    Cardiff, Barry and Swansea:
        Glamorgan R.G.A. (T.F.)
Armament:-

Milford Haven	4—9.2 inch
	6—6 inch
	8—12 pdrs.
Cardiff	4—6 inch
Barry	2—6 inch
Swansea	2—4.7 inch

(9) *Scottish Coast Defences.*

Defended Ports of Forth, Clyde, Tay, Aberdeen and Scapa.
    R.G.A. Garrisons:-
    Forth:
        No. 21 Coy. R.G.A.
        Forth R.G.A. (T.F.)
    Clyde:
        Clyde R.G.A. (T.F.)
    Tay and Aberdeen:
        North Scottish R.G.A. (T.F.)
    Scapa:
        Orkney R.G.A. (T.F.)

Armament:-

Forth	6—9.2 inch
	12—6 inch
	14—4.7 inch and 4 inch
	12—12 pdrs.
Clyde	4—6 inch
	4—4.7 inch
Tay	2—6 inch
	2—4.7 inch
Aberdeen	2—6 inch
Scapa Flow	—None yet mounted.

(10) *North Irish Coast Defences.*

Defended Ports of Lough Swilly and Belfast.
R.G.A. Garrisons:
    No. 15 Coy. R.G.A.
    Antrim R.G.A. (S.R.)
Armament:-

Lough Swilly	2—9.2 inch
	2—6 inch
Belfast	4—6 inch.

(11) *South Irish Coast Defences.*

Defended Ports of Queenstown and Berehaven
R.G.A. Garrisons:-
    Nos. 10, 43 and 49 Coys., R.G.A.
    Cork R.G.A. (S.R.)
Armament:-

Queenstown	4—9.2 inch
	6—6 inch
	8—12 pdrs.
Berehaven	2—9.2 inch
	6—6 inch
	2—4.7 inch
	8—12 pdrs.

(12) *Channel Islands.*

Coast Defences of Jersey, Guernsey and Alderney.
R.G.A. Garrisons:-
    Jersey:
        No. 20 Coy. R.G.A.
        R. Jersey Artillery (Ch. Is. Militia)
    Guernsey:
        No. 17 Coy. R.G.A.
        R. Guernsey Artillery (Ch. Is. Militia)
    Alderney:
        R. Alderney Artillery (Ch. Is. Militia)

Armament:-
Jersey	2—6 inch
	2—4.7 inch
Guernsey	2—6 inch
Alderney	2—6 inch
	2—12 pdrs.

OVERSEAS.

(1) *India.*

Defended Ports of Bombay, Calcutta, Karachi, Madras and Rangoon.

R.G.A. Garrisons:-
Bombay:-
Nos. 52, 77, 79 and 85 Coys. R.G.A.
Bombay Volunteer Artillery.
Calcutta:-
No. 62 Coy. R.G.A.
Calcutta Port Defence Volunteers.
Karachi:-
No. 69 Coy. R.G.A.
Madras:-
Madras Volunteer Artillery.
Rangoon:-
Nos. 64 and 75 Coys. R.G.A.
Rangoon Port Defence Volunteers.

Armament:-
Bombay	6—7.5 inch
	6—6 inch
	2—12 pdrs.
Calcutta	4—6 inch
Karachi	2—7.5 inch
	2—6 inch
Madras	2—4.7 inch (on field-carriages)
Rangoon	4—6 inch
	2—4.7 inch.

(2) *Gibraltar.*

R.G.A. Garrison:-
Nos. 4, 6, 7, 8, 9, 54 and 55 Coys. R.G.A.

Armament:-
14—9.2 inch
11—6 inch
7—4 inch.

(3) *Malta.*

R.G.A. Garrison:-
Nos. 1, 5, 63, 65, 96, 99, 100 and 102 Coys. R.G.A.
Royal Malta Artillery.

Armament:-
  16—9.2 inch
  14—6 inch
  12—12 pdrs.

(4) *Aden.*
 R.G.A. Garrison:-
  Nos. 61, 70 and 76 Coys. R.G.A.
 Armament:-
  4—6 inch.

(5) *Cape of Good Hope.*
 Defended Ports of Capetown, Simonstown.
 R.G.A. Garrisons:-
  Nos. 84 and 97 Coys. R.G.A.
  Cape Garrison Artillery.
 Armament:-

Capetown	2—9.2 inch
	2—6 inch
	2—4.7 inch.
Simonstown	3—9.2 inch
	4—6 inch.

(6) *Sierra Leone.*
 Defended Port of Freetown.
  R.G.A. Garrison:-
   No. 50 Coy. R.G.A.
   Sierra Leone Coy. R.G.A.
  Armament:-
   2—9.2 inch
   4—6 inch.

(7) *Bermuda.*
 R.G.A. Garrison:-
  Nos. 3 and 95 Coys. R.G.A.
  Bermuda Militia Artillery.
 Armament:-
   3—9.2 inch
   6—6 inch
   4—4.7 inch.

(8) *Jamaica.*
 Defended Ports of Kingston and Port Royal.
  R.G.A. Garrisons:-
   No. 66 Coy. R.G.A.
   Jamaica Militia Artillery.

Armament:-
    1—9.2 inch
    4—6 inch.

(9) *Mauritius*
 Defended Port of Port Louis.
  R.G.A. Garrison:-
   Nos. 56 Coy. R.G.A.
   No. 4 Coy, Hong Kong—Singapore Bn. R.G.A.
  Armament:-
    2—9.2 inch
    2—6 inch
    4—4.7 inch.

(10) *Ceylon.*
 Defended Port of Colombo.
  R.G.A. Garrison:-
   No. 93 Coy. R.G.A.
   Ceylon Garrison Artillery.
  Armament:-
    4—9.2 inch
    4—6 inch.

(11) *Singapore.*
  R.G.A. Garrison:-
   Nos. 78 and 80 Coys. R.G.A.
   No. 5 Coy., Hong Kong—Singapore Bn. R.G.A.
  Armament:-
    5—9.2 inch
    4—6 inch.

(12) *Hong Kong.*
  R.G.A. Garrison:-
   Nos. 83, 87 and 88 Coys. R.G.A.
   Nos. 1, 2 and 3 Coys., Hong Kong—Singapore Bn. R.G.A.
  Armament:-
    8—9.2 inch
    11—6 inch
    4—4.7 inch.

CHAPTER XIX

1914 to 1918. (concluded)

AS has been stated above, never has Coast Artillery in the British Army been so ready and prepared for a great war as they were in August 1914, and, as fate would have it never have they had less to do in the way of action during a major struggle than they had during the First World War. The naval superiority of the Allies over the Central Powers was so complete that once the German cruisers, which were overseas when the war started, had been dealt with, surface activity of the German fleet was restricted to spasmodic efforts in the North Sea and at the eastern end of the Channel, and a few far flung activities by disguised raiders. As the German armies failed to capture the Channel ports, there was never any real threat of invasion. In fact, when the Germans began their submarine-campaign against Britain's trade-routes in 1915, they virtually admitted they no longer hoped for any substantial results from their surface ships. The Germans however did not abandon their attempts to cause Britain serious damage by raiding the east coast with some of their larger warships until the war had been underway for some six months, until in fact after the Dogger Bank action on 24th January 1915 had proved to them what dangerous operations these hurried raids across the North Sea were. During those first six months, one of these raids provided the one and only classic example of a contest between modern warships and coast-defences, the attack on the Yorkshire coast on 16th December, 1914. However, it is necessary first to deal with the activities of the German cruisers which were at large overseas when war broke out.

The strongest and most dangerous group of these was the German Pacific Squadron under Admiral von Spee which consisted of two heavy cruisers (*Scharnhorst* and *Gneisenau*) and three light (*Emden*, *Nurnburg*, and *Leipzig*) and which on 4th August, 1914 was concentrated at the German owned Caroline Islands. Besides this powerful force, there were two light cruisers (*Karlsruhe* and *Dresden*) in the Caribbean Sea and one (*Konigsberg*) at large off the coast of East Africa. There was also a number of German armed merchant-cruisers ready to proceed to sea to prey on the British trade-routes. Having detached the *Emden* to attack British sea-borne commerce in the Indian Ocean, von Spee with his squadron made for South American waters and, having caused much alarm and damage among the Allied South Pacific Islands on the way, on 12th October reached Easter Island where he found the *Dresden* which had come from the West Indies to join him. He then made for the coast of Chile where on 1st November off the small Chilean port of Coronel he met and destroyed Admiral Cradock's cruiser squadron (*Good Hope Monmouth* and *Glasgow*—only the last escaped to fight another day) which had been on the look out for him. However, von Spee and his squadron did not survive long. On 8th December, when approaching Port Stanley, Falkland Islands, where they hoped to destroy the port and any shipping found lying at anchor in the harbour, they were surprised by Admiral Sturdee's squadron of two battle-cruisers, three heavy cruisers, and two light cruisers, and the whole force, with the exception of the *Dresden* which managed to escape, was sent to the bottom.

The German light cruiser *Emden* (3,600 tons, 24 knots, 10—4 inch guns), which von Spee had detached to attack British sea-borne commerce in the Indian Ocean, on 7th September, unseen and unheralded, entered the Bay of Bengal where she began to creat havoc, by capture and sinking, among allied shipping using the busy trade-routes in that area. She started her deadly work on the main Calcutta-Colombo track on 10th September, and, working northwards, arrived off the mouth of the Rangoon River on the evening of 18th September, having destroyed seven merchant-ships of various nationalities on her way. Here Captain von Müller heard that British cruisers were after him, so he doubled back south and at about 9.30 on the evening of 22nd September the *Emden*, appearing off the entrance to Madras harbour, began to bombard the Burma Oil Company's tanks which stood close to the sea-front. The port was taken completely by surprise, and very soon the tanks were ablaze, but the Madras Volunteer Artillery, quickly manning their two 4.7 inch Q.F. guns—which, though mounted on field-carriages, had been placed in position for coast-defence—opened fire on the raider and forced her to withdraw to the southward. As the result of this attack, five people were killed 15 wounded, and half-a-million gallons of kerosene utterly consumed.

The *Emden* then made for the focal shipping area outside the port of Colombo and between 25th and 29th September captured seven more merchant-ships, but, learning of the approach of allied warships, slipped away south and reached Diego Garcia on 9th October. At this island von Müller was able to coal ship and take in supplies at his leisure as the inhabitants of this small, isolated British colony were unaware that a state of war existed between Britain and Germany. Allied cruisers were now closing in on the *Emden* from all directions so that von Müller hurridly steamed away north and, audaciously returning to the scene of his previous exploits off Colombo, rapidly scooped in seven more prizes. The *Emden* then dashed eastwards, and, having coaled at the Nicobar Islands, made for Penang which von Müller had most daringly decided to raid. In Penang harbour, quietly at anchor, were the Russian light cruiser *Zhemchug*, the French Torpedo-gunboat *D'Iberville*, and three French destroyers, and there were no coast-defences whatsoever to protect them, not a single gun. At about five o'clock on the morning of 28th October the *Emden*, disguised as a British cruiser with a fourth dummy funnel, painted dark grey, and flying the white-ensign, entered the harbour, the piquet boat on guard allowing her to pass unchallenged. When within 800 yards of the *Zhemchug*, the *Emden* hoisted the German colours, and firing a torpedo which hit the unlucky Russian cruiser towards her stern, engaged her with her guns at point blank range of about 300 yards. The surprise was complete: the *Zhemchug* was ablaze and sinking before anyone realised what was happening, before anyone either aboard the *Zhemchug* or the French men-of-war, was able to man their guns, much less fire them. The *Emden*, taking no notice of the other warships, circled round the *Zhemchug*, gave her another torpedo which struck her amidships and caused a tremendous explosion, and firing a final contemptuous salvo, sailed out of the harbour unmolested and unscathed. The *Zhemchug*, shattered and in flames, went down in a cloud of smoke and debris within 15 minutes of the firing of the first shot with a loss of 91 men killed and 108 wounded. All of which goes to show what can happen when warships lie peacefully at anchor in a harbour which is not protected by coast-defences.

There is no need to follow the career of the Emden much longer. Having sunk the French destroyer *Mousquet* which she encountered outside Penang harbour, she steamed for the Cocos Islands, coaling on the way from the captured collier *Exford*, and, entering the harbour of Direction Island at dawn on 8th November landed a party to destroy the cable-station. But her fate was upon her. The telegraph officials had already recognised who their strange visitor was and had sent out an alarm message both by wireless and cable. Very shortly, before the landing-party had finished their work of destruction and could rejoin her, the *Emden*, having sighted on the horizon what was evidently the smoke of an approaching warship, had to raise anchor and steam out to meet her opponent which turned out to be *H.M.A.S. Sydney* (5,400 tons, 25.7 knots, 8—6 inch guns.) This was about 9 a.m. By 12 noon the *Emden* lay a shattered and blasted wreck upon the foreshore of North Keeling Island, her gallant captain and most of her crew dead or dying. She had put up a brave and spirited resistance, but the heavier metal of the *Sydney* had been too much for her, and she had gone to her final resting place with her flag flying and her few remaining guns still firing.

The German light-cruiser *Königsberg* (3,350 tons, 24 knots, 10—4 inch guns) had been preying upon the British trade-routes leading to Aden from the ports of East Africa since the outbreak of war and had managed to evade all attempts to bring her to action. At 5 a.m. on 20th September she was approaching the busy and prosperous little port of the British protectorate of Zanzibar where, anchored in the bay, quite oblivious of the proximity of the enemy warship, lay the British light-cruiser *Pegasus* (2,135 tons, 21 knots, 8—4 inch guns) and where there were no coast-defences to protect the harbour. Caught at anchor without steam up, the *Pegusas* never had a chance, and, after 30 minutes of unequal contest, turned over and sank with a loss of 24 killed and 55 wounded, the *Königsberg* steaming away unflurried and undamaged as soon as she was sure that her task was accomplished. Which again goes to show how dangerous it is for a warship to lie peacefully at anchor in a harbour which is unprotected by coast-defences. The *Königsberg* did not meet her final end until July 1915 when, embedded far up the estuary of the Rufiji River, she was blasted to pieces by the fire of monitors specially brought out from Britain to deal with her. The *Karlsruhe* and *Dresden* did not survive so long, the former mysteriously blowing up at sea when some 300 miles south of Barbados on 4th November 1914 and the latter being sunk by H.M.S. *Glasgow* at Juan Fernandez on 4th March, 1915. With the destruction of these enemy cruisers, the coast-artillery abroad were able to relax a little and were left only with their long and monotonous watch and ward. So complete indeed was our command of the seas, that even the great Mediterranean coast-fortresses of Malta and Gibraltar were never brought into action.

It is now necessary to return to home waters and the "Narrow Seas". In the opinion of the Admiralty the enemy was unlikely to attempt any raids except across the North Sea. The "Grand Fleet" was based on Scapa and Cromarty, special patrol forces of destroyers were based on the Forth, the Tyne, the Humber, Harwich, the Thames estuary, and Dover, and local defence flotillas were stationed at the various lesser ports along the East Coast. The coast-defences of the two bases of the Grand Fleet, Scapa Flow and Cromarty, were, when war broke out, extremely inadequate. The Orkney R.G.A. (T.F.) had been specially raised and organised to man the Scapa Flow defences, but by the end of July 1914 not a coast-defence gun had been mounted. The Admiralty naturally was much disquieted by this, and tremendous efforts were made at once to get guns—mostly naval—put into position to cover the various entrances to the anchorage. At first 12 pdr., Q.F. and 3 pdr. Q.F., landed from warships, were mounted, but these were gradually increased by 4 inch and 6 inch guns until adequate defences had been erected to protect the main base of the Grand Fleet. These defences were manned by a joint command of Royal Marines and Orkney R.G.A. (T.F.) under Lieut.-Colonel G. N. Harris, R.M.A. The coast-defences of Cromarty were under the sole control of the Royal Navy right from the beginning, and naval guns, mounted on shore and manned by Royal Marines, guarded this base.

The enemy did not make their first raid across the North Sea until the autumn. On the evening of 22nd November a squadron of three battle-cruisers (*Seydlitz*, *Moltke* and *Von Der Tann*) three heavy cruisers, three light cruisers and a flotilla of destroyers left the Heligoland Bight and made across the North Sea for the coast of East Anglia. At seven o'clock the next morning in the thick mist of early dawn the raiding squadron ran into the *Halcyon* (mine-sweeping gunboat) and two destroyers from the Yarmouth Local Defence Patrol some four miles south of the Cross Sands lightship. Giving the alarm immediately, these quickly made off into the mist under cover of a smoke-screen and managed to get away safely without loss. Meanwhile, the German battle-cruisers had opened fire on Yarmouth and Gorleston with their 11 inch guns, but in the early morning mist their observers could not see what they were firing at so that the rounds fell harmlessly either on the open beach or in the sea. The German warships did not wait for the mist to rise but, fearing they might be cut off from their base by superior forces, fled back across the North Sea as fast as they could go, losing one heavy cruiser, the *Yorck*, from mines on the way, and managed to return to harbour safely before any of the strong British forces, set in motion to intercept them, could catch up with them. In spite of the powerful force employed, this was merely a "tip and run" raid of the most useless kind, the only object gained being the sowing of mines under cover of the operation. The Germans subsequently reported "The bombardment of English coast towns was successfully carried out on 3rd November. Early that morning our battle-cruisers appeared off Yarmouth to bombard the harbour and its fortifications while mines were laid under their protection." (Scheer.)

The next operation was planned on a grander scale. Its object was "to bombard the fortified coast-towns of Scarborough and Hartlepool and to lay mines along the coast" (Scheer). The raiding force was to consist of four battle-cruisers (*Seydlitz*, *Moltke*, *Von Der Tann* and *Derfflinger*) one heavy cruiser (*Blücher*) four light-cruisers, and two flotillas of destroyers, all under the command of Vice-Admiral Hipper who flew his flag in the *Seydlitz*, and the High Sea Fleet itself was to come out of its harbour into the North Sea in support of the raiding squadron and to cover its retirement after it had struck its blow. Information that an operation across the North Sea was impending reached the Admiralty by the second week in December, and certain dispositions of the Grand Fleet and of the destroyer patrol and local defence forces were made with a view to heading off any enemy raiding-force from the coast and

1914 to 1918 (concluded)

intercepting and destroying it before it could return to its base. During the evening of 15th December, Hipper's force steamed out into the North Sea, followed later by the High Sea Fleet, and at about dawn next morning, having crossed the North Sea unobserved by the Grand Fleet's scouts, made contact about five or six miles south-east of Hartlepool with the destroyers of the Local Defence Patrol which, having given the alarm and after an exchange of shots, scattered and made off to the north-east. The heavy ships of the German raiding force then split into two divisions, two of them, the *Von Der Tann* and *Derfflinger*, turning southwards towards Scarborough, and the remaining three, *Seydlitz*, *Moltke* and *Blücher*, making for Hartlepool.

It was shortly after eight o'clock when the *Derfflinger* (26,180 tons, 26.5 knots, 8—12 inch guns) and the *Von Der Tann* (19,100 tons, 26 knots, 8—11 inch guns) suddenly appeared out of the early morning mist off Scarborough. This pleasant Yorkshire sea-side resort was of no military importance and quite undefended. The two German battle-cruisers steamed leisurely up and down the coast at a range of from 2,000 to 3,000 yards, firing their great 12 and 11 inch guns into the helpless town. They shelled the coast-guard station, the empty cavalry barracks, the Grand Hotel, the suburb of Falsgrave, and Scarborough Castle, that ancient coast-fort which has so often figured in these pages but which had now neither gun nor invalid gunner to defend its charge. After some 30 minutes of this very one-sided contest, the two enemy battle-cruisers disappeared to the north-east.

Hartlepool was a defended port with two coast-defence batteries:-

Heugh Battery	2—6 inch guns
Lighthouse Battery	1—6 inch gun

manned by the Durham R.G.A. (T.F.), seconded by a regular master-gunner with his small District Establishment R.G.A., all under the command of Lieut.-Colonel L. Robson R.G.A. (T.F.) one time mayor of Hartlepool and now commanding the Durham Territorial Garrison Artillery. The two batteries were close together, the Lighthouse gun being just on the south side of the Lighthouse, and Heugh Battery about 75 yards to the north of it. All three guns were sited low, their height being only about 50 feet above the sea. Some $3\frac{1}{2}$ miles to the south-east was South Gare Battery (2—4.7 guns), part of the Tees Defences and not under Hartlepool Fire Command of which Robson was the fire-commander with his F.C. post just in rear of Heugh Battery. Most of the men of the Durham R.G.A. (T.F.) manning the guns were natives of Hartlepool, many of them living quite close to the batteries. The total manning detail was 11 Officers and 155 Other Ranks. The Port War Signal Station, operated by a naval petty-officer and his assistants, was in the lighthouse itself, and in the harbour that morning were two flotilla-leaders, *Patrol* (2,940 tons, 25 knots, 9—4 inch guns) and *Forward* (2,850 tons, 25 knots, 9—4 inch guns) and a submarine *C9*.

During the night a warning message had been received that strong enemy naval forces were at sea, but, as the guns were always manned one hour before daybreak and preparation for action carried out, the coast-artillery were in no way perturbed. Dawn on that morning of 16th December broke cold and misty with little wind and a calm sea. Not long before eight o'clock firing was heard out to sea, and soon afterwards South Gare reported seeing large warships steaming north. Immediately on top of this the P.W.S.S. urgently warned the fire-commander that three unidentified warships were coming in at great speed. Their identity was not long in doubt for, hoisting the German colours, at 8.10 a.m. precisely the leading ship of the three opened fire on the coast-defence batteries. The German plan was for all three ships to concentrate for the first 15 minutes on the guns with the object of putting them out of action. Then, during the next 30 minutes, the *Seydlitz* (24,600 tons, 26.5 knots 10—11 inch guns) and the *Moltke* (22,640 tons, 27 knots, 10—11 inch guns) were to bombard the harbour works and the town while the *Blücher* (15,500 tons, 25 knots, 12—8.2 inch guns) contained the batteries if any of their guns were still firing. Finally, all ships were to withdraw together, rendezvous with the *Derfflinger* and *Von Der Tann* off Whitby, and then steam in company at full speed across the North Sea to their base.

The first shell from the *Seydlitz* fell between the two batteries —a bronze tablet now marks the spot—cutting all the fire-commander's telephone-lines and thus putting the control of the ensuing action out of his hands right from the start. The three enemy ships rapidly closed into to between 5,000 and 4,000 yards, their shells falling all round the guns, but, as they were using "armour-piercing" with delay action fuzes, many bounced, ricochetting into the unfortunate town behind. The coast defence guns hotly returned the enemy fire, the two guns at Heugh Battery engaging the battle-cruisers and the Lighthouse gun the *Blücher*. Unfortunately, this last gun was severely handicapped by the lighthouse itself blocking its arc of fire to the north and by much trouble with its electric firing gear which put the gun out of action for some time. Nevertheless, the fire of the coast-guns soon began to take effect, hits being registered both on the *Moltke* and *Blücher*. At 8.25 a.m. the enemy warships began the second phase of their programme, the two battle-cruisers turning their guns onto the town while the *Blücher* fired old powder-filled shell in front of the coast-defence batteries, hoping to blind them by a primitive form of smoke screen. The battle-cruisers by this time had passed northwards out of the arc of Heugh Battery so that both guns were able to turn on the *Blücher* which was now drifting practically stationary at a range of about 4,000 yards. Several hits were obtained, including one on the forebridge which put two guns of the secondary armament out of action and wounded some seamen.

Meanwhile the two flotilla-leaders and the submarine of the Royal Navy, which were inside the harbour when the action began, were attempting to clear the bar and reach the open sea so as to be able to attack the enemy. As soon as the *Patrol* appeared in the bay, she was engaged by the *Blücher*, hit twice by heavy shell, and driven aground. The submarine *C9*, which was following close behind the *Patrol*, in spite of the shallow water was forced to dive, and, bumping hard on the bottom, took no further part in the action, while the *Forward* was unable to clear the bar until the contest was over. In the meantime the *Seydlitz* and *Moltke* had turned south to rejoin the *Blücher*, once more engaging the batteries as they passed, and then all three ships turned away together, making for the open sea. Heugh Battery continued to fire at them as they disappeared into the mist, the last round being fired at a range of 9,200 yards at 8.52 a.m., 42 minutes after the commencement of the action. As the fire-commander's communications were cut right from the start, the battle was fought throughout by the battery-commanders under "standing-orders" which had previously been prepared to deal with all possible forms of attack. Most of the engagement was carried out at Case II firing H.E. shell except that Heugh Battery took on the *Blücher* with auto-sights and A.P. shell during the second phase: 123 rounds were fired in all by the three guns. In spite of many enemy shells falling close to the batteries, the detachments suffered only two casualties, both killed. The excellent behaviour of the Durham R.G.A. (T.F.) throughout this action was acknowledged by the award of the D.S.O. to Robson, the fire-commander, and one D.C.M. and two M.M.s to the three gun-captains. It is sad to relate that the whole of Hipper's raiding squadron, preceded by the High Sea Fleet, returned safely to base, escaping the British forces which were so eagerly searching for them. The German warships fired 1,150 shells into Hartlepool, killing 112 and wounding more than 200 civilians and doing extensive damage to the docks and property.

The next time the Germans were not so lucky. The impunity with which they had raided Scarborough and Hartlepool encouraged them to try again, and on the evening of 23rd January 1915 Hipper set out with the *Seydlitz, Moltke, Derfflinger,* and *Blücher,* four light-cruisers, and a flotilla of destroyers, his object this time being "to reconnoitre off the Dogger Bank and to destroy any of the enemy's light forces met there" (Scheer). The British Admiralty had received information of the impending movement early on the 23rd and at night-fall Admiral Sir David Beatty put to sea with his battle-cruiser squadron, a squadron of heavy cruisers, a squadron of light cruisers, and two destroyer flotillas. About a half-hour before dawn next morning, the British advanced destroyers made contact with the enemy near the Dogger Bank, and during the ensuing battle the *Blücher* was sunk, the *Seydlitz* crippled and the *Derfflinger* damaged, in fact the whole German force would probably have been destroyed but for an unlucky shell which penetrated the feed-tank of the *Lion*, Beatty's flagship, bringing her to a halt and so depriving the admiral of control at the critical moment of the battle.

This reverse caused grave misgiving in Germany and brought about a change of command of the High Sea Fleet and of general naval policy at the German Admiralty. The new commander-in-chief of the High Sea Fleet was Admiral von Pohl, and his instructions were "that the maintenance of the fleet intact at that stage of the war was a necessity. The intention was, by frequent and constant advances of the High Sea Fleet, to induce the enemy to operate in the North Sea, thus either assuring incidental results or leading to a decisive battle under favourable conditions to ourselves, that is to say, close to our own waters" (Scheer). So attacks and raids on Britain across the North Sea by surface ships came to an end, and coast-artillery had no more violent actions. The activities indeed of coast-artillery generally in the First World War may be said to have come to an end with the raid on Hartlepool. The watch and ward had still to be kept at every naval base and defended port from Hong Kong to Bermuda, and from Scapa Flow to Simonstown, but, except for occasional sweeps into the North Sea—one of which in the early summer of 1916 brought about the Battle of Jutland—and the operations of disguised raiders in distant seas, German naval activities for the rest of the war were all concentrated on her submarine campaign against the British trade-routes, and coast-artillery could take no serious part in this. So the artillery manning the coast-defences throughout the Empire remained constantly on the alert through the wearisome years until November 1918 but were never called upon again to spring into action.

During the first year of the war a great number of strong and robust men were tied up in coast-artillery, but as the war progressed, and the enemy's naval activities, except for his vigorous submarine campaign, became less and less, it was found necessary to reduce the strength and take away for more active service the fit and healthy. Trench warfare proved the need for more and more great guns and howitzers in the field, and these were all manned by the R.G.A. which had to find its manpower from the highly trained and well disciplined regular coast-artillery companies and Territorial units. Heavy and siege batteries were raised in great numbers, both from regular and Territorials, and were formed, organised, and generally mothered by the coast-artillery units from which they were constituted. This process of finding the manpower for heavy and siege batteries from coast-artillery went on both at home and abroad, the auxiliary artilleries playing their part, the Hong Kong—Singapore Battalion producing a mountain-battery for service in the Middle East, the Cape Garrison Artillery forming heavy batteries for France and Flanders, and the remainder willingly giving up their personnel for siege batteries. As aircraft assumed a more prominent part in the war, the demand for anti-aircraft guns, and men to man them, became urgent and imperative, and once again the R.G.A. produced the required manpower from its coast-artillery pool. All this of course left the coast-defences very short of "manning details" which had to be filled up by those who had been wounded with the heavy and seige guns overseas and were now recovering, by those who were either too old, or too young for active service, by those who were halt and lame, in fact by all those who were unfit for any reason for the field of battle. Yet the watch and ward had to be kept continuously and efficiently, and it was never let up for one minute until 11 a.m. on 11th November 1918.

Before finishing with the First World War, it is fitting to quote from a special order of the day, issued by a fortress commander to a Territorial coast-artillery unit which had been manning the coast-defences throughout the war, on the closing down of defensive precautions:-

"On the conclusion of all defensive arrangements in the garrison, the General Officer Commanding wishes to convey his thanks to all ranks of the Royal Artillery and Royal Engineers who have manned the coast-defences of the port for the period of four years and four months. The duty has been hard and monotonous, and has not been relieved by the interest of repelling a hostile attack, but the work has been most cheerfully and efficiently carried out, and the General Officer Commanding has no doubt that, if the enemy had attempted any hostile action against the port, it would have been defeated with the same success that has crowned the efforts of the British Army in the other theatres of war. The conduct of the troops under most disheartening conditions has been excellent, and the General Officer Commanding can never sufficiently express his thanks to all ranks engaged in manning the defences, for the support they have so willingly and efficiently given him during the very trying period of the war. He wishes that he could thank each man personally, and hopes that Commanding Officers will convey to all ranks his extreme satisfaction with the manner in which they have upheld the great name of the British soldier."

CHAPTER XX
1918 to 1939.

WHEN Great Britain emerged from the First World War, she found herself without a naval rival among the European powers. The ships of the High Sea Fleet of her late enemy lay at the bottom of Scapa Flow, and the Treaty of Versailles severely restricted the size and tonnage of any fleet Germany might wish to build in the future: Russia was in the midst of her shattering revolution, and the ships of the Imperial Navy had either been seized by Britain and France or lay useless at anchor in Russian ports: Austria-Hungary had ceased to exist, and France and Italy, the only two naval powers left in Europe, were too exhausted and impoverished to be capable of maintaining fleets, which even combined, could threaten the superiority of Great Britain. However, there were two great powers outside Europe who were not suffering from the effects of the First World War, who in fact had come out of that war far richer and more powerful than they had been in 1914, and whose vital interests moreover clashed with each other in the Far East and Western Pacific. The decade following 1919 saw the probability of a naval armaments race between the United States of America and Japan, each of whom had the resources to build up a great powerful modern fleet and wished to dominate Eastern Asia, especially China, and control the oceanic communications of the Western Pacific.

Great Britain herself had great interests in the Far East which, in case of a conflict between U.S.A. and Japan, she would have to protect against one or the other, so in 1920 she found herself having to decide which of the two Pacific powers she was prepared to side with and support with her foreign policy, and possibly in due course also assist with her battle-fleet. She was already in alliance with Japan by the Anglo-Japanese Treaty of 1902, but Japan's expansion into China during 1916-18 and her quite undisguised aggressive ambitions in Eastern Asia forced Britain to realise that her interests in the Far East were in much greater danger from Japan than from the United States. Moreover, friendship and co-operation with the latter became after 1919 one of the basic principles of Britain's foreign policy, the British Government fully appreciating that a naval armaments race between the two English-speaking peoples would be a moral disaster, and furthermore that the U.S.A. had the financial ability to outbuild Great Britain if she were determined to do so. So the Anglo-Japanese Alliance was not renewed, and Great Britain's policy in the Far East developed into resistance against the aggressive tendencies of Japan and co-operation with the United States as far as that was possible. However, whereas the Anglo-Japanese Alliance was a definite treaty, the "isolationist policy" of the U.S.A. made any such exact agreement between the two English-speaking powers impossible, and Great Britain had to rely upon a general similarity of interests and policy and a special feeling of friendliness and fraternity to produce the necessary assistance and provide an alliance when the crucial time came. In any case, her decision to be prepared to resist Japanese ambitions in Eastern Asia—with or without American assistance—involved a challenge to Japanese naval power and thereby necessitated, in case of war, the presence of a British battle-fleet in Far Eastern waters. Such a fleet would require a base—in 1920 no British battle-fleet base existed east of Malta—and, while a fleet could be moved from Europe to the Far East in some two or three months, a fleet-base with all its docks, dockyards, shore-installations, and defences required many years for its building. It was early in 1921 then that the British Government decided to construct such a base, and Singapore was chosen for its location.

It can be truly said that the Singapore base quite overshadowed everything else as far as Coast-Artillery and coast-defence were concerned during the period between the two World Wars. The development of the base unfortunately pursued a somewhat erratic career, and its progress was twice interferred with by Labour governments at home—1924 to 1925 and 1929 to 1931—but from 1932 onwards the British government was determined to construct a first-class base at Singapore and to protect it by a coast-fortress of superlative strength. No expense was spared, and great guns were mounted, most modern instruments installed, and large establishments constructed on a scale and with a lavishness which were formerly practically unknown to coast-defence and reflected the strong resolution of the British government to provide a base in the Far East which would furnish a safe haven and dockyard for the battle-fleet, securely defended against any efforts an enemy might make against it. The planning and construction of the Singapore coast-fortress during the period 1922-1939 prevented coast-artillery and the art and practice of coast-defence from sinking into the usual trough of neglect, penury and indifference which were their normal fate between wars. Novel systems and methods had to be thought out and put into practice, new and complicated instruments devised and tested, and more effective guns designed and mounted. All this kept coast-artillery very much alive, especially the School of Gunnery—renamed in 1919 the Coast Artillery School—at Shoeburyness which was the main source and testing ground for all these new conceptions.

The immediate post-war years were as usual a period of strict economy in all defence expenditure, and coast-artillery and coast defences both at home and abroad were left to lie fallow very much in the same state as they had entered the war. Retrenchment was the watchword and parsimony the rule so that all coast-defences—except perhaps the nascent coast-fortress at Singapore—tended to become forgotten and neglected. It will be recalled that when the First World War broke out in 1914, the coast-defences both at home and abroad were manned mainly by the R.G.A., a regular force organised in companies and consisting of a large number of highly trained and most efficient officers and men. In spite of the fact that Britain's foremost enemy, Germany, possessed a powerful fleet whose bases lay only just across the North Sea, and, at the start of the war, disposed of several enterprising cruisers positioned abroad ready to prey upon Britain's commerce and trade-routes, scarcely any of the R.G.A. was ever called upon to fight its guns in action against enemy ships and, as the war progressed, most of its man-power was drafted away to the field-army to provide the personnel for the heavy and siege batteries. In effect, the guns of the coast-defences proved most capable deterrents, and enemy commanders were very rarely prepared to risk their ships within range of them. Very naturally, when the whole problem of coast-artillery became to be considered after the War, it was seriously questioned whether a force such as the R.G.A., absorbing a large number of highly-trained regulars and above all costing a great deal, was really necessary for manning the coast-defences. After much deliberation and taking into consideration Britain's naval superiority in European waters and the paramount need for economy, it was decided that at home regulars were not required, and the coast-defences could safely be handed over to the Territorials.

This change over however did not take place immediately after the end of the War. It took some years before the War Office could decide on a definite policy for coast-artillery. In 1920, the regular R.G.A. on coast-defence at home was organised in Fire Commands, numbered from 1 to 36, each commanded by a field officer and composed of a "district establishment" and one or more coast-batteries which were designated by letters. In 1921 the Fire Commands were abolished, and there were now only coast-artillery stations, each commanded by a lieut-colonel or major according to its size and importance, with district establishments and coast-batteries. Abroad the pre-war organisation of stations and companies remained until 1923 when a start was made on altering the latter into lettered coast-batteries. This had not been completed before a major change took place in the whole organisation of the Royal Artillery. The War had shown that heavy and siege batteries were more akin to field than to garrison artillery, that anti-aircraft artillery would be an ever growing necessity, and that coast-artillery was not as important as it had been thought to be before the War, so that in the post-war organisation of the R.G.A. a large number of mechanised heavy and siege batteries—now called medium batteries—were retained, mobile anti-aircraft brigades were constituted, and coast-artillery only performed a minor part. In fact four-fifths of the whole Regiment were now on wheels, their training and outlook were similar, and it was therefore no longer necessary to maintain a separate corps for the now small and less important static portion. So in January 1924 the cleavage of 1899 was healed, and once more the Regiment was united as one single corps, the Royal Artillery. This re-union brought about a change in designation of coast-artillery units, the lettered coast-batteries now becoming numbered heavy batteries, this term being chosen to maintain the now all mobile appearance of the Regiment, and to show that, although coast-defence was the primary role of these batteries, they were also trained and ready to man the heavier guns and howitzers in the field if called upon to do so.

At last, towards the end of 1926, the War Office came to a firm conclusion concerning the future policy for coast-defence both at home and abroad. It was decided that at home, except in Ireland, the manning of the coast-defences should be handed over to the Territorials, whilst in Ireland—which was now divided into the Irish Free State and Northern Ireland—and abroad they should still be held by regular batteries. The general organisation laid down for Great Britain was that Coast Artillery should come directly under the Brigadier R.A. of each Command—except in Southern Command where there was to be a Brigadier, Coast Defences under the B.R.A.—and that each group of defences should have:-

(1) *Regulars.*
    Headquarters.
        Lieut.-Colonel commanding at Major Ports
        Major commanding at Minor Ports
            (to become Commander, Coast Defences on mobilisation.)
        with Headquarters Staff.
    District Establishment.
        Specialists (Range Takers, etc.)
        Master Gunner and Maintenance Staff.
(2) *Territorials.*
    Manning Establishment.
        Territorial C.O.s to be Fire Commanders.

This re-organisation, which was carried out early in 1927, entailed the disbandment of all the heavy-batteries on coast-defence in Great Britain, only one being retained (No. 21) at Shoeburyness as depôt battery to the Coast Artillery School. On the other hand the District Establishments were increased and included, not only the master-gunners with their maintenance staffs, but instrument specialists on a scale of one complete detachment per instrument as it was thought that Territorial specialists would not be sufficiently trained to take over the range-finders etc. immediately on mobilisation.

Now to consider the Territorials. The Territorial Force was resuscitated after the War in 1920 and in 1921 became the Territorial Army, the main difference being that the Territorial soldier was now liable to serve overseas in time of war instead of at home only as had been his previous terms of service. The Territorial coast-artillery units were reformed with the rest of the T.F. in 1920 and at the time of the change over to Territorial Army were:-

    Tynemouth
    Hampshire
    Devon
    Sussex
    Forth
    Cornwall
    Kent
    Clyde
    North Scottish
    Essex and Suffolk
    Lancashire and Cheshire
    Dorsetshire
    Glamorgan
    East Riding
    Pembroke
    Durham
    Orkney.

The decision in 1926 that the coast-defences of Great Britain should be manned by the T.A. only (it should be remembered that this also included the manning of the searchlights by Fortress R.E. of the T.A.) brought about some re-organisation, amalgamation, and reductions and by 1932, when the scheme for the coast-defences of Great Britain was put into its final form, the coast-artillery was organised as follows:-

(1) *Coast Defences, South Western Ports.* H.Q. Plymouth.
    The defended Ports of Plymouth and Falmouth.
        Headquarters and District Establishment, R.A.
        Devon and Cornwall Heavy Brigade, R.A. (T.A.)

(2) *Coast Defences, Southern Ports.* H.Q. Portsmouth.
    Defended Ports of Portsmouth and Portland.
        Headquarters and District Establishment, R.A.
        Hampshire Heavy Brigade, R.A. (T.A.)
        Dorsetshire Heavy Brigade, R.A. (T.A.)

(3) *Coast Defences, Eastern Ports.* H.Q. Sheerness.
  Defended Ports of Dover, Medway, Thames, Harwich and Newhaven.
    Headquarters and District Establishment, R.A.
    Kent and Sussex Heavy Brigade, R.A. (T.A.)
    Thames and Medway Heavy Brigade, R.A. (T.A.)
    Suffolk Heavy Brigade, R.A. (T.A.)
(4) *Coast Defences, Northern Ports.* H.Q. Tynemouth.
  Defended Ports of Tynemouth, Tees, Hartlepool and Humber.
    Headquarters and District Establishment, R.A.
    East Riding Heavy Brigade, R.A. (T.A.)
    Durham Heavy Battery, R.A. (T.A.)
    Tynemouth Heavy Battery, R.A. (T.A.)
(5) *Coast Defences, Scottish Ports.* H.Q. Leith.
  Defended Ports of Forth and Clyde.
  Headquarters and District Establishment, R.A.
    Forth Heavy Brigade, R.A. (T.A.)
    Clyde Heavy Battery, R.A. (T.A.)
(6) *Coast Defences, North-Western Ports.* H.Q. Liverpool.
  Defended Ports of Mersey and Barrow.
  Headquarters and District Establishment, R.A.
    Lancashire and Cheshire Heavy Brigade, R.A. (T.A.)
(7) *Coast Defences, Welsh Ports.* H.Q. Pembroke Dock.
  Defended Ports of Milford Haven and Severn.
    Headquarters and District Establishment, R.A.
    Pembroke Heavy Brigade, R.A. (T.A.)
    Glamorgan Heavy Battery, R.A. (T.A.)

The coast-defences of the Channel Islands—for reasons of economy—were dismantled and abandoned in 1929, and the artillery-corps of the Channel Islands Militia, after so many, many years of faithful service, were disbanded.

As has already been stated, the War Office decided that regulars should be retained in Ireland and abroad for manning the coast-defences. Three of the four coast-artillery stations in Ireland—Lough Swilly, Queenstown Harbour, and Berehaven—were in the Irish Free State which was now a dominion and clamouring to have the last British soldiers removed from its soil. The Antrim and Cork Militia Artilleries had not been revived after the War, so that the units of the Irish defences had to rely upon their own resources for manning the guns. Abroad the general situation had not altered very much since 1914. The companies had become heavy batteries, and, as the base progressed, Singapore gradually developed into the most important defended port abroad, but otherwise everything remained very much the same. The Cape of Good Hope—the defended ports of Capetown and Simonstown—was handed over to the Union of South Africa in 1920, the R.G.A.

being the last British troops to leave the Union. The Sierra Leone Company R.G.A.—for reasons of economy—was disbanded in 1922, and Sierra Leone itself ceased to be a defended port from 1929 onwards. However, the Royal Malta Artillery continued to flourish, and the Hong Kong—Singapore Royal Artillery provided invaluable additions to the regular coast-artillery of the two far-eastern ports. The Mauritius Company was not retained after the War, so that the H.K.S.R.A. consisted of one mountain-battery and three heavy-batteries at Hong Kong and one heavy-battery at Singapore. In India the native content of the coast-artillery had become so large during the War that a separate Indian Coast Artillery, stationed at Bombay and Karachi, had been formed, this becoming the Indian Heavy Battery in 1924, but it was disbanded in 1926, the Indian coast defences being left to the British heavy-batteries with Indian auxiliaries. In the western Atlantic, the coast-defences of Bermuda and Jamaica were in the hands of one regular heavy-battery stationed at Port Royal (Jamaica) and the Bermuda and Jamaica Militia Artilleries. In April 1933 the coast-artillery commands in Ireland and abroad were for the first time termed "heavy brigades" and numbered from two to eight (1st Heavy Brigade being a mobile field formation), and the general organisation was as follows:-

(1) *North Irish Coast Defences.* H.Q. Belfast.
    Defended Ports of Belfast and Lough Swilly.
        Headquarters and District Establishment, R.A.
            17th Heavy Battery, R.A.

(2) *South Irish Coast Defences.*
    Defended Ports of Queenstown Harbour and Berehaven.
        2nd Heavy Brigade, R.A.. H.Q. Queenstown Harbour.
            1st Heavy Battery, R.A.
            8th Heavy Battery, R.A.
            19th Heavy Battery, R.A.
            26th Heavy Battery, R.A.

(3) *Indian Coast Defences.*
    Defended Ports of Bombay and Karachi.
        5th Heavy Brigade, R.A. H.Q. Bombay.
            13th Heavy Battery, R.A.
            14th Heavy Battery, R.A.

(4) *Ceylon.*
    Defended Port of Colombo.
        6th Heavy Brigade, R.A. H.Q. Colombo.
            15th Heavy Battery, R.A.
            18th Heavy Battery, R.A.
            Ceylon Garrison Artillery (Volunteers.)

(5) *Singapore.*
        7th Heavy Brigade, R.A. H.Q. Singapore.
            11th Heavy Battery, R.A.

22nd Heavy Battery, R.A.
3rd Heavy Battery, Hong Kong—Singapore R.A.
(6) *Hong Kong.*
8th Heavy Brigade, R.A. H.Q. Hong Kong.
12th Heavy Battery, R.A.
20th Heavy Battery, R.A.
24th Heavy Battery, R.A.
Hong Kong—Singapore Brigade, R.A. H.Q. Hong Kong.
2nd Heavy Battery Hong Kong—Singapore R.A.
4th Heavy Battery, Hong Kong—Singapore R.A.
5th Heavy Battery, Hong Kong—Singapore R.A.
(7) *Gibraltar.*
3rd Heavy Brigade. H.Q. Gibraltar.
4th Heavy Battery, R.A.
27th Heavy Battery, R.A.
29th Heavy Battery, R.A.
(8) *Malta.*
4th Heavy Brigade. H.Q. Tigne.
6th Heavy Battery, R.A.
10th Heavy Battery, R.A.
Royal Malta Artillery. H.Q. Valetta.
1st Heavy Battery, R.M.A.
2nd Heavy Battery, R.M.A.
3rd Heavy Battery, R.M.A.
(9) *Aden.* H.Q. Aden.
Headquarters and District Establishment, R.A.
9th Heavy Battery, R.A.
(10) *Mauritius.* H.Q. Fort George.
Defended Port of Port Louis.
Headquarters and District Establishment, R.A.
25th Heavy Battery, R.A.
(11) *Jamaica.* H.Q. Port Royal.
Defended Ports of Port Royal and Kingston.
Headquarters and District Establishment, R.A.
22nd Heavy Battery, R.A.
Jamaica Militia Artillery.
(12) *Bermuda.* H.Q. St. George's.
District Establishment, R.A.
Bermuda Militia Artillery.

CHAPTER XXI

1919 to 1939 (concluded)

ALTHOUGH the British Government had realised as far back as 1921 the necessity for the construction of the naval base and coast-fortress at Singapore, by 1931 no very great steps had been taken to press forward with the work. There were three main reasons for this, firstly the Government assumption that there would be no major war for ten years, secondly the necessity for strict economy in the spending of money for defence purposes following the prodigal expenditure of the war years, and thirdly the presumption that the League of Nations and the system of collective security would effectively prevent any outbreak of war among the Great Powers. However by 1932 the British Government was convinced that Japan was pursuing an aggressive policy in Eastern Asia which would in the none too far future bring her into conflict with Great Britain—and possibly the United States as well—so in March of that year the Cabinet decided to cancel the ten year assumption and to give priority to re-armament in the Far East. It was laid down that fully protected fleet-bases should be completed at Singapore, Hong Kong and Trincomalee as soon as possible, Singapore being the most important of the three.

As has already been stated, up to 1931 not very much work had been done on the new Singapore coast-fortress. Previously the coast-defences of Singapore, mounted on the two islands of Blakang Mati and Pulau Brani, had protected the harbour of Singapore city only, but, as the new naval-base was being constructed on the northern shore of Singapore Island opposite the mainland on the Johore Strait, the principal coast-defences of the fortress had to be sited to cover the eastern entrance of the Johore Strait which meant that the actual batteries themselves had to be erected at the eastern end of Singapore Island *i.e.* in the Changi area, or on the small islands which obstructed the entrance to the Strait. As the result of the government decision to press on with the construction of the base and fortress, the War Office issued orders during the summer of 1932 for plans of all the proposed new coast-defence batteries and cantonments to house the troops to be prepared as soon as possible "so that the work will be commenced before the end of 1934." In actual fact work was started long before that, by August 1934 the guns of three batteries (Tekong Besar: 3—9.2 inch guns. Sphinx: 2—6 inch. Changi: 2—6 inch) were already in position—and from 1933 onwards until war with Japan broke out in December 1941 there was never any pause or lull in the

1919 to 1939 (concluded)

intense activity of preparing the coast-defences of Singapore Fortress.

Very little could be done at first to augment or improve the coast-defences of Hong Kong as that colony came within the bounds of Article XIX of the Washington Treaty for the Limitation of Armaments 1922 which forbade any increase, either in size or numbers, of the guns defending fortresses within its area. However work was set in hand to remount the guns already there on more modern carriages, thereby increasing their range, or to move them to better positions. However in December 1934 Japan denounced the Washington Treaty—a sure sign that she was meditating aggression—and the War Office at once called for a report on the state of the coast-defences of Hong Kong. This report (1935) stated that 3—9.2 inch guns, which were to be mounted on 35° carriages which would give them a range of close on 30,000 yards, were already being shifted to Stanley where a new battery was being constructed. The report went on to say that the whole range-finding system was inadequate and out of date, the counter-bombardment batteries—except for the new one at Stanley—possessed insufficient range, and the close-defence system lacked depth and its guns were wrongly sited. The report strongly recommended that every effort should be made to bring the coast-defences up to date, that the "Fortress Observation System" should be installed at once, that a new counter-bombardment battery should be erected at D'Aguilar, and the whole close-defence system reconstructed. These recommendations were accepted by the War Office, and work was begun to implement them as soon as money was made available. Trincomalee, the fine harbour on the east coast of Ceylon, had been abandoned as a defended port before 1914 and possessed no defences whatsoever, so it was decided that it should be reconstituted as soon as possible and armed with 2—9.2 inch guns and 3—6 inch. It was also considered necessary—to complete the Far Eastern and Indian Ocean design of naval-bases and defended ports—to defend Penang (4—6 inch guns) and Kilindini (2—6 inch guns).

All this expansion of coast-defence armaments in the East clearly required an increase in coast-artillery to man them. This necessity was never fully faced up to—for reasons of economy—and right up to the outbreak of war with Germany in September 1939—and even afterwards—there was continual acrimonious correspondence between the War Office and the various Commands abroad concerning Higher and Lower Colonial Establishments, Manning Tables, and "Rules for calculating Royal Artillery Garrisons". Even in June 1939 it had not been decided whether the extra man-power needed at Singapore to bring the coast-artillery up to war-establishment on mobilisation should come from the Chinese or Tamil settlers in the Straits Settlements. In actual fact, neither were found suitable, and a recruiting party had to be sent off hurriedly to India to raise additional H.K.S.R.A. personnel. However, to go back to 1934, a new Heavy Brigade, the 9th, consisting of 7th and 22nd Heavy Batteries, was constituted during that year at Singapore for the coast-defences at Changi, and in 1939 a battery of East African natives was raised from the King's African Rifles, thus forming the first artillery unit that famous colonial-corps ever included in its establishment, for the new defences at Kilindini. The Hong Kong—Singapore R.A.—their Indian personnel costing less than British gunners—was also called upon to expand to furnish extra units. In 1936 the H.K.S. Brigade at Hong Kong ceased to be coast-artillery altogether and was re-organised to provide the mobile artillery (two mountain batteries, two medium batteries) for landward defence, and in 1938 three new heavy batteries were raised, 7th H.K.S. for Singapore, 8th H.K.S. for Penang, and 'X' H.K.S. for Hong Kong; in the same year three new British heavy-batteries were also formed, 30th Heavy for Hong Kong and 31st and 32nd Heavy for Singapore.

A more detailed account of the expansion of the Hong Kong—Singapore Royal Artillery must now be given. In May 1934 this excellent Indian corps of artillery consisted of one mountain and three heavy batteries at Hong Kong, and one heavy battery at Singapore. By September 1939 it had grown into two mountain batteries, two medium batteries, and one heavy battery at Hong Kong, six anti-aircraft batteries and two heavy batteries at Singapore, and one heavy battery at Penang. Moreover, the extra man-power required to bring the British heavy batteries, both at Hong Kong and Singapore, from peace to war-establishments, had also to be produced from H.K.S.R.A. personnel, the local Asiatics—as has already been stated in the case of Singapore—proving unsuitable. The H.K.S.R.A. had always been recruited from the best class of Punjabi Mussalmans and Sikhs, and a very high standard of recruit was thus steadily maintained, but this rapid and large expansion forced the corps to lower its standards, to enlist Punjabi Mussalmans and Sikhs of inferior quality, and to extend their recruiting to Jats and other Hindu classes. The last group of 350 to be enlisted for Singapore in a great hurry in March 1940, because it had been found—after years of fruitless controversy—that Straits Settlements Chinese and Tamils were completely unsuitable to perform the duties of "higher gun numbers", were graded as Category III Personnel, only fit for unskilled labour! This tremendous enlargement meant not only a steep drop in the standard of the rank and file but also in the quality of the Indian officers and N.C.O.s which together had the most disastrous results on the morale, efficiency, discipline, and loyalty of the units. It was a shortsighted policy which ruined this fine corps of Indian artillery by such reckless and improvident expansion.

1919 to 1939 (concluded)

It was not only in the East that Great Britain found herself threatened by aggressive enemies. The rise to power of the two dictators, Hitler and Mussolini, the constitution of the Berlin-Rome Axis, and the failure of the League of Nations and the system of collective security forced Britain to reappraise her position in the "Narrow Seas" and the Mediterranean, with the result that coast-defences, which had been neglected and almost forgotten since 1919, were once more recalled to mind and given attention by the War Office. In 1937 the number of Territorial Army Heavy Brigades was increased, and the Antrim Heavy Battery, R.A. (T.A.), the first Territorial Army unit ever to be produced in Ireland, was formed to man the coast-defences of Belfast. During the next year Sierra Leone was revived as a defended-port, a special Sierra Leone Heavy Battery being raised for its defences, and at the eastern end of the Mediterranean, Haifa with its oil pipe-line terminus and installations was armed with 4—6 inch guns, 19th Heavy Battery, R.A. being sent there to man them. On the other hand, in July (1938) the coast-defences of Lough Swilly, Queenstown Harbour, and Berehaven were handed over to Eire (Irish Free State) as a gesture of trust and goodwill, and the regular heavy batteries withdrawn from the North and South Irish Coast Defences, a transaction which was to cost Britain dear in the near future.

In March 1938 the Regiment was once again divided into two parts, although on this occasion the split was not so drastic as it had been in 1899. There was no talk this time of two distinct corps, but only of two separate branches:-

(a) *Field Branch.*
    R.H.A.
    Field.
    Mountain.
    Medium.
    Field Army Heavy.
    Survey.

(b) *Coast Defence and Anti-Aircraft Branch.*
    Coast Defence Heavy.
    Anti-Aircraft.
    Anti-Tank.

The main reason for this division was the same as it had been before; a large part of the Regiment had become static and did not indulge in fire and movement on the field of battle. This was entirely due to the ever increasing threat of attack by enemy aircraft on the home-country and our naval bases and defended ports abroad, and the consequent rise in importance and great increase and expansion of anti-aircraft artillery. In addition to those already allotted for service with the field army, the whole of Great Britain and Northern Ireland had to be covered by a vast system of A.A batteries, provided by the Territorial Army, and every defended port had to have its quota of A.A. guns. Most of this A.A. artillery was substantially static in its function and had nothing in common in its duties, gunnery, tactics, and organisation with the artillery of the field-army. Therefore, as Coast Artillery had never been considered to have anything in common with the artillery of the field-army, the two were now joined together as the "Coast Defence and Anti-Aircraft Branch".

The Army Order, which announced the division of the Regiment, laid down that:-
- (a) There would be a combined list of officers of both branches for both promotion and posting up to and including the rank of Captain.
- (b) Majors would be on separate lists for promotion and posting in each branch.
- (c) Lieut.-Colonels would be on separate lists for posting in each branch.
- (d) Officers of the C.D. and A.A. Branch would be posted to any unit in that Branch.
- (e) Majors would be considered according to their recent experiences in, and suitability for, a particular branch, and as far as possible in accordance with their wishes. There would always be a limited number of exchanges of field officers between the two branches.
- (f) Other Ranks would be enlisted either into the Field or C.D. and A.A. Branch.

At the same time the term "Brigade", as denoting a Lieut.-Colonel's command of a group of batteries, was abolished, and "Regiment" introduced in its place.

In 1919 there were only three first-class naval powers left in the world, Great Britain, U.S.A. and Japan, and everything pointed to an intense warship building race between the three of them. However, the good sense and integrity of their statesmen and leaders in power at that time prevented them from entering this suicidal competition, and in February 1922 the Five Power Naval Treaty was signed at Washington between Great Britain, U.S.A., Japan, France, and Italy by which "the high contracting powers" agreed that:-
- (a) No new battleships should be built for 10 years and then none over 35,000 tons.
- (b) No cruisers of more than 10,000 tons should be built.
- (c) The ratio of naval strengths of Great Britain, U.S.A. and Japan should be represented by the figures 5: 5: 3.

This treaty effectively held up the naval building race until December 1934 when it was denounced by Japan with the result that the competition was once again thrown open to all entrants with Japan well in the lead and Germany and Italy joining in.

1919 to 1939 (concluded)

In the main the Coast Artillery problem did not change very much between the two World Wars from what it had been in 1914. The probable targets were still the battleship, heavy cruiser, light cruiser, blockship and destroyer. These of course had all increased somewhat in size and effectiveness since 1914, and added to them were the aircraft-carrier and fast motor torpedo-boat, but the former always kept well away out of range of any coast-defence gun while the latter replaced the destroyer as the threat to ships lying peacefully at anchor in harbour and demanded a special gun to deal with it. As Britain did not expect to meet a power armed with 35,000 tons (or over) battleships except in the East, the 9.2 inch gun was considered quite sufficient to contend with the Washington Treaty 10,000 ton, 8 inch gun cruiser, and the 6 inch with the light cruiser, blockship, and destroyer, but both had to be given extra range. For the 35,000 ton (or over) battleship, a 15 inch gun was considered necessary for Singapore, and for the motor torpedo-boat a special twin-barrelled 6 pdr. was designed to replace the old 12 pdr. Q.F. as the guardian of harbour entrances.

The guns intended for dealing with battleships, heavy cruisers, and light cruisers lying off a port attempting to bombard the ships at anchor, the dockyard, shore-installations, establishments etc., that is to say the 15 inch, 9.2 inch with 35° elevation mounting, and the 6 inch with 45° elevation mounting, were known as counter-bombardment guns. These guns had extreme ranges of 42,000 yards, 30,000 yards and 24,500 yards respectively, and the old system of range-finding with a Battery Depression Position-Finder was obviously no longer adequate nor effective. The Fortress System was therefore evolved to cope with these increased ranges. This system consisted of a series of Fortress Observation Posts, from four to ten thousand yards apart, sited to cover by observation all the water areas within range of the counter-bombardment guns. High sited O.P.s were equipped with a D.P.F., low sited ones with a Director for taking bearings and a Barr and Stroud range-finder for measuring ranges; all fed their bearings and ranges to the Fortress Plotting Room which was situated in a central position and strongly protected. Each O.P. also had a stereoscopic telescope for observing the fall of shot. At the Fortress Plotting Room was the Fortress Plotter, the function of which was to transform the bearings and ranges as they came in from the O.P.s into map co-ordinates and to chart the course of the target. The co-ordinates were then sent by telephone to the battery selected to engage the target where they were converted into battery bearings and ranges by a Co-ordinate Converter. Fall of shot reports received in the F.P.R. from O.P.s were recast into clock code by an Encoder, then passed to the engaging battery where the observations were decoded as plus or minus and left or right of the target. At the battery there was the Fire Direction Table which, on being fed continuously with battery bearings and ranges, calculated mechanically the necessary travel, ballistic and spotting corrections. Should a target appear within range of the batteries but out of sight of any O.P., it could be engaged by observation from an aeroplane, the observer reporting fall of shot by clock code.

The 6 inch guns on the old 15° elevation mountings for dealing with blockships, boom-smashers, cruisers, destroyers, indeed with any hostile ships that attempted to approach too close to a defended port or its coast-batteries, were known as "Close Defence" guns. These still depended on the old methods, their ranges being provided by the battery D.R.F., or, if the ranges were very short, the gun auto-sights. For night shooting close-defence guns required the collaboration of searchlights (Defence Electric Lights) which were used as observation lights for tactical control and as fighting lights for fire control.

The hostile warship, which it was feared might penetrate into a harbour at night and torpedo the men-of-war and other ships lying peacefully at anchor there, was no longer the destroyer—which had grown by this time far too large—but was now the motor torpedo-boat. This fiercesome little craft was much smaller and much faster than any destroyer, and the 12 pdr. Q.F. was far too slow in every way to deal with it. The Anti-M.T.B. 6 pdr. twin gun was therefore devised to replace the 12 pdr. This weapon, as its name implies, had two barrels set together on one mounting and could produce a rate of fire of more then 70 rounds per minute. The chief point of interest in the mounting was that corrections for line and elevation could be applied to each barrel independently of the cradle and therefore of the layers' telescopes. Owing to the high rate of fire it was found necessary to produce a displaced auto-sight and line sight, the Director No. 13, which was operated by two layers, line, elevation, and spotting corrections being transmitted direct to dials on the guns. The guns, which were loaded manually with one loading number and two ammunition numbers per barrel, had semi-automatic breeches. As with the 12 pdr. Q.F., the Anti-M.T.B. guns depended on an "Illuminated Area", produced by D.E.L. with dispersed beams, to illuminate their targets, the area normally being laid out so as to give the guns one minute's firing time at targets speeding at 40 knots.

Thus, when war with Germany broke out once again in September 1939, the standard coast-defence guns for the British Army were:-

15 inch (Singapore only)
9.2 inch with 35° elevation mounting } Counter-Bombardment.
6 inch with 45° elevation mounting
6 inch with 15° elevation mounting } Close Defence.
6 pdr. twin Anti-Motor Torpedo Boat

1919 to 1939 (concluded)

but it must be remembered that, such had been the general neglect of and utter lack of money for the coast-defences since the end of the First War, most of the 9.2 inch guns were still on their old mountings, and many 12 pdr. Q.F. were still to be found guarding the entrances to defended-ports. The ammunition for both 9.2 inch and 6 inch guns had been improved during the period between the two wars, for, whereas during 1914-1918 2 and 4 C.R.H. were the usual shapes for shell, 6 C.R.H. was introduced before 1939. Armour-piercing shell with base fuses, which had optional delay/non-delay plugs, and H.E. shell with nose fuses for unarmoured vessels or for firing landwards were now the normal coast-defence ammunition.

During the years from 1919 to 1939 the Coast Artillery School at Shoeburyness continued its most excellent work of holding courses for both officers and men, and of giving deep thought to and, more often than not, finding the solutions of the various problems with which coast-artillery was faced. (The Co-ordinate Converter was invented by an Assistant Instructor of Gunnery at the School.) However, the School, like the rest of Coast Artillery, suffered from the economy axe, and the instructional staff were reduced to a Commandant, one Major Instructor of Gunnery, two Captains Instructors of Gunnery, and four Assistant Instructors of Gunnery. Nevertheless all the regulation courses—long courses, short courses, refresher courses, and courses for Territorials—continued to be held. The development of the "fortress system of range-finding" and the production of all the new and improved coast-defence instruments ensured that the School kept not only up to date but well ahead in the study and practice of coast gunnery and on the technical application of modern scientific knowledge. Units, both Regular and Territorial, carried out annual or bi-annual practice seawards both by day and by night from the batteries they would have to man on mobilisation, and this practice was always assisted and supervised by I.sG. and A.I.s G. who had qualified at the long course (Gunnery Staff Course) at the School. With the help of the new systems and instruments and as the result of a general speeding up of methods and training, coast-gunnery steadily progressed and became most efficient. In the late 1920s the old, disarmed battleship *Centurion*, controlled by wireless, was used as a target by the Portsmouth (Isle of Wight) and Plymouth batteries with considerable effect: later the *Centurion* was taken to Malta where 4th Heavy Brigade and the Royal Malta Artillery scored so many hits and did her so much damage that she had to be withdrawn from target-work for major repairs. In one branch of coast-artillery training however there seems to have been no progress. In spite of the efficiency of modern mechanical means, the old "Repository Exercises" continued to be taught and much time was given to practising "gun shifts" with sheers, derricks, skidding etc. There was indeed a special unit, under Captain R. Shrive, R.A., which went around the coast-defences of Great Britain moving guns, carriages etc. when required by these methods.

As there was very little money available for coast-defence, except at Singapore, during the period between the two World Wars, there was no significant change in coast-fortifications or the genreal lay-out of coast-batteries. Even at Singapore, the new batteries were designed very much on the same lines as the old ones except that the 15 inch guns, controlled by the "Fortress System", were placed in positions back from the sea-shore and out of sight of the sea, the first coast-artillery guns ever to be mounted from whose sights the sea could not be seen. The 15 inch and new marks of 9.2 inch guns required "power" for their elevation, traverse, loading and ammunition supply, and this was supplied either by fluid pressure—the accumulators being initially charged by hand-pump—electric motors, or "elaulic" *i.e.* electric power used to provide oil under pressure. No great thought had been given to protecting coast-batteries against low-flying or bombing aircraft, indeed the guns and gun-detachments, standing exposed on the built-up battery positions, were very vulnerable to air attack. Except at Singapore, there was on the whole very little difference between the coast-defences of 1914 and 1939. In most cases the same guns were still there, only some of them had been changed to improved mountings to give them increased range. The most notable difference from 1914 was the fact that the whole of the Coast Artillery at home was provided by the Territorial Army instead of the regular R.G.A. The Territorial coast-regiments were all full of enthusiasm, had trained very hard, and, after some experience on their defences, were most efficient. The War of 1939 came too soon after the War of 1914 for it to be approached with any particular fervour, nevertheless, if most of the guns of the Coast Artillery in the British Army were somewhat out of date, the officers and men at home and abroad, Regular, Militia, Territorial, or Volunteer, were ready to man their defences and, if necessary, to fight them as long as their guns remained in action.

CHAPTER XXII

1939 to 1945.

HITLER and his Nazi Party came into supreme power and took over complete control of Germany in January 1933, and from that time onwards it became more and more apparent as each year went by that the German dictator intended war. It was the Munich crisis of the early autumn of 1938 which really opened the eyes of Great Britain to the warlike intentions of Hitler and his government. So threatening indeed was the German attitude at that time that the British Government ordered the coast and anti-aircraft defences of Great Britain to be manned, the Territorial Heavy Regiments mobilizing and taking up position in their defences where they remained until the end of September when the crisis for the time being faded out with the signing of the Four-Power Agreement on Czechoslavakia. This was a most excellent rehearsal for the following year, and many weak points in mobilisation and defence schemes were discovered and corrected during the succeeding 11 months. In March of the next year (1939) Hitler invaded and seized the whole of Czechoslavakia, and from that moment onwards it was obvious that war was imminent. During the summer Hitler turned his attention to Poland, working up the situation until there was no doubt that he intended to deal with that country as he had already dealt with Czechoslavakia. Both Britain and France had pledged themselves to defend the integrity of Poland, and during August the outlook became so menacing that on the 22nd the British Government ordered the "Precautionary Period" to come into effect. At once, both at home and abroad, the coast-defences were manned, the Territorial coast units in Great Britain and Northern Ireland mobilising at their drill-halls and going immediately to their defences. On 1st September the Germans invaded Poland, and on the 3rd both Britain and France went to war with them. The order of battle of the Coast Artillery and the guns they manned on that 3rd September 1939 were as follows:-

GREAT BRITAIN AND NORTHERN IRELAND.

(1) *Defended Port of Portsmouth (and Isle of Wight).*
    Garrison:-
        Hampshire Heavy Regt., R.A. (T.A.)
        Isle of Wight Rifles, Heavy Regiment, R.A. (T.A.)
    Armament:-
        6—9.2 inch
        16—6 inch
        8—12 pdrs.

(2) *Defended Ports of Plymouth and Falmouth.*
    Garrison:- Devonshire Heavy Regt., R.A. (T.A.)
    Armaments:-
       Plymouth 6—9.2 inch
                 4—6 inch
                 11—12 pdrs.
       Falmouth 2—6 inch.
(3) *Defended Port of Portland.*
    Garrison:- Dorsetshire Heavy Regt., R.A. (T.A.)
    Armament:-
       2—9.2 inch
       4—6 inch
       4—12 pdrs.
(4) *Defended Ports of Dover and Newhaven.*
    Garrison:- Kent and Sussex Heavy Regt., R.A. (T.A.)
    Armament:-
       Dover   2—9.2 inch
                 6—6 inch
                 2—12 pdrs.
                 2—6 pdrs.
       Newhaven 2—6 inch
                 2—12 pdrs.
(5) *Defended Port of Thames and Medway.*
    Garrison:- Thames and Medway Heavy Regt., R.A. (T.A.)
    Armament:-
       5—9.2 inch
       7—6 inch
       2—4.7 inch
       4—12 pdrs.
(6) *Defended Port of Harwich.*
    Garrison:- Suffolk Heavy Regt., R.A. (T.A.)
    Armament:-
       2—9.2 inch
       4—6 inch
       2—4.7 inch
(7) *Defended Port of Humber.*
    Garrison:- East Riding Heavy Regt., R.A. (T.A.)
    Armament:-
       2—9.2 inch
       4—6 inch
       2—4.7 inch.
(8) *Defended Port of Hartlepool.*
    Garrison:- Durham Heavy Regt., R.A. (T.A.)
    Armament:-
       1—9.2 inch
       4—6 inch.

(9) *Defended Port of Tyne.*
   Garrison:- Tynemouth Heavy Regt., R.A. (T.A.)
   Armament:-
      1—9.2 inch
      4—6 inch.

(10) *Defended Port of Milford Haven.*
   Garrison:- Pembrokshire Heavy Regt., R.A. (T.A.)
   Armament:-
      2—9.2 inch
      4—6 inch.

(11) *Defended Ports of Cardiff and Swansea.*
   Garrison:- Glamorgan Heavy Battery, R.A. (T.A.)
   Armament:-
      Cardiff 2—6 inch
      Swansea 2—4.7 inch.

(12) *Defended Ports of Mersey and Barrow.*
   Garrison:- Lancashire and Cheshire Regt., R.A. (T.A.)
   Armament:-
      Mersey 4—6 inch
      Barrow 2—6 inch.

(13) *Defended Port of Forth.*
   Garrison:- Forth Heavy Regt. R.A. (T.A.)
   Armament:-
      3—9.2 inch
      12—6 inch
      4—12 pdrs.

(14) *Defended Port of Clyde.*
   Garrison:- Clyde Heavy Battery, R.A. (T.A.)
   Armament:-
      4—6 inch.

(15) *Defended Port of Aberdeen.*
   Garrison:- Wallasey Battery, Lancs. and Cheshire Heavy Regt. R.A. (T.A.).
   Armament:-
      2—6 inch.

(16) *Defended Port of Scapa Flow.*
   Garrison:- Orkney Heavy Regt., R.A. (T.A.)
   Armament:-
      4—6 inch
      1—4.7 inch.

(17) *Defended Port of Belfast.*
   Garrison- Antrim Heavy Battery, R.A. T.A.
   Armament-
      4—6 inch.

Abroad.

(1) *Fortress of Gibraltar.*
   Garrison:-
   - 3rd Heavy Regt., R.A.
   - 4th Heavy Battery, R.A.
   - 26th Heavy Battery, R.A.
   - 27th Heavy Battery, R.A.

   Armament:-
   - 7—9.2 inch
   - 6—6 inch
   - 6—6 pdrs.

(2) *Fortress of Malta.*
   Garrison:-
   - 4th Heavy Regt., R.A.
   - 6th Heavy Battery, R.A.
   - 10th Heavy Battery, R.A.
   - 23rd Heavy Battery, R.A.

   Royal Malta Artillery
   - 1st Heavy Battery, R.M.A.
   - 2nd Heavy Battery, R.M.A.
   - 3rd Heavy Battery, R.M.A.

   Armament:-
   - 7—9.2 inch
   - 10—6 inch
   - 6—12 pdrs.
   - 9—6 pdrs.

(3) *Defended Port of Haifa.*
   Garrison:-   17th Heavy Battery, R.A.
   Armament:-
   - 4—6 inch.

(4) *Defended Port of Alexandria.*
   Garrison:-   Egyptian Artillery.
   Armament:-
   - 4—6 inch
   - 4—4 inch
   - 2—12 pdrs.
   - 2—6 pdrs.

(5) *Defended Ports of Port Said and Suez.*
   Garrison:-   19th Heavy Battery, R.A.
   Armament:-
   - 4—6 inch.

(6) *Defended Port of Port Sudan.*
   Garrison:-   Sudan Detachment, R.A.
   Armament:-
   - 2—6 inch.

(7) *Defended Port of Aden.*
    Garrison:- 9th Heavy Battery, R.A.
    Armament:-
        4—6 inch.
(8) *Defended Port of Kilindini.*
    Garrison:- King's African Rifles Coast Battery.
    Armament:-
        2—6 inch.
(9) *Mauritius: Defended Port of Port Louis.*
    Garrison:- 25th Heavy Battery, R.A.
    Armament:-
        2—6 inch.
(10) *Ceylon: Defended Ports of Colombo and Trincomalee.*
    Garrison:-
        6th Heavy Regt., R.A.
        15th Heavy Battery, R.A.
        18th Heavy Battery, R.A.
        Ceylon Garrison Artillery (Volunteers).
    Armament:-
        Colombo 2—9.2 inch
                4—6 inch.
        Trincomalee 2—9.2 inch
                3—6 inch.
(11) *Defended Port of Bombay.*
    Garrison:-
        14th Heavy Battery, R.A.
        Bombay Battery, Auxiliary Force (India).
    Armament:-
        2—7.5 inch
        3—6 inch.
(12) *Defended Port of Karachi.*
    Garrison:- 13th Heavy Battery, R.A.
    Armament:-
        2—7.5 inch.
(13) *Fortress of Singapore.*
    Garrison:-
        7th Heavy Regt., R.A.
        11th Heavy Battery, R.A.
        31st Heavy Battery, R.A.
        5th Heavy Battery, H.K.S.R.A.
        7th Heavy Battery, H.K.S.R.A.
        9th Heavy Regt., R.A.
        7th Heavy Battery, R.A.
        22nd Heavy Battery, R.A.
        32nd Heavy Battery, R.A.

Armament:-
- 5—15 inch
- 6—9.2 inch
- 18—6 inch
- 4—6 pdrs.

(14) *Defended Port of Penang.*
Garrison:- 8th Heavy Battery, H.K.S.R.A.
Armament:-
- 4—6 inch.

(15) *Fortress of Hong Kong.*
Garrison:-
- 8th Heavy Regt., R.A.
- 12th Heavy Battery, R.A.
- 30th Heavy Battery, R.A.
- 'X' Heavy Battery H.K.S.R.A.
- *12th Heavy Regt., R.A.*
- 20th Heavy Battery, R.A.
- 24th Heavy Battery, R.A.
- Artillery Group: Hong Kong Volunteer Defence Corps.

Armament:-
- 6—9.2 inch
- 10—6 inch
- 4—4.7 inch.

(16) *Sierra Leone: Defended Port of Freetown.*
Garrison:- Sierra Leone Heavy Battery, R.A.
Armament:-
- 2—6 inch

(17) *Bermuda.*
Garrison:- Bermuda Militia Artillery.
Armament:-
- 2—6 inch.

(18) *Jamaica: Defended Ports of Kingston and Port Royal.*
Garrison:-
- 2nd Heavy Battery, R.A.
- Jamaica Militia Artillery.

Armament:-
- 1—9.2 inch
- 4—6 inch.

(19) *Trinidad: Defended Port of Port o' Prince.*
Garrison:- Trinidad Volunteer Artillery.
Armament:-
- 4—6 inch.

(20) *Falkland Islands: Defended Port of Port Stanley.*
Garrison:- Falkland Islands Defence Corps.
Armament:-
- 2—6 inch.

The situation in the "Narrow Seas" around Britain the on outbreak of war with Germany was somewhat different from the coast-artillery and coast-defence point of view from what it had been in 1914. Then the enemy possessed a "High Sea Fleet" powerful enough to challenge the naval supremacy of Britain: in 1939 the German Navy had only just begun to recover from the restrictions laid on her by the Versailles Treaty. (It was not until 1935 that Hitler denounced the naval terms of that Treaty and declared he would build warships up to the limits set by the Washington Agreement). The main strength of the German Navy in surface ships in September 1939 was:-

Battle Cruisers	2
Pocket Battleships ...	3
Heavy Cruisers	3
Light Cruisers	6
Destroyers...	18

the battle-cruisers and pocket-battleships being armed with 11 inch guns, the heavy cruisers with 8 inch, and the light-cruisers with 5.9 inch. There was indeed very little danger of raids across the North Sea, and, besides maintaining constant vigilance, support of the Examination Service and routine duties were about all the Coast Artillery at home had to do during the first eight months of the war. Like the rest of Great Britain, the Coast Artillery had to take precautions against attack from the air, and much time was spent in practising "Passive Air Defence", the only offensive A.A. weapons held by coast-defence batteries being their light machine-guns.

In December 1939 it was decided that the Fortress Royal Engineers, who manned the searchlights and their engine-rooms, should be abolished, and the work taken over by the Coast Artillery units with whom the searchlights co-operated, most of the Sapper personnel being absorbed into the R.A. The "take over" was carried out in two stages, first the manning of the D.E.L. themselves, and then some six months later the engines and engine rooms. This was a wise move, the dual control being always somewhat unsatisfactory. On the other hand, Coast Artillery and Fortress Engineers had worked together happily for a good many years, and both were sorry that the partnership should now be dissolved. There was also difficulty about exposing the D.E.L. either fo operational or training purposes. It was feared that the searchlight beams would act as guiding lights for approaching enemy aircraft and mark the coast for them. It was finally decided that searchlights could only be exposed on the authority of the local combined H.Q. except when there was definite information that hostile surface craft were in the vicinity.

The successful German campaigns of the spring and early

summer of 1940 completely altered the war situation. By the end of June the enemy were in occupation of the western coast of Europe from North Cape to the Franco-Spanish frontier and a line of ports and harbours which directly threatened the British Isles. There once again was Antwerp in enemy hands, pointed like a loaded pistol at the heart of England, and the shores of Britain, from the Wash to Lands End, were laid open to the attacks of hostile light-craft swarming from the French and Flanders ports. Moreover, the Germans were now able to move their airfields close up to the French shore of the Channel, and, by determined use of their aircraft, make it extremely dangerous for British warships to operate during the hours of daylight in the narrow waters of the Channel, the Straits of Dover, and off the Thames Estuary. Every defended-port on the east and southern coasts of England was now under the constant menace of attack both by enemy aircraft and by light coastal motor torpedo-boats (known colloquially as E. Boats).

Hitler, like Napoleon before him, decided that, to win the war finally and decisively, he must defeat Great Britain, and that the surest and quickest way of doing so was by direct invasion across the Channel. But whereas Napoleon demanded command of the waters of the Channel for 24 hours as the fundamental condition without which he could not launch his assault, it was control of the air over the Channel that Hitler required before he would begin his attack. So on the 2nd and 16th of July, 1940 Hitler issued directions for the invasion of Britain which was to be preceded by an all-out assault by the German *Luftwaffe* on the British Air Force and its ground installations with the intention of utterly destroying the air-defences of Great Britain. The general outline of the plan of invasion, which was known as OPERATION SEA LION, was for a landing to be made on the southern shores of Britain between Folkestone and Brighton with a force of nine infantry divisions and two airborne divisions, supported by 250 amphibian tanks, 28 anti-aircraft ferries and 72 rocket-projectors, all carried in 155 transports, 471 tugs, 1,722 barges, and 1,161 smaller craft. This first wave was to be followed, as soon as possible, by a second of four armoured, two motorised and two infantry divisions. It was hoped to launch the invasion during the period 19th-27th September when moon and tides would be most favourable.

The remains of the British Expeditionary Force and the French northern armies had all been evacuated from Dunkirk to Great Britain by 5th June, and on the 22nd of the same month France gave up the unequal struggle and surrendered with the result that Great Britain was left all alone in her little group of islands—in which Eire was strictly neutral which did not make things easier—to face her victorious and powerful enemy. It was soon quite clear that the Germans across the Channel were preparing for invasion and, amongst the many designs and expedients set in hand to resist a sea and air-borne assault, it was decided to surround Great Britain from the Orkneys to the Outer Hebrides with a ring of coast-defence batteries which would cover every probable and possible landing-place, be it port, harbour, bay, cove, inlet, or open beach, very much as had been attempted in the days of Napoleon. This of course entailed a tremendous expansion of coast-artillery, an expansion beyond even the wildest dreams of anyone before the war. The first necessity and most formidable problem was the finding of guns. There was scarcely any reserve of coast-defence armament held by the Ordnance Department, but luckily and wisely the Royal Navy had kept in store a great assortment of guns and mountings from the ships which had been scrapped after the First War. From this providential supply, 6 inch 5.5 inch, 4.75 inch, and 4 inch guns were issued in great number for what were now named "Emergency Coast Batteries" and which were to protect the minor ports and cover every threatened beach of Great Britain. Ammunition for these guns was also produced from naval sources, but quantities were very meagre, averaging only about 50 rounds per gun.

The second problem was to equip these emergency batteries with the necessary searchlights (with their engines) and instruments, for without the former no firing could be done at night, and without the latter such firing as was done by day would be most inaccurate and ineffective. Searchlights were in very short supply and could only be issued to batteries being erected along what was considered to be the most threatened part of the coast, *i.e.* from Harwich to Portsmouth, but all kinds of substitutes were produced for the other batteries, ranging from portable lights on tripods to magnesium flares burnt in front of polished metal reflectors (Ryder Flares). A means of finding the range was the foremost requirement in instruments, for none of the naval guns were equipped with autosights, and there were very few spare D.R.F.s. However every available Barr and Stroud rangefinder of every possible size, large or small, was produced and hurried to the emergency batteries, and these, together with Dumaresques and rate-clocks, furnished the necessary instruments for engaging rapidly moving targets.

The third problem was the provision of officers and men to man and serve the emergency batteries. The first to be installed were manned by the Royal Navy and Royal Marines, but it was soon obvious that the Fleet could not spare the large number of men required so it was decided that the batteries should be taken over by Coast Artillery. The sources from which Coast Artillery could produce trained coast-gunners were slight, in fact there were only the Territorial regiments already on duty at the major ports and the staff and recruits under training at the Coast Artillery Training

Unit at Plymouth. G.H.Q. Home Forces finally laid down that personnel for the Emergency Batteries should be provided as follows:-
(a) Officers: by posting from the Territorial Coast Artillery regiments at the major ports and by calling up officers of the "Officers Emergency Reserve" and the "Territorial Army Reserve".
(b) Senior N.C.O.s: by posting from the Coast Artillery Training Unit.
(c) Junior N.C.O.s: by posting from the Territorial Coast Artillery regiments at the major ports.
(d) Search-light Personnel: by posting from the School of Electric Lights at Gosport.
(e) Specialists: from recruits under training as such at the Coast Artillery Training Unit.
(f) The main Body of Gunners:-
  (i) from Medium and Heavy Regiments which had returned from Dunkirk having lost their guns.
  (ii) from Recruits who had done one month's basic training at any R.A. training unit.

The establishments for the Emergency Coast Batteries varied according to the size and number of the guns, ranging from about four officers and 135 men for a three gun 6 inch battery to three officers and 90 men for a two gun 4 inch battery.

The mounting of the guns at the selected positions, the construction of concrete gun-houses to protect them from hostile dive-bombers, normal air bombing, and shell fire, the building of command and observation posts, searchlight emplacements, and engine houses, and the erection of accommodation for officers and men were no light task, but by employing a mixture of special gun-mounting parties, experts from the Coast Artillery School, sailors from the Royal Navy, gunners who were to man the guns when they were in position, and the Royal Engineers Services with contract labour, and by using vast quantities of quick-setting concrete, the job was done with surprising speed. It was carried out in seven instalments:-

During June 1940, 1st instalment	47 batteries
2nd instalment	24 batteries
During July 1940, 3rd instalment	8 batteries
During August 1940, 4th instalment	28 batteries
During September 1940, 5th instalment	5 batteries
During November 1940, 6th instalment	15 batteries
During January 1941, 7th instalment	26 batteries

The batteries were grouped into regiments under lieutenant-colonels, and a brigadier was appointed in each Corps defending the coast to command all the Coast Artillery in the corps area. It was during this period (Summer 1940) that at last Coast Artillery throughout

1939 to 1945

the British Army was given its proper name and title. Batteries were called "Coast Batteries", regiments "Coast Regiments" and the brigadier on corps headquarters, the Commander Corps Coast Artillery. The complete list of the coast-defences of Great Britain, when they were at about their greatest extent (Autumn 1941) is given below. It must be remembered that the Channel Islands had neither coast-guns nor coast-gunners when war broke out in 1939 and were unfortunately occupied by the enemy in July 1940.

Place.			Guns.				
	14″	9.2″	6″	5.5″	4.7″	4″	12 pdrs.	6 pdrs.
(a) Major Ports								
Orkneys			11		3		17	9
Invergordon ...			6					
Aberdeen ...			4					
Dundee			4					
Forth		3	16				4	6
Blyth			4					
Tyne		1	6			1	2	
Sunderland ...			4				2	
Tees & Hartlepool		1	6				2	
Humber		2	6		4	3		6
Yarmouth ...			4				2	
Lowestoft ...			6				2	
Harwich		2	4				2	3
Thames & Medway		2	12	4				5
Dover	2	6	16			2	3	3
Newhaven ...			4				2	
Portsmouth & Southampton			6	14			8	8
Portland		4	4				4	
Dartmouth ...					2			
Plymouth ...		6	6				11	
Falmouth ...			4				2	2
Avonmouth ...			2					
Cardiff			2					
Swansea			2					
Barry			2		2			
Newport							2	
Milford Haven ...		2	4					
Mersey			4					
Barrow			4					
Clyde			4		2		1	
Belfast			6					
Londonderry ...			2				1	

(b) Minor Ports	6"	5.5"	4.7"	4"	12 pdrs.
Shetlands	2			4	
Wick	2				
Montrose	2				
Inverness				2	
Berwick	2				
Amble	2				
Seaham Harbour	2				
Whitby	2				
Scarborough	2				
Boston	2				
Kings Lynn	2				
Brightlingsea			2		
Burnham	2				
Ramsgate	2				2
Shoreham	2				
Littlehampton	2				
Poole			2		
Exmouth			2		
Torquay			2		
Brixham			2		
Salcombe				2	
Looe				2	
Fowey			2		
Penzance				2	
Padstow				2	
Appledore			2		
Llanelly				2	
Fishguard	2				
Carnarvon				2	
Holyhead	2				
Preston				2	
Fleetwood				2	
Whitehaven				2	
Workington				2	
Stranraer				2	
Loch Ewe	2				
Stornoway				2	
Larne	2				
(c) Beach Batteries					
Lossiemouth	2				
Peterhead	2				
Fidra	2				
South Shields	2				
Seaton Carew	2				
Filey	2				

1939 to 1945

	6"	5.5"	4.7"	4"	12 pdrs.
Hornsey			2		
Paull				2	
Stallingborough			2		
Grimsby	2				
Mablethorpe	2				
Skegness	2				
Hunstanton	2				
High Cape	2				
Cley Eye	2				
Sheringham	2				
Cromer	2				
Happisburgh	2				
Winterton				2	
Kessinghand	2				
Covehithe	2				
Southwold	2				
Dunwich				2	
Thorpeness	2				
Minsinere	2				
Aldeburgh	2				
Felixstowe	2				
Frinton	2				
Clacton	2				
Mersey Island			2		
Shoeburyness	2				
Coalhouse		2			
Shoremead		2			
Shellness	2				
Herne Bay	2				
Margate	2				
Kingsgate				2	
Joss Bay		4			
Sandwich Bay	2				
Bethlehem	2				
Deal	4				
Kingsdown	2				
St. Margaret's Bay		4			
Folkestone	4				
Mill Point		4			
Hythe	2				
Dumpton Point		2			
Dymchurch	4				
Littlestone	3				
Dungeness	3		2		
Jury's Gut	2				
Greatstone	2				

	6"	5.5"	4.7"	4"	12 pdrs.
Winchelsea	2				
Pett Level	2				
Hastings	2				
Bexhill				2	
Normans Bay			2		
Pevensey		2			
Eastbourne	2				
Seaford	2				
Brighton	2				
Worthing	2				
Bognor		2			
Hengisbury Head				2	
Swanage				2	
Abbotsbury				2	
West Bay		2			
Lyme Regis			2		
Seaton				2	
Sidmouth				2	
Dawlish				2	
Teignmouth			2		
Par				2	
Newquay				2	
Hayle				2	
Barnstaple				2	
Ilfracombe				2	
Minehead				2	
Portishead		2			
Port Talbot				2	
Lytham				2	
Maryport				1	

Thus Great Britain, from the Orkneys to the Outer Hebrides, was ringed by coast-defence batteries. Nevertheless it was not only the British Isles that had to be defended. The Germans had already occupied the Channel Islands, and precautions had to be taken to prevent them from seizing Iceland and the Faroe Islands as air and submarine bases. So defending forces were sent to these islands, and included in them were coast-batteries, nine for Iceland and three for the Faroes. Iceland finally mounted:-

      4—6 inch guns
      2—4.7 inch
      10—4 inch
      2—12 pdr.

and the Faroe Islands:-

      4—5.5 inch guns
      2—12 pdr.

As it turned out, the German *Luftwaffe* was decisively defeated by the R.A.F. during the Battle of Britain, Hitler therefore did not obtain command of the air over the Channel as he had demanded, and the great invasion, Operation SEA LION, was never launched. On 17th September Hitler ordered postponment of the invasion until spring of the next year (1941), but he never really intended preparations for it to be restarted. His mind was already turned towards the East, and in June 1941 he invaded Russia where, like his great predecessor Napoleon, he finally found nothing but defeat and disaster. So once again the coast-artillery and their batteries were baulked of their prey, and long hours of anxious looking out across the sea in all kinds of weather were never rewarded by the sight of the enemy flotillas making towards the shore. Nowhere was the constant watch more anxious than at Dover which of all places was closest to the Germans, a mere 22 miles from the enemy port of Calais. Part of Operation SEA LION had been the closing of the Straits of Dover by the fire of heavy guns emplaced along the French Coast, and before the end of June (1940) the Germans were busy mounting batteries of the largest calibre from Cap Gris Nez to Calais, the most famous of which during the next few years were:-

Batterie Lindemann	3—16 inch guns
Batterie Todt	4—15 inch
Batterie Grosser Kurfurst	4—11 inch
Railway Batteries	4—11 inch.

Soon the German guns were firing on British ships—mostly cargo ships—passing through the Straits, on minesweepers at work clearing enemy mines, and on Dover itself, its defences, harbour, docks and township.

The British had no guns in the Dover area capable of dealing with the hostile batteries. As early as May the Prime Minister, Mr. Winston Churchill, had suggested the mounting of long-range guns at Dover to command the Straits, but the best that could be produced were two 14 inch naval guns, manned by the Royal Marines, which, known locally as Winnie and Pooh, did what they could to counter the enemy fire. However, the Coast Artillery were quite determined they were not going to be mastered by the German artillery along the French coast. If the enemy could engage British shipping passing through the Straits, shell the coast defences, and bombard Dover, the Coast Gunners were equally resolved to stop German ships moving through French coastal waters between Dunkirk and Boulogne and to bring counter-battery fire to bear on the German guns. So it was decided in September (1940) to construct three new batteries in the Dover area specially for these purposes:-

Fan Bay Battery: 3—6 inch guns, extreme range 25,000 yards
South Foreland Battery: 4—9.2 inch guns, extreme range 31,000 yards
Wanstone Battery: 2—15 inch guns, extreme range 42,000 yards.

Fan Bay Battery was ready in February 1941, South Foreland in October 1941, and Wanstone in June 1942.

Meanwhile the threat of invasion had passed away and the main task of these batteries was to close the Straits of Dover to enemy shipping. The Germans, to lessen the strain on the French railways, continually sent small convoys of storeships and tankers through the Straits from north to south to supply the ports on the Channel and Atlantic coasts of France. Very soon the new Dover batteries put an end to this practice, and the Germans were reduced to slipping single ships through by night, hugging the French coast, even these being liable to engagement and sinking by the Dover guns, for there had appeared a wonderful new invention which enabled coast-batteries to pick up, follow, and fire at targets without them ever being seen from any observation post or by any range-finding instrument. Radar had originally been developed for spotting and giving warning of the approach of aircraft, but it worked equally well over water and could pick-up and follow the course of surface craft. Basically it consisted of a wireless beam, sent out and traversed like the beam of a searchlight but having a much longer range, which, on striking a solid object, was reflected back to its source, the time taken for the beam to go out and return to its origin being measured, and thus the range to the object could be calculated; its direction when following the object was also registered so that the bearing was known. Thus range and bearing were recorded and continued to be whilst the target proceeded on its course. It can therefore now be understood how the Dover batteries were able to spot and pick up enemy ships in the Straits or hugging the French coast in thick weather and by night.

The first Radar sets for Coast Artillery began to appear in February 1941. These were CD/CHL or Early Warning Sets designed to give warning of the approach of surface ships or low flying aircraft. Such a set could pick up a battleship or cruiser at 40,000 yards—at even greater ranges should conditions be particularly favourable—and smaller warships or a collection of barges at 12,000 yards. These were distributed around the coast of Great Britain in a "Coast Watching" cordon, being housed in special emplacements, manned by specially trained "Coast Observer Detachments" of the Coast Artillery, and having direct communication with the nearest coast-artillery plotting-room. Later specific "Coast-Artillery" Sets were designed and issued to counter-bombardment fire-commands for use with their batteries. These sets were for two purposes: one set tracked the enemy ship, recording its bearing and range, while the other reported fall of shot, so much right or left, plus or minus of the target. CD/CHL Sets were also issued to most of the coast-fortresses and defended ports abroad during 1941, the first ones being sent to Malta in March, but special C.A. Sets did not reach them until 1944.

To return now to the counter-bombardment batteries at Dover. Their first important action was against the German battle-cruisers *Scharnhorst* and *Gneisenau* and the heavy cruiser *Prinz Eugen* when they sailed up Channel and through the Straits on 12th February 1942. The three enemy warships had been sheltering for many months (the battle-cruisers since March and the cruiser since June 1941) in Brest where they had been continuously bombed by the R.A.F., so Hitler decided it was time they were brought back to Germany, for, he said, they were like "a patient with cancer who is doomed unless he submits to an operation." Hitler further decided that the return could best be carried out as a surprise move up the English Channel, through the Straits of Dover, and along the eastern verge of the North Sea. To ensure that the ships should remain unseen as long as possible they were ordered to leave Brest on 11th February as soon as it was dark enough to hide their movements. Nevertheless the *Scharnhorst*, *Gneisenau*, and *Prinz Eugen* did not sail until 10.45 p.m., but by dawn they were east of La Hogue, steaming up Channel covered by 16 fighters overhead and screened by six destroyers and ten torpedo-boats. The morning of the 12th was thick, and visibility was poor, and it was not until 10.40 a.m. that the enemy squadron was sighted off Le Touquet by an R.A.F. fighter patrol, and almost immediately afterwards (10.50 a.m.) it was picked up by the CD/CHL Radar Set at Fairlight, just east of Hastings, at 67,000 yards, at that time a record range for that type of set. By 11.45 the German warships were passing Boulogne, by 12 noon they were off Cap Gris Nez, and by 12.15 were entering the narrowest part of the Straits. Here they were attacked by a division of five motor torpedo-boats from Dover, but these small craft were unable to penetrate the hostile screen of destroyers.

Meanwhile at his headquarters at Dover Castle, the C.C.C.A. 12th Corps (Brigadier C. W. Raw) had received further radar reports of the approach of the German squadron in spite of vigorous efforts being made by the Germans to jam the British sets. At 11.30 the set at Lydden Spout picked up and followed the "plot" and at 12.15 it was taken up by the set at South Foreland. Raw decided to engage the enemy at once with South Foreland Battery (4—9.2 inch)—the 15 inch battery not yet being in action—although the targets were almost at extreme range and could not be seen from the Fortress O.P.s on account of the thick weather, visibility being down to five miles. At 12.19 the battery opened fire, using the South Foreland radar set to provide ranges and bearings, and continued to fire at approximately 30,000 yards for 17 minutes until the targets passed out of range; 34 rounds of armour-piercing shell were fired, three direct hits being claimed. As soon as the German warships emerged from the Straits they were attacked by six Swordfish torpedo-bomber aircraft of the Royal Naval Air Service escorted by 10 fighters of the R.A.F. In spite of the attack being driven home with the greatest gallantry, it failed completely, none of the torpedoes finding their marks and all six Swordfish being shot down into the sea. After this the enemy squadron was left to steam on its way until about 3.30 p.m. when off the coast of Holland it was again attacked by destroyers from Harwich and by R.A.F. bombers but once more with little success. The German admiral must now have thought he was safely in the home run, but between seven and nine p.m. he ran into a minefield, laid unbeknown to him by the R.A.F. off the Frisian Islands, and both the *Scharnhorst* (twice) and the *Gneisenau* were struck, the former very seriously. In spite of these mishaps all three ships managed to make the mouth of the Elbe safely during the 14th, but both the battle-cruisers were severely crippled.

After this contest the Dover guns had no more major targets of this type, but they were kept busy throughout 1942 and 1943 and until September 1944, when the Germans evacuated the French Channel coast, preventing the Germans from running their small convoys of storeships and tankers through the Straits by night. The enemy vessels would attempt to creep through, hugging the French coast to keep out of range and to deceive the British radars, but once the 15 inch guns of Wanstone Battery were in action, they were almost always detected and engaged, many ships being sunk or crippled. The batteries also co-operated whenever possible with the light coastal-forces of the Royal Navy when the latter were attacking enemy vessels in the Straits. An instance of this took place during the night 4th/5th July 1943, and the contemporary report from Coast Artillery Headquarters, Dover, is given below in full:-

"(1) At 0130 hours on 5th July, 1943 a [radar] plot was picked up off Dunkirk which from its size and speed was correctly estimated to be two destroyers known to be in that port. It was decided that M.T.B.s should attack under cover of the fire from C.A. guns.
(2) By 0240 hours the enemy was in range of the Eastern Counter-Bombardment Fire Command, and Wanstone Battery opened fire followed by South Foreland Battery. Wanstone Battery fired 20 rounds and South Foreland Battery 40 rounds 8 c.r.h.

1939 to 1945 237

(3) Range finding was by Wanstone Battery [radar] set through the Fortress Plotter. Fall of shot reports were good.
(4) The enemy was steaming at 25 knots and began snaking turns as soon as fire was opened.
(5) Striking Force 'A', consisting of three M.T.B.s, had proceeded to a position about five miles N.N.E. of Cap Gris Nez and at 0251 hours, when moving E.S.E. sighted the enemy one mile away on the port bow.
(6) At 0252 hours, after consultation with the Naval Staff, the Operations Room, who were plotting our own and the enemy's forces, ordered the batteries to add 1,000 yards followed two minutes later by a further add 1,000 yards. Owing to the long time of flight this did not immediately take effect on the water, and the M.T.B.s closed the target with shells bursting round them and the enemy vessels.
(7) Between 0253 hours and 0254 hours M.T.B.s 235 and 240 fired their torpedoes but missed. M.T.B. 202 did not not sight the target. At 0254 the enemy altered course sharply to port and this brought him very close to two salvoes fired with the first correction of add 1,000 yards. He then altered course to starboard and, as the second add 1,000 yards had now taken effect, salvoes began to fall a long way plus. Shortly after the firing of the torpedoes a black cloud of smoke was seen to come up from sea level by the second enemy ship. This cannot be claimed as a hit as no flash was seen but may have been caused by shell fire. The M.T.B.s then withdrew apparently undetected by the enemy and at 0301 hours C.A. Operations ordered the guns to resume fire for effect.
(8) Meanwhile Striking Force 'B' (3 M.T.B.s), who were somewhat out of position, had got very close to the enemy and fire was stopped from 0307 hours to 0311 hours to allow them to get clear. Fire was then continued until 0317 hours when the target was out of range. At 0321 hours N.T. 284 was jammed for nine minutes.
(9) There were no casualties to guns or equipment or to the striking forces.
(10) Enemy batteries opened fire at 0309 hours but their fire was not very heavy. Only about 10 rounds were fired, falling in Dover, Folkestone, and Ramsgate, and in the sea in front of South Foreland Battery.
(11) This is the first time such close co-operation between C.A. guns and light coastal-forces has been attempted and, in so far as M.T.B.s were able to make their attack undetected and unscathed, was entirely successful. The mean range of the action was about 32,000 yards."
As can be seen from paragraph 10 of the above report the enemy

did not take these activities of the Dover guns lying down. The opening of fire either by Wanstone or South Foreland Battery was usually the signal for the German batteries of heavy guns around Cap Gris Nez to reply. They would normally choose the battery which was firing as their primary target but would always throw some salvoes into the British coast towns for good measure. Dover was their favourite victim, and the good citizens of that long suffering town objected strenuously and vociferously to the casualties and damage caused by the enemy shelling. They did not understand that the Coast-Artillery guns were engaging enemy ships in the Straits but thought they were merely indulging in an artillery-duel with their opposite numbers around Cap Griz Nez, and, as the British guns always seemed to open fire first, the inhabitants of Dover blamed them for starting a contest in which they, the inhabitants, though non-participants, received most of the hard knocks. These complaints were brought up in Parliament by the member for Dover, and as the result of this steps were taken to explain both to the Press and to the citizens of Dover that the Coast Artillery guns did not open fire just to begin a light-hearted tussle with the enemy batteries on the other side of the Channel but to sink German ships which were trying to pass through the Straits. However, Dover and the other coast-towns were not the only targets for the enemy's return fire. He more often than not began with counter-battery fire on the offending Coast Artillery battery, causing casualties both to men and equipment, on one occasion (1st September 1944) one man being killed, five wounded, and much damage done to guns, instruments, and buildings. During the period the Dover counter-bombardment batteries were in action:-

(1) They had 50 engagements against enemy shipping and convoys.
(2) They had six engagements against hostile batteries on the French Coast.
(3) They sank 26 ships of varying types and sizes and damaged many more.
(4) They fired the following number of rounds:-
    15 inch— 1,243 rounds
    9.2 inch—2,248 rounds
    6 inch— 73 rounds.

While the Dover guns were kept very busy, once the threat of invasion was past, the rest of the Coast Artillery defending the shores of Great Britain was left with no particular objective. There was of course always the chance of a sudden enemy sea-borne raid but, as time went on, this became less and less likely. Thanks to radar however, the batteries aligned along the coast from the Wash to the Lizard did get the opportunity from time to time of engaging the enemy, for German "E Boats" would prowl off shore at night, patrolling the British coast and seeking what they could destroy.

These would be picked up by radar and reported to the nearest coast-battery which, getting ranges and bearings from the radar, would then open fire on them. Such happened at Falmouth during the night of 29th February/1st March 1944:-

"At 0058 hours Fire Commander informed that Navy had identified plot as hostile and had allotted number E11F. Alarm was sounded and all batteries in Fire Command put at immediate readiness. Target was now at 12,600 yards and was coming in rapidly at 35 knots. Response on Radar set was equivalent to M.T.B. target.

At 0109 hours F.C. allotted radar to 6 inch battery for fire control. Bearing and range were read every 15 seconds and passed by head and breast-set to P.F. detachment in Battery O.P. thence being transmitted electrically by magslip to guns. At 0120 hours target was stationary at 6150 yards, and first salvo was fired. Immediately target began to move away from battery rapidly, increasing speed to 35 knots.

At 0126 hours, immediately after radar had reported round as range, flash was seen on water which lasted for 15 seconds and was followed by whitish cloud of smoke. At 0127 searchlights were exposed for 3 minutes but no targets were picked up.

At 0132 hours battery again opened fire. Target was now at 13,800 yards rapidly increasing to 18,000 yards. Radar set continued to follow target until well out of range. Echo split into three distinct echoes, indicating at least three vessels.

At 0258 hours batteries were ordered to resume normal routine. During action sea was smooth and visibility good except for occasional rain shower. There was no moon. 23 rounds of 6 inch H.E. fuze 230 fired. After each salvo from 2 gun battery, correction was made on radar fall of shot reported before next was fired. No failure of equipment occurred."

The Dover guns had their last action during September 1944. The British and American armies had won a great victory in Nomandy and were now pursuing the defeated enemy across northern France, the 2nd Canadian Corps advancing northwards along the French Channel coast with objectives Boulogne—Calais—Dunkirk. Coast Artillery H.Q. at Dover was asked to co-operate by neutralising, and if possible destroying, the German batteries around Cap Griz Nez which had for so long bombarded Dover and other coast towns. Raw, Coast-Artillery commander at Dover, decided to use both Wanstone Battery—(2—15 inch) and the Royal Marine Siege Battery (2—14 inch), observation being carried out from aircraft (Air Observation Post) provided by the Canadians. On 16th September both batteries registered, and on the 17th the onslaught began. The targets on this day were Batterie Lindemann and the Railway Batteries, Wanstone firing 75 and R.M. Siege 102 rounds, 11 direct hits being reported. On the 19th R.M. Siege engaged Grosser Kurfurst Batterie, firing 72 rounds and obtaining 25 direct hits. On the 20th the last target, Blanc Nez Batterie, was attacked by both batteries, Wanstone firing 85 rounds, R.M. Siege 198, but on this occasion only six direct hits were recorded. This three day bombardment did not completely silence the enemy batteries, which continued to shell Dover and Folkestone intermittently, but on 30th September the Canadians overran Cap Gris Nez and the German heavy guns fired no more.

The tremendous expansion of Coast Artillery in 1940, due to the threat of invasion, had tied up a very large number of men in coast-batteries. By the end of 1942 it was quite obvious that invasion was no longer a possibility, and in January 1943 it was determined to reduce coast-artillery strengths at home in order to provide more men for the armies fighting overseas. It was agreed that the only threat to the harbours and coasts of Britain was prowling "E-Boats" and "U-Boats" which might raid harbours or attack convoys within range of the coast-batteries, so in July it was decided to consign certain batteries to care and maintenance, and to replace regular troops by Home Guard at others. The general plan was that:-
(1) Counter Bombardment Batteries were only to be retained where they were capable of closing the Straits of Dover. Elsewhere all counter-bombardment batteries were to be relegated to care and maintenance.
(2) Close Defence Guns on a scale of two guns per port and Examination Batteries were to be retained fully manned by regulars. Remainder of close defence guns were to be relegated to care and maintenance.
(3) Anti-Motor Torpedo Boat Guns were to be retained in action fully manned by regulars.
(4) Beach Defence Batteries were either to be manned by Home Guard or to be relegated to care and maintenance.

The permanent regular detachment of a Beach Defence (Emergency) Battery manned by Home Guard was one officer and seven men, the Home Guard establishment four officers and 150 men, the latter coming for their training and practice during week-ends and in the evenings after they had completed their days work. The following is an extract from an instruction issued to regular battery-commanders of Home Guard manned batteries in February 1944:-

"The introduction of Home Guard Batteries and the withdrawal of regular personnel has greatly increased the importance of Home Guard training. Although much of the training will have to be done by Home Guard officers and senior N.C.O.s in the same way as in a pre-war Territorial unit, the Battery Commander and his seven Other Ranks must be regarded as the permanent staff.

Battery Commanders faced with this have been heard to remark 'How shall I fill in my time? It will be dreadful living alone with only 7 men!' Such an attitude is of course ridiculous —an officer of the correct temperament to be B.C. of a Home Guard Battery should be very busy indeed. Admittedly he will lack the companionship of a full mess, but with supervision and participation in the maintenance of equipment under his charge (as frequently he himself will have to take off his jacket and get down to it) and with organisation of Home Guard training, he will find his time fully occupied.

Already there have been quite a few cases of regular Battery Commanders failing to take the necessary interest in Home Guard parades; indeed failing to turn up at them or even failing to see that a junior officer or senior N.C.O. was there to carry out or supervise the instruction. How often have business executives, civil servants, solicitors, bankers etc. been left in the care of a Lance-Bombardier with no idea at all on how to instruct? The most shocking case was one where for several successive Sunday mornings the Home Guard Parade consisted of an improvised football match between Home Guard and regular personnel.

The Home Guard parade must be a period of interesting well-planned and well-rehearsed training without a single moment of delay or dullness during which interest is allowed to lag."

In October 1944 the number of batteries in action was further reduced, and it was no longer compulsory for Home Guard to attend training parades: in December work was started on dismantling the coast-defence batteries which were in care and maintenance and were unlikely to be required for the future defence of Britain. The war in Europe was now rapidly coming to an end, and by the early spring of 1945 the Allied Armies were biting deep into Germany. There was no longer any need for Coast-Artillery to defend the ports, harbours, and shores of the United Kingdom, and after five and a half years of constant watch the Coast Artillery were able to stand down. Their last task, organised as infantry, was to form the relief force for the Channel Islands which were surrendered by the Germans immediately after the cease-fire in Germany on 5th May.

Before closing the account of Coast Artillery at home during the Second World War, it is necessary to say something concerning the training establishments. The Coast Artillery School moved from Shoeburyness during September 1940. The danger of invasion, the continual obstruction of the sea-range, the constant bombing

R

of the Thames Estuary with the resulting interference with instruction made it obvious that the School could no longer remain in its old home. After considerable reconnaisance Llandudno on the coast of North Wales was selected for its new site. As soon as it was safely settled in, the School was greatly expanded, Searchlight and Radar Wings being added to the already old established Gunnery Wing. A "War Gunnery Staff Course" was organised, and all Coast Artillery Cadets came to the School for their final five weeks training before being commissioned. 14 different types of course were held for officers and other ranks, ranging in length from 18 to two weeks, and the instructional capacity of the School was increased to 150 Officers, 115 Cadets, and 445 Other Ranks.

The training of recruits was carried out by the Coast Artillery Training Centre which was formed on mobilisation at Plymouth and which consisted of three Training Regiments and an Officer Cadet Training Unit. During the period of most rapid expansion further training regiments were raised in the Isle of Wight and at Milford Haven. A recruit on joining spent his first month with the Basic Training Regiment and then moved on for two months specialist and gunnery training with one of the other training regiments. If considered suitable, he then passed on to the O.C.T.U. which, besides taking recruits who had completed their training, received recommended candidates from service batteries. In spite of having to move its establishments several times on account of enemy bombing, the C.A.T.C., which was finally distributed between Brixham, Paignton, Ivybridge, Whitsand Bay, Bovisand and Staddon Heights, carried on its most essential and valuable work with efficiency throughout the War and never failed to turn out well trained recruits for the Coast Artillery.

CHAPTER XXIII

1939 to 1945 (continued)

ON 10th June 1940 Italy declared war on Great Britain and France, and the area of the struggle was at once extended to the Mediterranean, Red Sea, and East Africa. This expansion brought into the theatre of operations the great fortresses of Gibraltar and Malta and the defended-ports of Alexandria, Haifa, Port Said, Suez, Port Sudan, Aden and Kilindini. The Italian Fleet consisted of:-

      6 Battleships
      7 Heavy Cruisers
      12 Light Cruisers
      60 Destroyers

the *Vittorio Veneto* (battleship) carrying 15 inch guns, and the *Trieste* (heavy cruiser) 8 inch guns. The British Mediterranean Fleet was weak, its strength rarely rising above:-

      5 Battleships
      1 Aircraft Carrier
      8 Light Cruisers
      25 Destroyers

and very often being much less. The British in the Mediterranean did not fear the challenge of the Italian fleet for supremacy in that sea and felt itself well able not only to maintain its position but to defeat the Italians should it come to a fight. However it was not the Italian fleet which made it difficult, and for long periods almost impossible, for the British Fleet to rule the waves of the Middle Sea, but the German and Italian air-forces which, with their land-based aircraft and airfields surrounding the central and, after the fall of Greece and Crete, the eastern Mediterranean, made it highly dangerous and often suicidal for British ships to cruise in daylight on the surface of the sea.

Crete was taken by the Germans during the latter part of May 1941, and its loss was almost entirely due to German air superiority in the eastern Mediterranean at that time. The Coast Artillery assisting in the defence of the island were the 15th Coast Regiment R.A. (three batteries) and two batteries of the Royal Marine Mobile Defended Base Organisation, the ports to be defended being:-

      Suda Bay ⎫ 4—6 inch
      Canea   ⎬ 4—4 inch
                 ⎭ 2—12 pdrs.
      Maleme 2—4 inch
      Georgeopolis 2—4 inch
      Heraklion 2—4 inch.

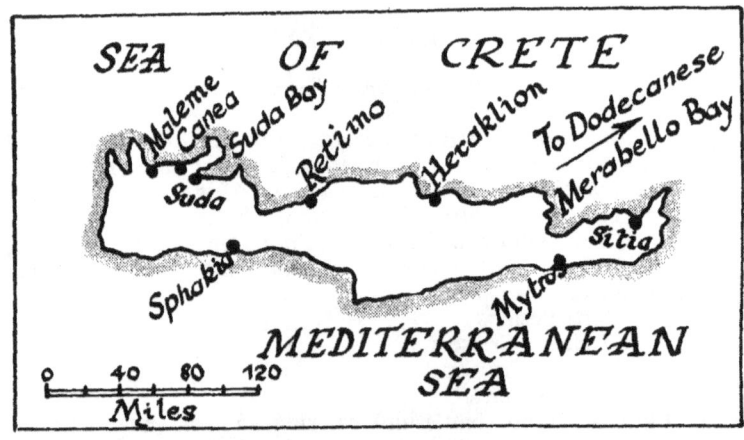

Suda Bay was the most important of these being an advanced base for the fleet. The German plan for the capture of Crete was to land from the air by parachute, glider and, when the airfields had been captured, transport aircraft some 16,000 men of the 7th Airborne Division, to be followed up by sea convoys bearing 6,000 more infantry with guns, tanks, ammunition and supplies. The airfleet to support and carry this great air-borne invasion consisted of 228 bombers, 205 dive-bombers, 233 fighters, 500 transport-planes, and 72 gliders, all based on the airfields of Greece. The British-Greek defending force was about 27,500 strong, but it was very weak in arms, especially guns of all kinds, most of the troops having only lately been evacuated from Greece, and it was distributed in a series of defended-localities along the northern coast road covering the ports and the three airfields.

After a heavy and violent air bombardment, the main German air-landings were made on 20th, 21st and 22nd May in two groups, a western and larger group between Canea and Maleme with first objective the airfields at the latter place and Retimo and second objective the ports of Suda and Canea, and an eastern and smaller group with objectives the airfield and port of Heraklion. The western attack succeeded, the loss of Maleme airfield on the 21st proving fatal to the defenders for it enabled the Germans to pour in airborne reinforcements. At Heraklion the Germans were less successful, failing to capture either of their objectives. The British-Greek forces were soon split up into widely separated pockets fighting desperately for their lives and unable to support or assist each other. At sea the Royal Navy dealt decisively with the follow-up, either sinking or driving back the convoys of small steamers and caiques carrying the German reinforcements and defeating their Italian escorts, but these fine efforts cost the Navy dear, three destroyers being sunk, and one battleship (*Warspite*) and two cruisers being severely damaged by air attack.

On 26th May General Freyberg, the commander of the Allied forces, came to the conclusion that it was no longer possible to hold on to Crete. It was evident that both Suda and Canea would shortly fall into the hands of the enemy, and the continuous and effective air-attacks on the British ships made it increasingly clear that the Navy would be unable much longer to prevent sea-borne reinforcements from reaching the island. It was therefore decided to withdraw the troops from the Suda-Canea area across the mountains to Sphakia on the south coast where the Navy would collect them, and to embark these around Heraklion at that port itself. The retreat from the Suda-Canea area began on the 27th, but the enemy was unrelenting in his attacks and gave the retiring troops no respite with the result that large numbers were cut off and forced to surrender. Rear guard actions were fought at Babali Hani and Askifou, and the retreating columns were under persistent and violent air attack the whole way as they struggled along the one mountain road. At Sphakia the Navy took off as many as they could throughout the 30th and 31st, but again enemy aircraft made embarkation extremely difficult, and at times almost impossible, attacking both the beaches and the ships at sea. After the 31st, no more attempts could be made. The force at Heraklion was safely embarked during the night 28th/29th, but, as soon as it was daylight, the ships were attacked by German aircraft which continued their assault throughout the day, two destroyers being sunk and one cruiser and two destroyers damaged. Of the 4,000 troops embarked at Heraklion, 800 were casualties before the day was out.

So Crete was relinquished to the enemy. The British casualties had been heavy. The Navy had lost three cruisers and six destroyers sunk and had two battleships, six cruisers, and seven destroyers severely damaged, besides suffering over 2,000 casualties to officers and ratings. The British and Dominion Armies, the Royal Marines, and Royal Air Force sustained some 15,000 casualties, 3,600 being killed and wounded and the remainder abandoned as prisoners. It was no glorious campaign for the 15th Coast Regiment either. Heavily bombed both before and during the German attack, they had no opportunity to use their guns against enemy ships, though one battery did get some shooting at land targets. During the night 26th/27th May the personnel of the batteries, having destroyed their guns, were withdrawn from the north coast and on the 27th set out on their long trek across the mountains to Sphakia. On account of the congestion on the roads regimental headquarters and the batteries soon lost touch with each other, and it was only as separate groups that the regiment reached the port of embarkation on the morning of the 29th. Owing to the mass of troops in the area and lack of cover from bombing, it was necessary to spread the men out among the surrounding hills, and it soon became very evident that only a small proportion of the regiment would be able to get away before embarkation was closed down. During the night 31st May/1st June, nine officers and 120 other ranks were embarked in *H.M.S. Kimberley* and returned safely to Egypt, the remainder being left behind to become prisoners of the enemy. The detachment at Heraklion was safely taken off during the night 28/29th May, but unfortunately the majority were lost with *H.M.S. Hereward*.

Malta, the formidable fortress placed right in the middle of the Mediterranean, had been a famous Coast Artillery station ever since its occupation in 1800. It had served as the chief British naval base in the Mediterranean throughout the Napoleonic and First World Wars and as a main advanced base for all purposes during the Crimean War. In the Second World War the island proved once again its high value as a base, showed outstanding bravery under attack and privation, but also demonstrated clearly that the advent of air power had quite altered its status and potentiality. When Italy entered the war in June 1940, the Coast Artillery and defences of Malta were:-

4th Heavy Regiment, R.A.—Outer Fire Command
 6th Heavy Battery, R.A. Madalena Battery—2—9.2 in.
             Binjemma Battery—1—9.2 in.
 10th Heavy Battery, R.A. Leonardo Battery—2—9.2 inch
             Benghaisa Battery—2—9.2 inch

1st Heavy Regiment, Royal Malta Artillery—Harbour Fire Comd.
 1st Heavy Battery, R.M.A. Rocco Battery—3—6 inch
             Ricasoli Battery—3—6 pdrs.
 2nd Heavy Battery, R.M.A. Tigne Battery—3—6 inch
 3rd Heavy Battery, R.M.A. St. Elmo Battery—6—6 pdrs.
 4th Heavy Battery, R.M.A. Delimara Battery—2—6 inch
             Campbell Battery—2—6 inch.

During 1941 additional guns were mounted at:-
 Isola Battery   2—4.7 inch
 Taxbiex Battery   2—4 inch
 St. Angelo Battery 2—4 inch
 Manoeldone   1—4 inch
 Bogebba   2—12 pdrs.
 Delimara   2—12 pdrs.

There was a major-general C.R.A. of all the artillery on the island with brigadiers under him commanding the Anti-Aircraft and Coast Defences, the latter being Brigadier G. C. Gatt, late of the Royal Malta Artillery and the first of that famous corps ever to attain that exalted rank. Gatt was a well known name in the R.M.A., many of the brigadier's forebears, relations, and sons either having served or were still serving in the regiment. When Gatt retired towards the close of 1941, he was succeeded as Commander Coast Artillery by Brigadier C. J. White.

Once Italy had entered the war, and especially so after Rommel's Africa Corps had crossed into Libya, the sea routes between Sicily, Italy, and the Libyan ports became absolutely vital to the Axis powers, for it was only by these communications that the North African campaign could be supplied, nourished, and supported. Malta was ideally placed in the midst of these essential life-lines for seriously interfering with them, and, from the moment war was declared, surface warships, submarines, and aircraft based on Malta proceeded to attack the convoys of storeships and tankers passing across to Tripoli, Benghazi, Derna, and Tobruk. So severe were these depredations that the Axis powers soon determined that Malta must be put out of action, and that it could and should be done by air-power alone. The Axis air-bases in Sicily and North Africa were so close to Malta that this could be done by two methods, or, as the Axis powers finally decided, by a combination of both.

Malta produced practically nothing for itself and was absolutely dependent for most of its food, and all of its petrol, ammunition, spares, expendable military stores, and replacements of arms, equipment, machinery, heavy appliances etc. upon outside sources, and all these commodities had to be brought to Malta by ship through the Mediterranean. The Axis powers therefore hoped to put Malta out of the war firstly by a continuous air bombardment aimed direct at its heart which would make it impossible for any warship to use Valetta harbour or any aircraft the airfields, and which would destroy the dockyards, air installations, military establishments and seats of government and command: and secondly by preventing, again by the use of air power, any convoys of British ships from reaching Malta and so slowly but surely forcing the island to expend all her irreplaceable resources from aeroplanes to tinned milk, and from petrol to potatoes, thereby forcing her to surrender from sheer lack of food and munitions.

The Coast Artillery at Malta suffered equally with the rest of the garrison and inhabitants. The bombardment from the air began almost as soon as war broke out with Italy and continued, with fluctuating intensity, until December 1942. The Germans joined in with the Italians in January 1941, and from that time onwards the attack was maintained both by day and night, reaching its greatest violence during the first six months of 1942. The garrison and inhabitants of Malta had not only to stand up to the rain of bombs from the air but also to endure, on account of the blockade, many shortages of which the most serious were lack of food, petrol, and ammunition. By April 1942 Malta was living under true siege conditions, and the troops' rations—as were those of the inhabitants —were cut to the bone, the main items being but ¼lb. of bully-beef and ½lb. of bread per man per day, not much for soldiers standing to their guns throughout the 24 hours. The Coast Artillery as usual suffered from a chronic shortage of man-power, it being quite impossible to find sufficient men for the watches on the guns, the necessary reliefs, local defence, and all the other multitudinous duties. Nevertheless this did not prevent the Coast Artillery from contributing strong parties to assist in the rapid unloading of such ships as did run the blockade successfully. However the Coast Artillery had two opportunities of action against the enemy during these long months of air assault and siege, firstly on the night 25th/26th July 1941 and secondly on the night 16/th17th May 1942. Both attacks were carried out by the Italians.

The object of the first attack was to penetrate into Grand Harbour Valetta and sink and destroy the ships of a convoy which were known to be unloading there at that time. The force detailed for the operation was:-
    (1) One Motor Boat Carrier, which carried:-
    (2) Ten baby Motor Explosive Boats—one 2-man, and nine 1-man boats.
    (3) One large Motor Launch, which carried:-
    (4) Two 2 men Human Torpedoes with detachable explosive bows.
    (5) Two Motor Gunboats for escort and rescue work.

This force left the port of Augusta (Sicily: about 90 miles north east of Malta) at about six o'clock on the evening of 25th July and travelled during the night to a rendezvous some 14 miles from Valetta where the M.Ex.B.s were lowered into the sea. The plan of attack was:-
    (1) Human Torpedoes to blow a breach in the viaduct connecting St. Elmo Breakwater, which covered the entrance to Grand Harbour, with the shore.
    (2) M.Ex.B.s then to pass through the breach and attack the shipping in the harbour.
    (3) The whole operation to be covered by an air attack lasting 30 minutes which would drown the sound of engines and divert the attention of the defenders.
    (4) After the operation, the crews of the Human Torpedoes and M.Ex.B.s to be collected and rescued by the Motor Gunboats and Motor Launch.

From the rendezvous the M.Ex.B.s and the Motor Launch carrying the Human Torpedoes moved towards Malta, and soon after 4 a.m., at a point some 900 yards from St. Elmo Breakwater, the Human Torpedoes were launched. At a quarter past four the air attack began, and about 30 minutes later the leading submarine blew the breach in the viaduct.

1939 to 1945 (continued)

The Malta radar had picked up the Motor Boat Carrier as early as 9 p.m. on the previous evening, but no tracks of approaching vessels were later indicated, they were probably too small. During the early hours of the 26th the sound of motor engines far out to sea was heard by several look-outs on the batteries, the guns were manned and searchlights exposed for a short time, but no targets could be seen. At 4.15 a.m. the enemy air attack started and continued until just before 4.45, the last Italian aeroplane having scarcely left when the look-out at St. Elmo Battery (6—6pdr.) saw the track of a fast-moving boat close in shore and almost simultaneously the viaduct went up with a loud explosion. The alarm was at once given, the searchlights exposed (Illuminated Area), and the twin 6 pdrs. of St. Elmo and Ricasoli Batteries immediately poured a stream of tracer shells at point blank range into the line of baby M.Ex.B.s clearly seen in the beams of the searchlights making straight towards the gap in the viaduct in line ahead. In less than two minutes the action was over, five of the M.Ex.Bs. being sunk or disabled and the remainder in flight. During the next hour various vessels attempting to escape were picked up by searchlights and engaged, the 6 inch battery at Rocco firing a salvo at a target, probably the carrier of the Human Torpedoes, at a range of 8,000 yards and destroying it, and St. Elmo and Ricasoli re-opening fire at about 5.50 on two M.Ex.Bs at a range of 3,000 yards and sinking them. As soon as it was light the scattered survivors were attacked by fighter aircraft from the Malta airfields and finally dispersed. The total number of victims were :-

(a) Sunk by Coast Artillery —1 Motor Launch Carrier of Human Torpedoes.
6 Motor Explosive Boats.
(b) Sunk by R.A.F.— 2 Rescue Motor Gunboats.
2 Motor Explosive Boats.
(c) Self-Destroyed 1 Human Torpedo.
(d) Run-Aground 1 Human Torpedo.

Some 580 rounds of 6 pdr. and two rounds of 6 inch ammunition were fired. All three batteries, St. Elmo, Ricasoli and Rocco were manned by the Royal Malta Artillery.

The second action took place during the night 16th/17th May 1942. Four Italian "E-Boats" (Motor Torpedo or Gunboats) were patrolling off Malta at about 1 o'clock on the morning of the 17th. when they were picked up by radar, and at 1.20 a.m. seen clearly in the beams of the searchlights of Campbell Fort which had been exposed. They were at once engaged by the battery at 10,000 yards but moved off north at high speed and were soon out of range. At 1.45 a.m. the enemy E-Boats were again picked up by radar, this time they were seen to be approaching Valetta Harbour. As soon as the enemy vessels were within range, the searchlights were exposed showing four E-Boats close together which were at once engaged by Tigne, Rocco, Madalena and Leonardo batteries at about 10,000 yards. The targets immediately took avoiding action by zig-zagging and making smoke but were firmly held in the searchlight beams. They then turned away at their fastest speed, retiring towards the north under cover of their smoke screen until well out of range. One enemy E-Boat was sunk during this action, and one found disabled on the surface at a range of about 11,000 yards at daylight. This was engaged and hit by Rocco battery but was finally sunk by fighter aircraft. The amount of ammunition fired in this action was three rounds of 9.2 inch and 39 rounds of 6 inch.

The Royal Malta Artillery was greatly expanded during the siege, producing anti-aircraft, and beach-defence units besides a second coast regiment. This last was the 5th Coast Regiment R.M.A. consisting of the 11th, 12th and 13th Coast Batteries. Once the battle of El Alamein was won in October 1942, and the British had advanced westwards well into Libya, the siege of Malta was raised, and in the New Year of 1943 convoys started to come in regularly from Alexandria. During the spring of that year the Axis powers were driven completely out of North Africa, in July and August the great island of Sicily was attacked and taken, and in September Italy was successfully invaded from the south. It was now possible to reduce substantially the coast-artillery at Malta, and in November 1943 all the British coast batteries (4th Coast Regiment R.A.) were sent home to Great Britain, the coast-defences being handed over entirely to the R.M.A., and the counter-bombardment guns left unmanned. The 1st Coast Regiment R.M.A. (1st, 2nd and 3rd Coast Batteries) continued to man the close-defence and anti-motor torpedo boat guns, and so it remained until the end of the war in Europe in May 1945.

The extension of the war to the eastern Mediterranean and North Africa brought about a great increase in the coast-defences of those areas, and by December 1941 the defended ports and guns were as follows:-

Alexandria	4—6 inch.
	2—4 inch.
	2—12 pdr.
	2—6 pdr.
Haifa	7—6 inch.
	2—12 pdr.
Famagusta	2—4 inch.
Port Said	2—6 inch.
Suez	2—6 inch.
Mersa Matruh	3—6 inch.

1939 to 1945 (continued)

The coast-guns at Tobruk during the siege (April-December 1941) were manned by the Nottinghamshire Yeomanry Coast Battery—yeomen turned coast-gunners—the French coast-defences of Syria were handed over to British Coast Artillery (August 1941) and, during the advance of the Eighth Army into Libya in the winter of 1942/43, the defences of the Italian ports (Tobruk, Benghazi and Tripoli) were taken over as each town was occupied.

At the other end of the Mediterranean Gibraltar stood on guard throughout the war, but the coast-artillery there were never called upon to open fire in anger. The principal threat with which the Gibraltar defences had to deal was attacks by Italian Human Torpedoes on shipping in the roadstead and harbour. These weapons employed an Italian tanker berthed in the Spanish port of Algeciras as their base and mother ship, and this fact was not discovered by the *Rock* authorities until after the armistice with Italy in September 1943. The Human Torpedoes would move across Algeciras harbour under water and only surface when close to their targets thus being almost impossible to discover. They were very rarely able to penetrate the harbour-defences but several times managed to sink merchant-ships at anchor in the roadstead. On one of the few occasions when they successfully reached the harbour entrance, two of them were spotted by a coast-artillery look-out and subsequently sunk by depth-charges. The Coast Artillery garrison of the *Rock* at its greatest strength (Spring 1943) was:-

    3rd Coast Regiment R.A.—4th and 41st Coast Batteries R.A.
    19th Coast Regiment R.A.—26th, 27th and 40th Coast Batteries R.A.

and the guns:-
    8—9.2 inch
    8—6 inch
    6—4 inch
    6—6 pdrs.

The closing of the Mediterranean sea communications by the Axis air-power enhanced the importance of the Cape route and brought about the establishment of the trans-African Air-route from Takoradi (Gold Coast), by way of Kano, Fort Lamy, El Obeid and Khartoum, to Cairo. This rendered necessary the arming of new ports and an increase in the number of guns at old ones, so that by the end of 1941 the defended-ports on the route to the Cape were:-

Bathurst (Gambia)	2—5.5 inch.
	1—4 inch.
	1—12 pdrs.
Freetown	2—9.2 inch.
	4—6 inch.
	1—4 inch.
Takoradi	6—4.5 inch. (dual purpose C.D./A.A.)

Lagos	2—6 inch
	1—4 inch.
	2—12 pdrs.
Ascension Isl.	2—5.5 inch.
St. Helena	2—6 inch.

During the latter half of 1942 and the first months of 1943 formations of mobile coast artillery were organised in the United Kingdom for providing the defences of ports abroad when captured from the enemy. Their armament was mixed, consisting chiefly of 6 inch, 4 inch, and 12 pdr. guns with transportable platforms which could be rapidly installed in position, and their equipment included radars, searchlights, and range-finding and fire control instruments. Each consisted of a headquarters Fixed Defences and two coast-regiments with coast-batteries, fire-command post detachments, coast-observer detachments, and maintenance units. The first of these formations to go into action was 201st Fixed Defences which followed close on the heels of the Allied landings in North Africa in November 1942 and quickly established itself and its guns in the ports of the Algerian coast as they were occupied by the British First Army. The next one was 203rd Fixed Defences (Colonel S. C. Tomlin) which arrived at the port of Augusta in Sicily on 19th July 1943. Its first task was to provide coast-defences for Syracuse and Cantania. This was done early in August, the former port being defended by Italian guns remade serviceable soon after capture and six British searchlights, the latter by a mixture of British (8—12 pdrs.) and Italian guns and again British searchlights.

The armistice with Italy was signed on 3rd September and on the same day British troops crossed the *Straits of Messina* and invaded the toe of Italy. 203rd Fixed Defences were ordered to take over and be responsible for the coast-defences of Taranto as soon as possible. This important naval base had been surrendered by the Italians practically intact, and by the last week in September it was possible to organize the coast-defences with complete Italian equipment, guns, searchlights, and personnel, all under British command and supervision. Brindisi was the next port to become the responsibility of 203rd Fixed Defences, and this was treated in a similar way to Taranto, the Italian armament etc., with the soldiers who manned them, being encorporated *en bloc* into the British coast-defence scheme.

Meanwhile the British and American armies were advancing up the leg of Italy, and on the east coast the ports of Bari, Barletta, Manfredonia, and Vasto were captured from the Germans, the coast-defence of each becoming in turn the charge of 203rd Fixed Defences. The same procedure was carried out in each case. A reconnaisance-party was sent forward as soon as possible after allied occupation to examine the port and get in touch with the naval authorities. A rapid inspection was then made of the Italian coast-defence batteries and searchlights to see if they were—or could soon be put—in working order and whether they were mounted in suitable positions. Next a map of the port was produced with the docks and anchorages marked upon it, and to these were added the most probable directions and areas of attack. It was then settled how best to prevent the enemy from reaching these areas, and on this the searchlight and gun scheme was based. Finally it was decided what Italian equipment already in *situ* to use—and also what Italian troops already on the spot to employ—and what British personnel, guns, searchlights, radar etc. should be brought forward to the port. For instance Bari was defended by Italian guns, British searchlights, and British radar, all manned by British coast-gunners.

On the west coast Naples had been occupied by the Allies on 1st October, and, as this was to become the main allied base for the Italian campaign, it was decided to concentrate all the remaining resources of 203rd Fixed Defences on its defence. Now that the Italian navy had been eliminated, the scale of possible German attack was considered to be no greater than destroyers, E-boats, human-torpedoes, and such like. The coast-defences of Naples, when finally deployed, consisted of a mixture of British and Italian guns and searchlights, British radars, and British and Italian personnel. The guns were:-

Ischia Batteries	4—90/53 guns
	4—76/40 guns
	2—90/53 guns.
Monte di Procida Batteries	2—76/40 guns.
	4—12 pdrs.
	2—4 inch guns.
Nisida Battery	2—4 inch guns.
Posillipo Battery	2—76/40 guns.
Torre Battery	4—76/40 guns.
Orlando Battery	2—76/40 guns.
Campanella Battery ...	2—76/40 guns.
Capri Battery	4—90/53 guns.

These guns, and the searchlights, radars etc. were manned by the 574th Coast Regiment which had an establishment of:-

British—23 officers—501 Other Ranks.
Italian—12 officers—819 Other Ranks.

Most of the Italians had remained with their batteries after the armistice and turned straight over to the British on the arrival of the Allies.

The Allies made their landing at Anzio late in January 1944 and in May recommenced their successful advance northwards up the Italian peninsula. 203rd Fixed Defences supplied the coast-artillery for the Anzio anchorage (June) for Civitavecchia (June) Piombino (July) Ancona (July) and Leghorn (August) and continued to hold these ports throughout the last winter of the war (1944/45). During the spring of 1945, when the war was rapidly coming to its violent end, there was much "last minute" activity amongst enemy light coastal craft, several attempts being made to attack shipping in ports occupied by the British and Americans and guarded by 203rd Fixed Defences. At Ancona during the night of 14th/15th April at about 8.15 p.m. the coast-artillery radar picked up a plot approaching the port and considered to be hostile. Lighthouse Battery (2—4 inch and 1—90/53) was ordered to open fire at 8.50, using the radar to provide ranges and bearings. The target was seen on the radar to slow up, circle round, and then fade out. Next morning four Italians were collected from rubber-dinghies by the Navy and R.A.F., these prisoners admitted they were crews of two midget M.T.B.s which had attempted to enter Ancona harbour with the object of attacking shipping at anchor there. Unfortunately for them they had run into the accurate fire of Lighthouse Battery which put their little vessels out of action and forced them to take to their dinghies.

During the night of 23rd/24th April a last desperate attempt was made by the enemy to cause havoc among allied shipping lying at anchor in Leghorn roadstead and to damage the harbour and port installations. The attacking force was a German assault-flotilla consisting of five M.T.B.s and 12 M.Ex.B.s, all manned by fanatical members of the Hitler Youth Corps and based on Porto Fino. At 2.15 a.m. the coast-artillery radar at Leghorn picked up a plot moving at speed towards the port which was soon identified as hostile and as a large group of very small vessels. At 2.50 Trotta Battery (2—4 inch: 2—76/40: 1—90/53) opened fire at a range of 15,000 yards, closing to 9,500 opening again to 11,300; and finally to 14,000. At this stage, about 3.45 a.m., the enemy sent up a series of red and green verey-lights and golden rain rockets, and the Royal Navy despatched two motor-launches at speed to investigate. Almost immediately radar plots again showed enemy craft closing the roadstead, so Trotta Battery once more opened fire with its 4 inch. This was the enemy's last attempt, for the hostile vessels now turned away, made off at about 30 knots, and then dispersed. During the next few days several survivors of the Hitler Youth Corps suffering from exposure were picked up from rubber dinghies, and much wreckage was seen. Trotta Battery fired 30 rounds of 4 inch, 100 rounds of 90/53, and 36 rounds of 76/40. None of the shipping in Leghorn roadstead was either attacked or damaged. Shortly after this the war in the Mediterranean came to its end.

CHAPTER XXIV

1939 to 1945 (continued)

WITH the surprise attack on Pearl Harbour in the Hawaiian Islands on 7th December 1941, Japan entered the war of intent and the United States by force of circumstance. For the past twelve months the British government had been expecting the Japanese to go to war with Great Britain and her empire and realised it was very possible that the eastern struggle—at least for the opening months—might have to be fought without the help of the U.S.A. which were still showing at that time strong isolationist tendencies. Without American assistance and with her hands full in the Narrow Seas, Atlantic and Mediterranean, it was quite evident that Great Britain would be much inferior to Japan in eastern waters, and the latter would have clear superiority at sea. The Japanese fleet in December 1941 was a formidable force consisting of:-

12 Battleships
10 Aircraft Carriers
18 Heavy Cruisers
18 Light Cruisers
18 Light Cruisers
115 Destroyers

and her two newest battleships, the *Musashi* and *Yumato*, were the most powerful in the world, each being of 64,000 tons burthen and armed with 9—18 inch guns. The British, foreseeing the probability of their inferiority at sea, during 1941 set about strengthening their coast-defences in the Pacific and Indian Oceans, and in the China Sea. Japanese naval supremacy in the Far East would not only threaten the great dominions of Australia and New Zealand, but would also endanger all the British possessions in Asia, in east Africa, in the Pacific, and even possibly in the south Atlantic, and would most certainly completely disrupt the sea-communications in those areas. The chief fleet bases and defended ports in the East were Singapore, Hong Kong, Trincomalee, Bombay, Colombo, Penang, and Karachi, and, in the southern Indian Ocean, Mauritius. With the Japanese menace ever growing more ominous, it was necessary to augment and expand these coast-defences, and by the time of the Pearl Harbour disaster they had increased and spread as shown in the list below. It is to be noted that some of the guns may not have been actually ready and in action by 7th December but work upon getting them mounted was in progress:-

Singapore	5—15 inch
	6—9.2 inch
	18—6 inch
	8—12 pdrs.
	10—6 pdrs.
Hong Kong	8—9.2 inch
	14—6 inch
	2—4.7 inch
	4—4 inch
Penang	4—6 inch
Rangoon	2—6 inch
Trincomalee	2—9.2 inch
	3—6 inch
Colombo	2—9.2 inch
	4—6 inch
Calcutta	2—6 inch
Vizagapatam	2—6 inch
Madras	2—6 inch
Cochin	2—6 inch
Bombay	4—7.5 inch
	3—6 inch
	2—4.5 inch C.D./A.A.
Karachi	2—6 inch
Lutong (Sarawak)	2—6 inch
Christmas Island	1—6 inch
Cocos Keeling	2—6 inch
Addu Atoll (Maldive Islands)	6—6 inch
	1—4 inch
Diego Garcia (Chagos Islands)	2—6 inch
Seychelles	2—6 inch
Mauritius	2—6 inch
Aden	4—6 inch
	2—12 pdrs.
Kilindini	2—6 inch
Tanga	1—6 inch
Zanzibar	1—6 inch
Dar-es-Salaam	1—6 inch

1939 to 1945 (continued)

Fiji	... ...	5—6 inch
		2—4.7 inch
Noumea	... ...	2—6 inch
Fanning Islands	...	1—6 inch
Falkland Islands	...	2—6 inch
		1—4 inch
		2—12 pdrs.
South Georgia	...	2—4 inch

The coast-artillery for manning all these extra guns were produced in many ways. A coast branch of the Indian Artillery was raised which, besides finding the coast artillery for all (six) of the Indian defended-ports, sent a regiment to Addu Atoll and a battery to Diego Garcia. The Ceylon Garrison Artillery was expanded into two regiments and, in addition to defending Colombo and Trincomalee, despatched one battery to the Seychelles and another to Cocos Keeling. The detachments for Lutong and Christmas Island were composed of a mixture of British and H.K.S.R.A. ranks, with 10 British and 41 H.K.S.R.A. at Lutong and five British and 26 H.K.S.R.A. at Christmas Island. A Mauritius Coast Regiment was raised from the local Territorial Force, and at Rangoon the 6 inch guns were manned by the Rangoon Coast Battery, Burma Auxiliary Force. At other places volunteer batteries were organised, South Georgia even producing two of such units. There was no shortage of men to man the guns, but only a shortage of guns.

The Pearl Harbour disaster placed the Allies in very definite inferiority at sea, and gave the Japanese almost a free hand in the western Pacific, China Sea, and Indian Ocean. The Japanese plan was to establish a deep defensive zone by pushing out east into the Pacific and south into the Indian Ocean, capturing the naval bases of Hong Kong, Manila and Singapore on the way, and occupying the Philippines, Indo-China, Siam, Malaya, Java, Sumatra, Burma, Borneo and New Guinea. The Philippines, Siam, and Malaya were attacked on the same day as the surprise air-assault on Pearl Harbour was launched, and on the next day (8th December) the first Japanese patrols crossed the British-Chinese border into the the mainland territory of Hong Kong. The Japanese were well practised in the art of attacking and capturing coast-fortresses. They had in the not far distant past taken Port Arthur from the Chinese in 1894, again from the Russians in 1905, and Tsingtao from the Germans in 1914. They had pursued the same general plan in all three cases. They had landed the main-body of their attacking forces well away from the coast-fortress itself, beyond striking distance of any field-army that might be covering the fortress. They had then closed in from the landside, attacking the back-door

s

of the fortress. They never risked tackling the coast-defences with their warships nor attempted to land their troops within range of the coast-artillery guns. The Chinese, Russian, and Germans, realising that the landside of a coast-fortress was its most vulnerable part, blocked that back-door entrance with strong and formidable fortifications which in 1904-05, the only occasion on which the assailants came up against really powerful and sustained resistance, the Japanese found most difficult to overcome and penetrate. The whole story of the siege of Port Arthur, when it was held by the Russians, shows what great strength and added security stout forts and fortifications on the landside give to an isolated coast-fortress and its garrison.

Hong Kong was a very isolated coast-fortress, it being some 2,500 miles from Singapore its nearest neighbour. It had not had so much money spent on it as Singapore as it was really too far forward to serve as a safe fleet base and for most of the between-war years had lain within the area of the Washington Agreement. Morover, loss of command of both the sea and air by the Allies made its final fall certain once it was invested, and its landward defences lacked depth and solidity. Nevertheless there was no reason why it should not put up a spirited and vigorous resistance provided it was adequately defended. The garrison of Hong Kong in December 1941 consisted of:-

Coast Artillery—8th and 12th Coast Regiments, R.A.
Mobile Artillery.—Hong Kong Regiment H.K.S.R.A.
                    965th Defence Battery, R.A.
Anti-Aircraft Artillery—5th Anti-Aircraft Regiment, R.A.
Royal Engineers—2 Companies.
Infantry—1st Brigade— 1st Bn. Middlesex Regt.
                              1st Bn. Winnipeg Grenadiers
                              1st Bn. Royal Rifles of Canada.
            2nd Brigade— 2nd Bn. Royal Scots
                              2nd Bn. 14th Punjab Regt.
                              5th Bn. 7th Rajputana Rifles.
Volunteers.—Hong Kong Volunteer Defence Corps
                          (Coast Artillery
                          Engineers
                          Signals
                          Infantry
                          Machine-Gunners.)
Ancilliary Corps.

in all about 11,000 men, with a small port-defence naval flotilla of one destroyer, two gunboats and eight M.T.B.s and a flight (six aircraft) of Vildebeeste Torpedo-Bombers R.A.F. at Kai Tak airfield.

The coast-defences were a mixture of new and old, and great efforts had been made since 1935 to bring both the fortress coast-defence system and the armament up to date. The distribution of units and guns were:-

Eastern Fire Command
12th Coast Battery, R.A.
30th Coast Battery, R.A.
36th Coast Battery, R.A.

Hong Kong Volunteer Defence Corps

Western Fire Command
24th Coast Battery, R.A.
20th Coast Battery, R.A.

Hong Kong Volunteer Defence Corps

8th Coast Regiment, R.A.
Stanley Battery—3—9.2 inch
Bokhara Battery—2—9.2 inch
Collinson Battery—2—6 inch
Chung-am-Kok—2—6 inch
Pakshawan Battery—2—6 inch
D'Aguilar Battery—2—4 inch
Bluff Head Battery—2—6 inch
12th Coast Regiment, R.A.
Mount Davis Battery—3—9.2 in
Stonecutters Battery—3—6 inch
Jubilee Battery—3—6 inch
Aberdeen Defences—2—4 inch
Belchers Battery—2—4.7 inch.

The C.R.A. was Brigadier T. Macleod.

The British plan was to defend Kowloon Peninsula (mainland) with the 2nd Brigade whilst the 1st Brigade held Hong Kong Island against landings from the sea and, if not already engaged, would also be available for employment as a central reserve. The mainland defences were the "Gindrinkers Line" (inner line), 10½ miles long, consisting of a series of partially constructed redoubts and defended localities, and having little or no depth. This compares somewhat badly with the complex system of fortifications protecting Port Arthur on the landside in 1904 although the Russian Empire of those days was not considered to be a model of efficiency.

The Japanese attacking force consisted of three divisions under Lieut.-General Sakai supported by some 80 aircraft and a powerful naval squadron. This army-corps was already concentrated on the Chinese mainland around Canton and during the first week in December moved towards the British-Chinese frontier. At dawn on 8th December the forward Japanese patrols crossed the border, and at the same time the enemy air-force made a devastating low-flying attack on Kai Tak airfield, destroying Hong Kong's little R.A.F. flight before it could get off the ground. From that moment onwards the whole colony was subjected to continuous and accurate high level bombing which seriously interfered with operations, causing many fires and much destruction, and spread distress and sometimes panic amongst the large Asiatic population. The advance of the enemy army-corps southwards towards the Gindrinkers Line was carried out with skill and speed, and by dusk on the 9th their leading patrols were probing the forward defence-posts. This

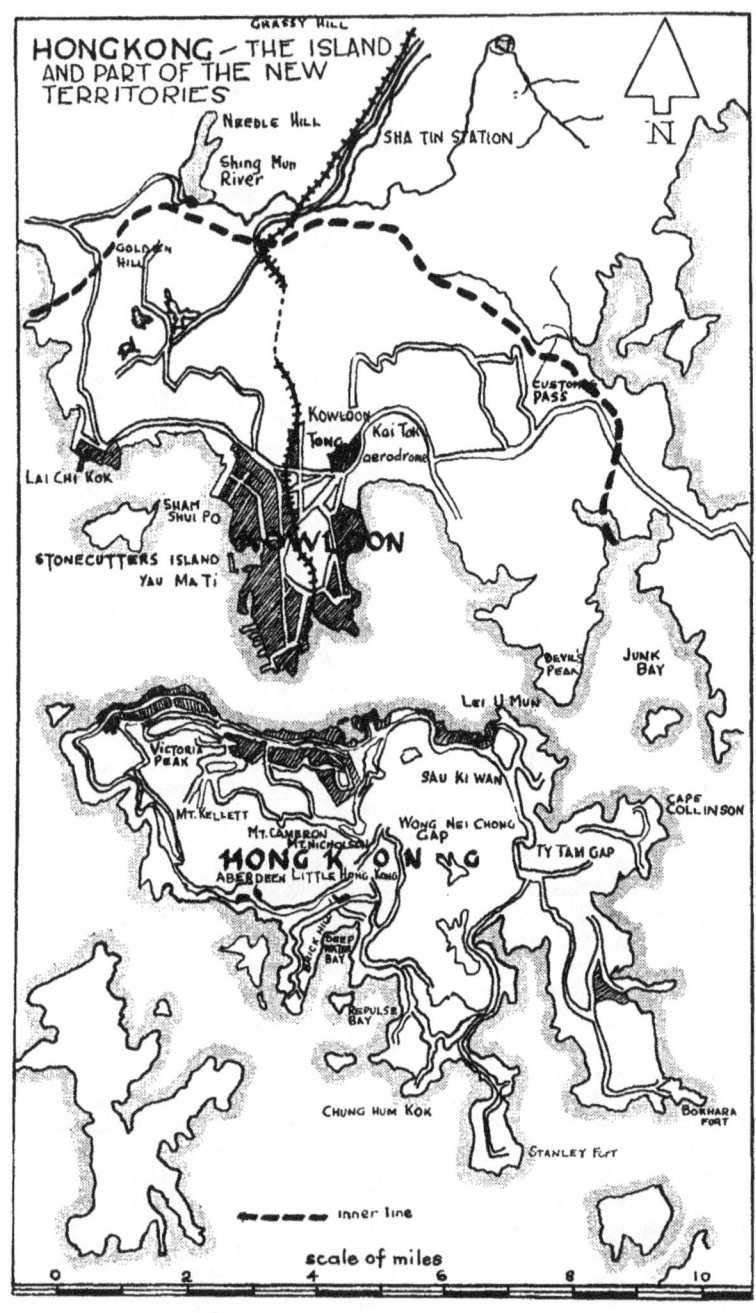

1939 to 1945 (continued) 261

position, as has already been stated, was 10½ miles long, indifferently fortified, and held by the 2nd Brigade of three battalions which were quite insufficient to hold so extended a line. During the night 9th/10th the Japanese captured Shingmun Redoubt (left centre), which was the key to the whole position, by a surprise assault, thus rendering it impossible to hold the remainder of the line and indeed any part of the Kowloon Peninsula at all. After some heavy fighting the 2nd Brigade abandoned the Gindrinkers Line during the 11th, and during the following night all the British forces on the mainland, except for the 5/7th Rajput Regiment holding the Devils Peak area, were safely withdrawn to Hong Kong Island.

The Coast Artillery up to the present had had little to do in their primary anti-ship role. On December 8th Bokhara Battery (2—9.2 inch), which had only just been mounted, opened fire on an enemy mine-sweeping trawler, and two days later Stanley Battery (3—9.2 inch) engaged a Japanese destroyer, straddling the target but unfortunately failing to hit it. During the forenoon of the 11th a small enemy party attempted to land on Lamma Island—off the south-west corner of Hong Kong—using captured junks, and these were fired upon by Jubilee and Aberdeen Batteries. However, right from the start of operations, the 9.2 batteries at Stanley and Mount Davis were used for firing on land targets, being employed on counter-battery work and on delivering harassing fire against the enemy communications on the mainland. The amount of land service ammunition held in the magazines (25 rounds per gun) was woefully insufficient, and much of this small supply was useless on account of age (manufactured in 1918), so that soon the 9.2 batteries were reduced to firing armour-piercing shell with non-delay fuzes but of course their fragmentation was very poor. The Japanese soon brought their heavy howitzers (6 inch and 9 inch) into action and on the 10th began an intense bombardment of Stonecutters Island which was continued throughout the 11th, West Fort receiving more than 40 direct hits, and many installations and buildings being completely wrecked. During the night 11th/12th, in conformity with the general withdrawal from the mainland, Stonecutters Island was evacuated, the 6 inch guns and all movable stores being destroyed.

Between the 12th and 18th December there was a pause in operations while the Japanese prepared for the crossing from the mainland to Hong Kong Island. By the 13th they had brought forward their artillery, and on the afternoon of that day the terrors of shelling were added to those of air bombardment which the garrison and unfortunate inhabitants of Hong Kong were already suffering. Mount Davis, Belchers, and Pakshawan Batteries came in for particular attention on the 14th, one 9.2 inch at Mount Davis being badly hit, the two 4.7 inch at Belchers being knocked out and the battery set on fire, and at Pakshawan the searchlights being put out of action and the fortifications much damaged. The bombardment continued during the next few days, Mount Davis being subjected to intense and accurate shelling on the 16th during which the battery plotting-room was hit and destroyed.

During the late evening of 18th December the enemy carried out successfully their crossing from the mainland, landing along the eastern section of the north shore of Hong Kong Island. The defence of the island had been re-organised in two brigade areas, 2nd Brigade holding the East Brigade Area and 1st Brigade the West Brigade Area, and every possible preparation had been made to put up an effective and vigorous resistance. The enemy crossed the eastern end of the harbour and the narrow Lyemun entrance in a mixture of ferry-boats, launches, lighters, junks, and sampans. Pakshawan Battery was attacked by three companies of Japanese infantry who had crossed Lyemun Strait on small rubber boats and petrol tin crafts right in the face of the fixed beam of the searchlight illuminating the entrance. The enemy drove out the searchlight-detachment with hand-grenades—the detachment left the light exposed and the door of the emplacement locked—and then attacked the battery which, after a stout resistance, was overrun. By morning of the 19th the Japanese were firmly established on the island, and during the day drove relentlessly southwards towards the southern shore which they reached at Repulse Bay at dawn on the 20th, thus splitting the island in two portions and successfully isolating one part from the other.

Throughout the 20th the enemy continued pouring troops and material across the harbour, and made every effort to expand outwards the wedge they had driven athwart the island. The majority of the British garrison was in the western larger part, and desperate counter-attacks were made to try and join up with the eastern area, but unfortunately none were successful. On the eastern side the Stanley Peninsula became the main centre of resistance, and the surviving troops retired towards its defences with the result that Bokhara, Collinson and D'Aguilar Batteries had to be abandoned, their guns destroyed, and the personnel withdrawn to Stanley. Nevertheless the 9.2 battery at Stanley continued to do excellent work, engaging the enemy as they pushed forward through the gaps in the hills of the central massif. The situation however was rapidly becoming hopeless. On the east side the remnants of the eastern brigade were hanging on desperately to the Stanley Peninsula. On the west the main-body was being driven back steadily from position to position, ammunition was fast running short and was most difficult to replenish, and the troops were becoming worn out and exhausted. The enemy continued their heavy shelling and bombing, and by the morning of what must have been Hong Kong's saddest Christmas Day ever, it was obvious that successful resistance could be carried on no longer, so at four o'clock in the afternoon the colony and its garrison were surrendered to the enemy. The troops, with such little assistance as the Royal Navy and Royal Air Force had been able to give them, had fought to their utmost limit, none could have done more or better. The ultimate fate of Hong Kong was settled when the Japanese found themselves supreme on the sea surrounding the colony and in the air over it: there could be no remote chance nor hope of relief. Its immediate fate was sealed when the Japanese captured Shingmun Redoubt and penetrated the Gindrinkers Line: there was then no longer any possibility of a successful defence. So on that Christmas Day 1941 the pleasant colony of Hong Kong went down in blood, slaughter, and misery.

CHAPTER XXV

1939 to 1945 (concluded)

JAPAN began operations against Malaya during the night 7th/8th December 1941. It has been related in a previous chapter why the British Government decided it was necessary to build a great naval base at Singapore and how, to protect that base, a powerful coast-fortress had grown up in the period between the two World Wars. This coast-fortress guarded against assault from the sea not only the naval-base, but an air-base which had been constructed beside it, and the important and considerable commercial port and city of Singapore itself. However, the lay-out and armament of the coast-fortress had been designed only to protect the naval-base and its adjuncts against attacks by enemy warships and against sea-borne landings on Singapore Island and the northern shore of Johore Strait. Now Singapore lies at the southern end of a long, mountainous, and jungle-clad peninsula which stretches away to the northwards for some 1,200 miles, and, to prevent an enemy army from advancing down that peninsula and capturing Singapore and its naval base from the rear, a field force was organised and held in readiness to fight and defeat any hostile land forces which might attempt such a move. The coast-fortress itself had no permanent defences of any kind on the landward side. During the years preceding the outbreak of war some field defences had been prepared, beginning with the Mersing area, subsequently being extended further north, and on the eastern front stretching as far back as Kota Tinggi, but the estimated cost was so ruthlessly cut down by the War Office that the work had had to be abandoned and so was never completed. In some sort of vague way it was hoped that the jungle would provide the necessary protection by proving impenetrable to troops burdened with modern heavy equipment. In any case the lessons of Sevastopol and Port Arthur, both of which held out for many months after the field-armies covering them had been defeated, had long been forgotten. There was to all practical purposes no further defence of Singapore Fortress once the British field-force operating in the Malayan Peninsula had been defeated. Yet even the tottering Manchu Empire of China furnished Port Arthur in 1895 with forts to protect the landward side.

The British forces available for the defence of Malaya in December 1941 were:-

(1) *Royal Navy (Admiral Sir Tom Phillips)*
  (a) "Z" Force
        Battleship ...   ...   H.M.S. *Prince of Wales.*
        Battle-Cruiser   ...   H.M.S. *Repulse.*
        6 Destroyers.
  (b) *Eastern Squadron*
        3 Light Cruisers
        4 Destroyers
        3 Gunboats
        40 Minesweepers, Patrol Boats etc.
(2) *Army (Lieut.-General A. E. Percival)*
  (a) *Field Force*
        3rd Indian Corps—9th Indian Division.
                         11th Indian Division.
                         Federated Malay States Volunteer Force.
        8th Australian Division.
  (b) *Singapore Fortress*
        Coast Artillery (Brigadier A. D. Curtis)
        7th Coast Regiment, R.A.
        9th Coast Regiment, R.A.
        16th Defence Regiment, R.A. (for beach defence).
        Anti-Aircraft Artillery
        3rd Heavy A.A. Regiment, R.A.
        1st Heavy A.A. Regt., H.K.S.R.A.
        2nd Heavy A.A. Regt., H.K.S.R.A.
        3rd Light A.A. Regt., H.K.S.R.A.
        1st Heavy A.A. Regt., Indian Artillery.
        5th Searchlight Regiment, R.A.
        Royal Engineers
        2 Companies.
        Infantry
        2nd Bn. Loyal Regiment.
        2nd Bn. Gordon Highlanders.
        2nd Bn. 17th Dogra Regiment.
        1st Bn. Malay Regiment.
        1st. Bn. Manchester Regiment. (Machine-Gun Bn.)
        Volunteers
        Straits Settlements Volunteer Force.
        Ancilliary Corps
  (c) *Penang Fortress*
        8th Coast Battery H.K.S.R.A.
        36th Fortress Company, R.E.
        5th Bn. 14th Punjab Regiment.
        Province Wellesley Bn. Straits Settlements Volunteer Force.

1939 to 1945 (concluded)

(3) *Royal Air Force (Air Vice-Marshal C. W. Pulford.)*
    4 Fighter Squadrons—43 Buffaloes.
    1 Night Fighter Squadron—10 Blenheims.
    3 Bomber Squadrons—34 Blenheims.
    2 Torpedo Bomber Squadrons—27 Vildebeestes.
    3 General Reconnaisance Squadrons—3 Catalinas: 15 Hudsons.
    The Coast Artillery at Singapore were distributed as follows:-

Changi Fire Command—9th Coast Regiment, R.A.

7th Coast Battery, R.A.	Johore Battery: 3—15 inch
	Betang Kush Battery: 2—6 inch.
22nd Coast Battery, R.A.	Tekong Battery: 3—9.2 inch.
	Sphinx Battery: 2—6 inch.
	Ladang: 1—12 pdr.
	Pulau Ubin: 1—12 pdr.
	Pulau Sajahat: 2—6 pdr.
	Calder Harbour: 2—6 pdr.
32nd Coast Battery, R.A.	Pengerang Battery: 2—6 inch.
	Changi Battery: 2—6 inch.
	Changi Outer: 2—6 pdr.
	Changi Inner: 2—6 pdr.

Faber Fire Command—7th Coast Regiment, R.A.

11th Coast Battery, R.A.	Connaught Battery: 3—9.2 inch.
	Serapong Battery: 2—6 inch.
	Siloso Point: 1—12 pdr.
	Pulau Hantu: 2—12 pdr.
	Behala Reping: 2—6 pdr.
31st Coast Battery, R.A.	Buona Vista Battery: 2—15 inch.
	Tanjong Tereh: 1—12 pdr.
	Batu Berlayer: 2—12 pdr.
5th Coast Battery, H.K.S.R.A.	Silingsby Battery: 2—6 inch.
	Siloso Battery: 2—6 inch.
7th Coast Battery, H.K.S.R.A.	Labrador Battery: 2—6 inch.
	Pasirlaba Battery: 2—6 inch.

and at Penang:-

8th Coast Battery, H.K.S.R.A.	Cornwallis Battery: 2—6 inch.
	Batumaung Battery: 2—6 inch.

The Japanese had long been preparing for their assault on Singapore. They had to avoid at all costs having to land in Malaya with an extended and highly vulnerable line of communication stretching all the way back by sea to Japan. It was necessary for them first to obtain a forward base closer to their objective, and this they did by occupying French Indo-China in two grabs, the first in September 1940 when they occupied the northern portion, the second in July 1941 when they seized the southern half. This latter grab gave them the large and safe harbour of Camranh Bay, only some 800 miles from the east coast of Malaya. Furthermore on

7th December 1941 the Japanese began to occupy Siam—without resistance—which brought them even closer to Malaya. The force detailed for the invasion of Malaya consisted of four divisions with two regiments of tanks under the command of Lieut.-General Yamashita, supported by some 250 aircraft and the Southern Fleet of the Japanese navy under Vice Admiral Kando.

During the first week in December (1941) the Japanese convoys sailed westward across the Gulf of Siam, and during the night 7th/8th began to land troops on the open beaches at the extreme north-east corner of Malaya, at Singora and Patani just north of the Malayan-Siamese border, where the main Japanese force (2 divisions) was put ashore, and at Kota Bharu just south of it. The Japanese plan was for their stronger force to drive straight down the western side of Malaya astride the main railway line, while the smaller body moved down the east coast containing as many British troops as possible. The Japanese obtained air-superiority right from the start of the campaign, violently bombing the British airfields and wrecking many of the R.A.F. machines before they could become airborne, most of the remainder soon being shot down in air-combat. On 10th December the British suffered a terrible disaster. As soon as Admiral Phillips at Singapore learnt of the presence of Japanese troop convoys moving westwards across the Gulf of Siam, he determined to attack them with 'Z' Force with a view to destroying them and cutting their line of supply. Having asked the R.A.F. to provide his ships with fighter air-cover —a request which the R.A.F. were most unfortunately unable to fulfil—the admiral set out on the evening of the 8th with *H.M.S. Prince of Wales* and *Repulse* and four destroyers. Throughout the 9th 'Z' Force sailed northwards seeking for Japanese convoys, but, lacking aircraft for reconnaisance and so being unable to find the enemy, the admiral turned south at about 8 p.m., having determined to go back to Singapore. The small British squadron however had not carried out its sweep unseen, having been sighted by a patrolling Japanese submarine during the afternoon. This submarine reported its discovery to its base at Camranh Bay, and during the morning of the 10th, whilst moving south off the east coast of Malaya where the admiral had deflected it to look for a reported enemy landing, 'Z' Force was heavily attacked by some 80 Japanese bombers and torpedo-bombers from the airfields of southern Indo-China. Without fighter cover the two British capital ships had very little chance of surviving, and, after being repeatedly struck by bombs and torpedoes, sank about one o'clock afternoon. The Japanese now had not only command of the air over Malaya but the sea surrounding it.

Meanwhile the Japanese forces, which had been successfully landed in northern Malaya, had begun their drive southwards

1939 to 1945 (concluded)

towards Singapore. Opposed by Third Indian Corps, which was strung out on both sides of the central mountain backbone of the peninsular, they were able to concentrate their main blow on the western side and, assisted by their tanks of which the British had none, were soon across the Malayan-Siamese border and, after some stiff fighting, took Alor Star on 15th December. Having defeated the 11th Indian Division, the Japanese pushed on and by the 16th were already threatening Penang. This port and township was by now in a parlous condition. Without any anti-aircraft artillery whatsoever for its protection it had been mercilessly bombed by the Japanese for four days. Most of the inhabitants had fled to the hills, leaving the town shattered and ruined, peopled only by the garrison and the dead, and a prey to disease owing to the fouling of the water supply and the breakdown of the sanitary services. It was therefore decided to evacuate the garrison and the remaining European residents by sea during the night 16th/17th. Among the garrison was 8th Coast Battery H.K.S.R.A. which was manning the two 6 inch batteries, Batumaung and Cornwallis. The guns of Batumaung were destroyed by exploding gelignite in the chambers, and those of Cornwallis, the battery now being in full view of the enemy on the mainland, put out of action by burring the breech-screws which were then broken with sledges and jammed home. The searchlights and engines were smashed, and the ammunition (1,500 rounds) blown up by delayed action charges. The battery reached Singapore by sea safely on the 18th.

Closely supported by their air-force the Japanese pressed on southwards, and by 24th December Third Indian Corps were behind the Perak River, while on the eastern side the enemy were moving down the coast of Trengganu. The Japanese did not relax their pressure, and, Third Indian Corps having been defeated at the battles of Kampar (1st January 1942) and Slim River (7th January), it was decided to withdraw all British troops into the state of Johore. The British forces in Johore consisted of Third Indian Corps (the remnants of two Indian Divisions), 8th Australian Division, and 45th Indian Brigade which had just arrived as reinforcements from India. On 13th January, in spite of Japanese air-superiority, an important convoy reached Singapore safely, bringing the 53rd Infantry Brigade (part of the 18th Division) two anti-aircraft regiments, one anti-tank regiment, and 50 Hurricane fighters with their crews. An attempt was now made to wrest the initiative from the enemy, but in the Battle of Muar (15th/18th January) in northern Johore the British forces were again badly beaten—mainly due to the Japanese ability to ship troops by small boats down the coast and thus land in rear of the defending force—and once more had to retreat, fighting unsuccessful rear-guard actions at Batu Pahat (22nd/23rd January) and Kluang (24th January). On the eastern side the Japanese continued their progress southwards and on 26th January landed a force at Endau in conjunction with this advance, the British "Eastforce" only with difficulty managing to escape the enemy pincers. This landing brought on the biggest air-battle of the campaign, the convoy carrying the enemy troops being twice attacked by British bombers and torpedo-bombers escorted by Hurricanes and Buffaloes. On each occasion a large force of Japanese Naval fighters was encountered, and in the subsequent fighting 14 aircraft were lost by each side.

So rapid and successful had been the enemy advance down the Malay Peninsula that there was now no alternative left to the British commander but to withdraw all his forces into Singapore Island, and this was carried out between 27th and 31st January, but unfortunately not without serious loss, the 22nd Indian Brigade being almost surrounded and destroyed at Layang on the 28th. Up to the present the strong and powerful coast-defences of the fortress had had no opportunity of coming into action with their guns. No Japanese warships had offered themselves within range as targets and except for some high level bombing which had caused a few casualties, the batteries had not suffered as yet from the effects of war. Now however, with all the British forces retiring into Singapore Island, the battle was on their very door step. It was evident that every possible coast gun must now be prepared to engage targets on the land front, but this was none too easy, as many of them, due to their positions, could not be brought to bear in a landward direction. However it was thought that the two 15 inch batteries (Johore and Buona Vista) and Connaught (3—9.2 inch) and Pasirlaba (2—6 inch) Batteries would be able to bring down fire on the northern approaches to the island, and the 6 inch batteries at Labrador, Serapong, and Changi on targets on the island itself should the enemy penetrate so far. Unfortunately land-service ammunition was very short, only 30 rounds per gun for 9.2 and 6 inch, and none at all for the 15 inch guns. Nevertheless, a counter-bombardment scheme was worked out, and arrangements for observation of fire landwards prepared.

With the defending troops withdrawn into Singapore Island, some of the coast-artillery batteries found themselves practically in the front line, or even beyond it. This was especially so in the Changi Fire Command, Pengerang Battery (2—6 inch) being on the mainland, and Tekong (3—9.2 inch) and Sphinx (2—6 inch) Batteries and the A.M.T.B. guns at Ladong, Calder Harbour, and Pulau Ubin on islands in the entrance eastern to Johore Strait. It was decided however to include them all within the defences, Pengerang being garrisoned by a battalion of the Indian State Forces, Tekong Island by the 2/17th Dogras, and Pulau Ubin by detached infantry posts. All coast-gunners were expected to fight as infantry if necessary to protect and defend their own works and guns. The main object of the whole defence plan was that the Japanese should be prevented at all costs from landing on the island or, if they succeeded in landing, should be stopped near the beaches and destroyed or driven out by counter-attack. For this purpose the defences were organised into three sectors:-

(1) Northern 11th Indian Division.
            18th Division. (The remainder of which had arrived by convoy on 29th January.)
(2) Southern —1st and 2nd Malayan Brigades.
            Straits Settlements Volunteer Force.
(3) Western —8th Australian Division.
            44th Indian Brigade.
(4) Central Reserve—12th Indian Brigade.

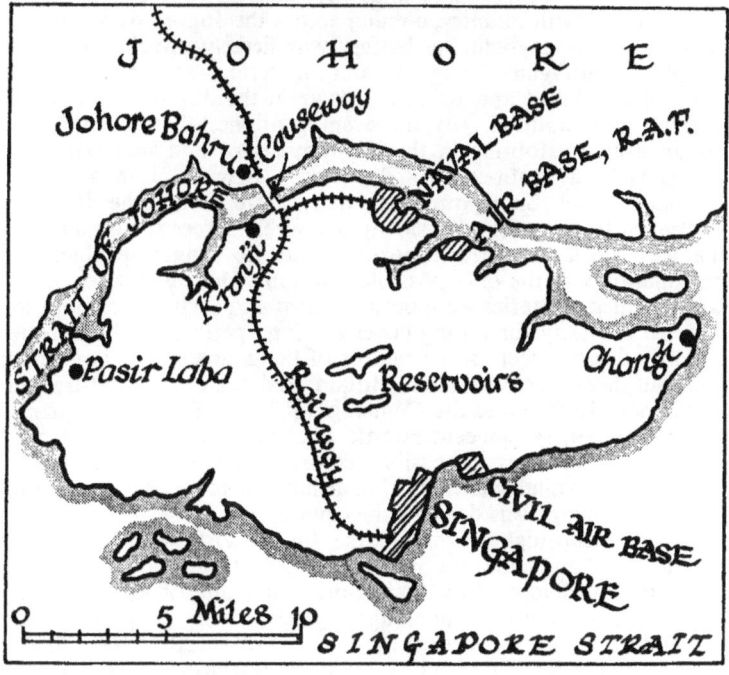

As soon as the British withdrew from the mainland, the Japanese began making preparations to cross over the Johore Strait to Singapore Island. Whilst engaged in this, they continued to carry out heavy attacks from the air, their main targets being Singapore docks, Kallang airfield, Singapore city, and the coast-defences in

Changi area, the Johore 15 inch battery being specially picked on for attention. Soon the Japanese artillery joined in, shelling the hastily thrown-up defences and the roads and tracks leading to them. The Johore and Connaught Batteries together with the field-artillery answered with counter-bombardment and harassing fire. During the afternoon and evening of 8th February the enemy put down an intense bombardment on the Western Sector, and during the night his infantry crossed the western arm of the Johore Strait in every sort of small craft and assaulted the forward defences of the 8th Australian Division.

In spite of stubborn defence the enemy quickly drove in the Australian forward posts, and by dawn were pushing forward towards the south-east. Pasirlaba Battery (2—6 inch) manned by 7th Coast Battery H.K.S.R.A., was right in the front line of the sector under assault. It was soon busy engaging the enemy assault craft, packed with infantry, coming across the Johore Strait, but as soon as it was daylight the battery was fiercely attacked by dive-bombers, both guns being put out of action, and many of the detachments killed and wounded; later in the day the battery was overrun and captured. By the evening of the 9th the enemy had obtained a firm footing in the Western Sector and was exploiting his gains in a south-easterly direction. Between 10th and 15th February, when Singapore was finally surrendered, the Japanese continued to press forward slowly but surely across the island from the west until the British position was no longer tenable, and capitulation was the only possible solution. During this period the coast-artillery batteries were both bombed and shelled continuously, and the necessity for taking cover for long periods in slit trenches had an adverse effect on the morale of the gunners, many of whom had but lately arrived from Britain and were but half-trained. The Johore 15 inch and the Connaught 9.2 inch Batteries frequently engaged Japanese concentrations, but as observation was very difficult and at times impossible, communications most uncertain, and the only available ammunition armour-piercing, it is feared that not much damage was done to the enemy.

During the night 10th/11th the Buona Vista 15 inch Battery was surprised and captured by the advancing Japanese. The detachments managed to destroy their guns, but many of the men were taken prisoner, the remainder fighting desperately in the dark and joining hands with the nearest infantry. On the 12th it was decided to draw the defence lines closer towards Singapore City with the result that the Changi area had to be evacuated. During the night 12th/13th the guns and batteries of the Changi Fire Command were blown up and destroyed, the men being safely withdrawn behind the new defence perimeter where they fought as infantry. All that were now left of the coast-defence batteries were:-

1939 to 1945 (concluded)

(1) On Singapore Island, opposite to Blaking Mati Island:-
    Labrador Battery   ...   2—6 inch.
    Batu Berlayer ...   ...   2—12 pdr.
    Pulau Hantu   ...   ...   2—12 pdr.
    (a small detached islet)
(2) On Blaking Mati Island:-
    Connaught Battery   ...   3—9.2 inch.
    Serapong Battery   ...   2—6 inch.
    Siloso Battery ...   ...   2—6 inch.
    Siloso Point   ...   ...   1—12 pdr.
    Behala Reping ...   ...   2—6 pdr.
(3) On Pulau Brani Island:-
    Silingsing Battery   ...   2—6 inch.
    Tanjong Tereh ...   ...   1—12 pdr.

During the 13th Blaking Mati, Pulau Brani, Labrador, and Batu Berlayer were heavily shelled and bombed, and in reply Connaught, Serapong, Siloso, and Labrador Batteries engaged the enemy both on land and at sea. At about three o'clock afternoon, the Japanese set about putting Labrador Battery out of action with a violent artillery concentration, both guns being quickly knocked out, and the detachments (7th Coast Battery H.K.S.R.A.) forced to evacuate the position, many being killed and wounded. During the night 13th/14th the guns and searchlights at Batu Berlayer and Pulau Hantu were destroyed and the men withdrawn to Blaking Mati as the enemy forward patrols were closing in on Labrador. On Blaking Mati Island, which had been for so many years the happy home of the Coast Artillery stationed at Singapore, the last remnants of that Coast Artillery now prepared to continue the struggle until they could fight no longer. During the day (14th February) whilst the enemy persisted with their bombing and shelling, defence positions were strengthened, extra sand-bags filled and stacked, and water containers replenished from the well at Connaught as all mains had been broken. However, the night passed without the enemy attempting to land on the island, and next morning (Sunday 15th February) it was evident the Japanese were concentrating all their efforts on Singapore itself. About six o'clock in the evening information was received that Singapore with all its defending forces had been surrendered unconditionally, so the remaining guns were destroyed, and those who wished to set about preparing for escape in any boat they could find, while the remainder waited patiently to be made prisoners-of-war.

So Singapore fell, probably the greatest disaster the British Empire had ever suffered. The war with Japan however did not end there, and soon the enemy were stretching out in all directions, southwards into Indonesia, northwards into Burma, and westwards across the Bay of Bengal towards India and Ceylon. The small

T

coast-artillery detachments at Lutong (Sarawak) and Christmas Island were soon mopped up. Burma had been invaded towards the end of January (1942) and immediately after the fall of Singapore, the Japanese Southern Fleet moved into the Bay of Bengal. The chief port of Burma was Rangoon, its coast-defences being a battery of 2—6 inch guns manned by the Rangoon Coast Battery, Burma Auxiliary Force. The Japanese advance into Burma met with only slight resistance, and by the first week in March the enemy were threatening Rangoon from three sides. It appeared quite impossible to hold the city, which had been frequently and heavily bombed, so it was decided to evacuate the place at once. On 8th March therefore Rangoon was abandoned to the enemy, all the British troops retiring northwards to Prome while the Coast Battery, having blown up the guns, was withdrawn by sea to India.

The Japanese Southern Fleet having moved into the Bay of Bengal, the whole of the east coast of India and the island of Ceylon were threatened. In the latter were the naval base of Trincomalee— 2—9.2 inch, 3—6 inch guns manned by the 6th Coast Regiment, Ceylon Garrison Artillery—and the defended port of Colombo— 2—9.2 inch, 4—6 inch guns manned by the 1st Coast Regiment, Ceylon Garrison Artillery. The enemy began operations by occupying the Andaman Islands on 23rd March, and during the first week of April the main Japanese naval striking-force:-

    4 Battleships
    5 Aircraft Carriers
    3 Cruisers
    8 Destroyers

sailed toward Ceylon. A few days later detached units of the British Eastern Squadron, which was based on Addu Atoll, were caught by enemy naval aircraft, two cruisers and an aircraft carrier being sunk, and two heavy assaults were made on Ceylon itself, Colombo being attacked by 115 aircraft on 5th April (Easter Sunday) and Trincomalee by a similar number on the 9th. Meanwhile Japanese light naval forces had been running riot in the Bay of Bengal, 28 allied merchant ships being sunk in the space of a few days. The position indeed appeared desperate, and it looked as if the invasion of Ceylon was imminent. Actually the situation was not so black as it seemed, for the Japanese had reached the limit of their westward drive, their battleships and aircraft-carriers were soon withdrawn to the Pacific for service against the Americans, and their fleet was no more seen in strong force in the Bay of Bengal with the result that never again throughout the rest of the war did the British Coast Artillery come into action against the Japanese.

CHAPTER XXVI

Conclusion.

THERE is not very much more to say about Coast Artillery in the British Army. During the Second World War it reached its greatest strength and its greatest extent. The United Kingdom was ringed with a perimeter of coast-batteries, there was not a port of any size throughout the whole Empire—on which indeed in 1939 the sun never set—which did not have its quota of coast-guns, and British coast-artillery were defending—or had been—the harbours of Italy, French North Africa, Libya, Syria, Iceland and the Faroe Islands. The two great coast-fortresses of Hong Kong and Singapore had indeed, been captured, by the Japanese which proved that, no matter how strong the front-door may be, if the back-door is left unbolted, the stronghold is insecure and endangered. The invention of radar had enabled coast-artillery to discover and engage targets which were invisible to searchlights, telescopes, and the human eye. In fact British Coast Artillery had reached its highest eminence just before it was to become extinct for ever, for science and invention were fast catching up with the business of defending ports, harbours, and coasts with guns. Just as the invention of cannon had made the mediaeval castles, which had stood so long on guard over the creeks and inlets, of no further use, so were the bomber and torpedo-bomber aeroplane, and the development of rocket-driven ballistic and guided missiles steadily usurping the role of coast-artillery and replacing the guns which had for four centuries protected the shores of Great Britain and her dominions and dependencies. Coast-artillery indeed was rapidly becoming out of date and no longer needed.

However, Coast Artillery in the British Army took ten years to die. When the Second World War came to an end in 1945, and the armies were demobilised, there were a very large number of guns in situ and a vast amount of equipment to be cared for. Though most of the "emergency batteries", which had been installed around the coasts of the United Kingdom, had been dismantled, both at home and abroad all the major ports had many guns in excess of their pre-war numbers, and most minor ports were still defended. The first necessity was to get all this equipment put into care and preservation, so once again "District Establishments" of Regulars, with their armament officers, master-gunners, and district-gunners, were organised at Coast Artillery stations to maintain and preserve the guns, instruments, equipment, and stores. By 1946 the District Establishments were at their work, and by 1948 five regular regiments of coast-artillery had been formed, three at home with their headquarters at Dover, in the Isle of Wight, and at Donibristle (Scotland) and two abroad at Gibraltar and Malta. The Royal Malta Artillery was still carrying on in its own island, but the Hong Kong Singapore Royal Artillery had been disbanded, India now being a free dominion and not prepared to provide recruits for the British Army.

In May 1947 the Territorial Army was reconstituted, and it was decided at first to raise 31 coast-regiments, but, with the usual post war-economy, this was considered far too many ,and the number was reduced next year to 20 (later to 18) regiments, organised into three coast-brigades. At the same time an interim scale of armament for the naval-bases and defended-ports both at home and abroad was agreed to, and, as this scale amounted to a considerable reduction in existing armament, a large number of guns had to be dismounted and removed, a special unit being available for this purpose "245th Armament Battery". Also for the first time in all their long history, indeed since the foundation of coast-artillery in 1540, the district-establishments were grouped into units called "Maintenance Batteries," one such battery being stationed in each Command. The Coast Artillery School, which had been at Llandudno when the war ended, was moved to Plymouth, being installed in the ancient and famous Royal Citadel. However in 1949 further reductions had to be carried through, two of the regular coast-regiments at home being disbanded, and the one remaining—47th Coast Regiment—being transferred from Dover to Plymouth where it controlled the training of recruits and provided the depot establishment for the School. As in the years before the war, the Territorial Army was now once more entirely responsible for the coast-defences of the United Kingdom.

By 1950 the post-war organization of Coast Artillery had been settled. The naval-bases and defended ports both at home and abroad were adequately protected with their guns and equipment in a state of "care and preservation". At home the three coast-brigades, consisting of the eighteen Territorial regiments and the six regular maintenance-batteries, covered the shores of the United Kingdom. The Coast Artillery School combined with 47th Coast Regiment— together now called the Coast Artillery Training Centre—was responsible for the training and instruction of all grades, from recruits to instructors of gunnery. Abroad Gibraltar and Malta retained their regular coast-regiments (plus the Royal Malta Artillery at the latter place) while at all other coast-artillery stations cadres maintained the armament, equipment and stores. But air-power, scientific progress, invention, and even possibly economy had caught up with Coast Artillery. On 17th February 1956 the Minister of Defence announced in the House of Commons that Coast Artillery was to be abolished. He stated that in the light of modern weapon

Conclusion

development it was clear that there was no longer any justification for maintaining coast-artillery. The seaborne threat could be countered more effectively by the Navy and Air Force, and other types of artillery could, if needed, be used for seaward defence. The Government had therefore decided that all existing coast-artillery units would either be converted to new roles or become inactive. It would, however, be the aim to preserve so far as possible the identity of all Territorial Army Coast Artillery Regiments in some other role. This was the death sentence, and after 31st December 1956 Coast Artillery in the British Army ceased to exist.

For more than four centuries Coast Artillery had defended the important ports and harbours of Great Britain and her overseas possessions, and thus provided secure and safe bases and anchorages for the Royal Navy wherever the fleet was operating. Coast Artillery from the very reason for its existence, was always very closely linked to the Royal Navy. It was at the great naval bases that Coast Artillery was to be found in force, and its guns, projectiles, instruments, tactics, and methods depended entirely upon naval developments. Coast Artillery was very proud to perform the honoured task of providing safe havens of refuge for the Navy and to prevent such happenings in harbours defended by its guns as befell *H.M.S. Pegasus* at Zanzibar and the Russian light cruiser *Zhemchug* at Penang in 1914. Until the appearance of the aeroplane, Britain's warships could feel safe when anchored under the protecting guns of Coast Artillery. This function will now be left to some other organisation, an organisation probably armed with torpedoes, rockets, and guided missiles, all with nuclear war-heads. It is hoped the duties of watch and ward will be carried out as efficiently as they were in the days of Coast Artillery.

Beginning with the coast-forts of King Henry VIII, coast-defence reached its zenith with the great fortress of Singapore. It is notable how many of those castles of King Henry continued all down the centuries to be the central core of the coast-defences of the United Kingdom. Several were still in use as coast-artillery headquarters during the Second World War, and as late as 1956 a maintenance battery was stationed in Southsea Castle. The Gunners of King Henry chose their sites well, and his sappers built even better. The history of Coast Artillery at home is not one of battles and violent actions. The command of the sea, established at the time of the *Armada* by Elizabeth I's seamen and maintained throughout the centuries, deterred the fleets and warships of Britain's enemies from attacking her shores, and invasion, although often threatened, was never launched. No continental pontentate would risk his transports and landing-craft upon the waters of the Channel without having even temporary command of that narrow sea—and in 1940 of the air over it as well. Nevertheless the well protected and defended ports prevented many a hostile squadron from attempting a raid on the British coast. It is extraordinary how few down the ages ever attempted to do so.

Overseas Coast Artillery was kept much busier. Temporary loss of command of the sea in certain areas enabled Britain's foes to assail her possessions, and it has been seen how during the War of American Independence the great Mediterranean fortresses of Gibraltar and Minorca were attacked, the latter even being captured, and how during the Second World War, when Britain lost command of the Indian Ocean and the China Sea, Hong Kong and Singapore were assaulted and taken. Moreover, during the eighteenth century, many of Britain's oversea possessions were isolated, and an enemy, even when inferior at sea, could make a sudden raid upon an unsuspecting coast-fort. Above all it was in the West Indies, where British and French islands lay so close alongside each other, that Coast Gunners were continually in action, and where so many found their graves, both from the blows of the enemy and the deadly effects of yellow fever. So we come to the end of the story, a story which has lasted for more than four centuries. It is a story of which Coast Gunners can be proud. Britain's greatness was securely founded on her sea-power, and Coast Artillery played no small part in providing a sure base for that power. There could have been no *"Rule Britannia"* without Coast Artillery safely keeping goal behind the Royal Navy.

INDEX

INDEX

Abbotsbury, 232.
Abercromby, General Sir Ralph, 113, 114.
Aberdeen, 98, 166, 187, 188, 221, 229; Volunteers, 103, 164.
Aberdeen (Hongkong), 259, 261.
Acre, 120.
Acul, Fort 'L' (San Domingo), 110
Additional Gunners, 8, 21, 25, 26, 30, 33, 34, 36, 51, 52, 53, 56, 63, 66, 73, 81, 82, 83, 88, 100, 101, 104, 117, 119, 131, 138, 159, 160, 212.
Addu Atoll (Maldive Islands), 256, 257, 274.
Aden: first occupied by British, 135; garrison, 135, 156, 190, 209, 223; armament, 190, 223, 256; general, 149, 195, 143.
Admiralty (British), 33, 74, 101, 139, 170, 195, 196, 199.
Admiralty (German), 200.
Adriatic Sea, 125.
Aeroplanes, see Aircraft.
Africa,
 East, 192, 195, 212, 243, 255.
 North, 247, 250, 251, 252, 275.
 South, 207, 208.
 West, 36, 39, 129, 160.
Africa Corps (German), 247.
Africa, Trans-, Air-Route, 251.
Air, Command of the, 226, 258, 263, 268, 277.
Airborne Division, 7th German, 244.
Aircraft, 201, 213, 216, 218, 226, 234, 236, 239, 243, 247, 258, 263, 267, 268, 275, 277.
Aircraft-Carriers, 215, 242, 255, 274.
Airforce, German (Luftwaffe), 226, 233, 235, 243, 244, 245, 247.
Airforce, Italian, 243, 247-9.
Airforce, Japanese, 259, 268, 269, 270, 272, 274.
Air Force, Royal, 226, 233, 235, 236, 245, 249, 250, 254; in defence of Hongkong, 258, 259, 263; in defence of Malaya, 267, 268, 270; replaces Coast Artillery, 277; Squadrons, 267.
Air Observation Post, 239.
Air-Route, Trans-African, 251.
Alamein, El (Battle), 250.
Alberoni, Cardinal, 26.

Aldeborough, 95, 100, 231; Volunteers, 103.
Alderney, 91, 99, 107, 136, 155, 163, 188; Militia, 107, 163, 188.
Alexandria, 156, 222, 243, 250.
Algeciras, 68, 70, 74, 251.
Algeria, 252.
Alliance, Anglo-Japanese, 202, 203.
Alor Star (Malaya), 269.
Amble, 230.
America, North, British and French Colonies in, 27; rebellion of British Colonies in, 48-74, 75; difficulties of campaigning in, 48; British Army in, 49; Coast Artillery in, 60-2; lines of communication to, 63; in First World War, 202, 203.
America, South, 192.
America, United States of, founding of, 74; victory in War of Independence, 74; declares war on Great Britain, 1812, 129; at war with Great Britain, 129-31; situation after First World War, 202, 203; and Japan, 210; a first-class naval power, 1919, 214; agrees to Washington Treaty, 214; enters Second World War, 255; Navy (fleet), 55, 129-31, 202, 274.
Americans, operations in War of Independence, 60, 61; victory in War of Independence, 74; on Great Lakes, 1812-14, 129-31; victory in Normandy, 239; advance in Italy, 252-4.
American War of Independence, 42, 44, 45, 48-74, 75, 87, 119, 278.
American War of 1812, 129-31.
Amherst, General Lord; 53.
Amherst, Lieut-Colonel William, 38.
Amherst Battery (Dover), 96.
Amherst Fort (Guernsey), 99.
Amherstberg (Lake Erie), 129.
Amiens, Peace of, 86, 92, 93, 102, 114, 119, 123, 128.
Ammunition, round-shot, 3, 4, 6, 12, 20, 21, 65, 74, 140; shell, 4, 65, 69, 73, 74, 140, 167; red-hot shot 71, 72, 73, 110; projectiles, 4, 167, 168, 172, 173, 217; responsibility for and care and preserva-

Index

tion of, 8, 21, 22, 23, 24, 25, 30, 76, 77, 135, 138; expended in Hartlepools engagement, 198, 199; between 1919 and 1939, 217; shortage of 1940, 227; fired against German warships, 236, 238; fires from Dover Batteries, 238; fired from Falmouth, 239; expended at Malta, 247, 249, 250; expended at Leghorn, 254; at Hongkong, 261; at Penang, 269; at Singapore, 270, 272.
Anacapri (Capri), 124.
Ancillary Corps, 256, 258.
Ancona, 254.
Andalusia, 126. 128.
Andaman Islands, 274.
Anderson, Captain, R. A., 42, 78.
Anderson, George, Master Gunner, 136.
Anglo-Japanese Alliance, 202, 203.
Anholt (Denmark), 125, 131.
Annapolis (Nova Scotia), garrison at, 17, 18, 27, 29, 36; armament of, 26, 28; fort at, 28; French attack on, 27-9; abandonment of, 80.
Anne, Queen, 16.
Anne's Battery, Queen, (Gibraltar), 66.
Anstruther Redoubt (Fort St. Philip, Minorca), 35.
Anti-Aircraft (Artillery and Guns), 201, 204, 212, 213, 214, 225, 246, 250, 258, 266, 269.
Anti-Aircraft Regiments, R.A., 3rd Regiment, 266; 5th Regiment, 258; Indian Artillery, 266; H.K.S.R.A., see H.K.S.R.A.
Antigua (West Indies), 36, 56, 58, 59, 108, 113, 114, 118, 119.
Anti-Tank (Artillery and Guns), 213, 269.
Anti-Torpedo Craft Guns, method of fighting, 1914, 183, 184, 215, 216, 240, 250, 270.
Antrim, Militia Artillery, 162, 163, 188, 207; Territorial Artillery, 213, 221.
Antwerp, 85, 91, 226.
Anzio, 254.
Appledore, 230; Volunteers, 103.
Appley House (Portsmouth Defences), 178.
Approaches, Western, 48.
Arbuthnot, Major-General, C. G. (D.A.G.R.A.), 152.

Archliff Fort (Dover), 96.
Archirondelle (Jersey), 98.
Arcon, Chevalier d' (French Engineer), 70, 72, 73.
Argyll, Militia, 162.
Agyll Redoubt (Fort St. Philip, Minorca), 35.
Armada (Spanish), 8, 9, 10, 277.
Armament, of Coast Artillery, 3, 6, 7, 11, 12; responsibility for, 8, 20, 24, 25, 30, 77, 135, 138, 157, 275, 276; reduction of, 19, 25; of sea-batteries, 32; in Ireland, 45, 46, 105, 106; of warships, 46, 75, 167-70; of coast defences, 1770, 46, 47; carronades, 47, in 1779, 50-3; of floating batteries, 72; in 1805, 95-99; of Marcouf Islands, 121; in 1815, 132; changes in, 133; care of, 138; strengths adjusted to, 151; of Volunteers, 164; R.B.L. and R.M.L. Guns, 167-9; in 1856, 1867, and 1881, 169, 170; position in 1899, 171; rearming of Coast Defence, 1904-14, 171, 172; superiority over warships, 172; movable, 184; at outbreak of war, 1914, 185-91; at Hartlepools, 1914, 197; of Hongkong and Singapore, 1934, 210, 211; between 1919 and 1939, 215-7; at home and abroad, 1939, 219-24; of Emergency Coast Batteries, 227, 229-32; at Malta, 246; in East, 157; at Hongkong, 259, 262; at Singapore, 264, 267; after Second World War, 276.
Armament Battery, No. 245, 276.
Armament Officers, 158, 275.
Armed Neutrality, 48.
Armour (Warships), 140, 167, 169, 170.
Armstrong, Sir William, 167.
Armstrong Gun, 167.
Army of England (French), 91, 92, 115.
Army, British, (Forces), expenditure on coast-defences, 2, 13; in North America, 49; strength of in 1797, 91; in Ireland, 104, 106; expedition to San Domingo, 108; in Guadaloupe, 112, 113; at Dominica, 115-118; at Malta, 122; in Calabria, 123; in Sicily, 123, 124; at Anholt, 125; at Cadiz and Tarifa, 126-8; at

Barrosa, 127; in Canada, 130; in the Crimea, 142; and Cardwell Reforms, 149; its victory in Normandy, 239; in Crete, 244-6; campaign in North Africa, 250-2 advance in Italy, 252-4; in defence of Hongkong, 258-63; in defence of Malaya and Singapore, 266-74; in Burma, 274.
Army, Eighth (British), 251.
Army, First (British), 252.
Army, Grand (French), 92, 93.
Arsac, Ternay d', (French Naval Captain), 37-39.
Arsenal, Woolwich, 141.
Arthur, Port (North China), 171, 257, 258, 259, 264.
Artibunite, River (San Domingo), 110.
Artificers, Royal Military, 69, 80, 120, 121.
Artillery, Anti-Aircraft, 201, 204, 212, 213, 214, 246, 250, 258, 266, 269.
Artillery, Anti-Tank, 213, 269.
Artillery, Cape Garrison, 161, 190, 201.
Artillery, Ceylon Garrison, 161, 191, 208, 223, 257, 274.
Artillery, Coast, functions of, 1; on loss of command of the sea, 3; at times of invasion, 3; command, administration, and organisation, 3, 7, 8, 12, 20, 24, 25, 26, 30, 76, 133, 135, 148, 149-54, 159, 204-9; methods, 3, 4, 7, 12, 20, 47, 71, 132, 141, 152, 173, 174, 181, 182-4, 204-9, 215-7, 277; armament, 3, 6, 11, 12, 19, 25, 32, 46, 47, 95-9, 133, 169, 170, 219-24, 229-32, 246, 256, 257, 259, 262, 267; garrisons, 7, 8, 10, 11, 16, 17, 20, 24, 25, 30, 36, 37, 39, 40, 41, 42, 43, 44, 45, 46, 51, 58, 77, 78, 134, 136, 137, 155-7, 185-91, 206-9, 211, 212, 213, 219-24, 246, 250, 257, 258, 259, 266, 267; and Spanish Armada, 8, 9; in Great Civil War, 10, 11; and Commonwealth, 11; first sent abroad, 16; during War of Spanish Succession, 17-9; on formation of Royal Artillery, 24; in the '45 Rebellion, 30, 31; during the Seven Years War, 32-40; in Newfoundland, 37-9; stations, 1764, 41; on formation of Invalid Companies, R.A., absorbed into R.A., 42; in Ireland, 44, 45, 46, 79, 104-6, 131, 134, 135, 188, 205, 207, 208, 213; in crisis of 1779, 49-53; volunteers in, 52, 101-4, 131, 164, 165; in West Indies, 56-60, 108-19, 134, 135, 155, 156, 160, 161; at St. Lucia, 57, 58, 156; at St. Kitts, 59; in North America, 60, 61; in Florida, 61, 62; in West Africa, 62; at Minorca and Gibraltar during War of American Independence, 63-74; role during French Revolutionary and Napoleonic Wars, 75, 131; at Home, 1790, 76-9; abroad, 1790, 79, 80, 81; at Toulon, 82, 83, 85; at Fishguard, 88, 89; and provision of gun detachments, 100, 102; during 1795 to 1804, 100, 101; in Channel Islands, 107, 134, 155; in Guadeloupe, 112, 113; at Dominica, 115-8; in Corsica, 120; at Marcouf Islands, 121, 122; at Malta, 122, 131, 134, 156, 246-50; at Minorca, 123; at Madeira, 123; in Sicily, 123, 124, 252; at Capri and Scylla, 123, 124; in Adriatic islands, 125; at Corfu, 125, 134, 156; at Cadiz and Tarifa, 126-8; at Genoa, 128; stations gained during Revolutionary and Napoleonic Wars, 128; at Capetown, 128, 131, 134, 156, in Ceylon, 128, 134, 156, 274; at Mauritius, 128, 131, 134, 156, 257; at St. Helena, 128, 134, 156; in Coast Forts on Great Lakes, 129-31; and landward defence, 132, 258; changes during 19th Century, 133; at Gibraltar, 134, 156, 251; and Master Gunners, 138; and Repository exercises 139; during the Crimean War, 142; transferred to War Office, 142; separated from mounted branch of R.A., 148, 151-4; reorganised 1860, 148-9; takes over coast-defence in India, 149, 156; reorganised 1877-91, 150, 151, 161; Royal Garrison Artillery formed, 151-4; organisation, 154, 155; withdrawn from Corfu

Index

Ionian Islands Newfoundland, and New Brunswick, 156; at St. Helena and Sierra Leone, 156, 157; leaves Canada, 157; the Coast Brigade, 157, 158; District Establishments formed, 158; formed from local natives abroad, 160, 161; Territorial Force in, 165, 166; first issue of R.B.L. guns, 171; formation of Fortress Companies, R.E., 174; becomes scientific, 174; training in "Scotch Up", 177; object of, 177; ready for war in 1914, 181; distribution on outbreak of war 1914, 185-91; during 1915 to 1918, 200, 201; at end of First World War, 201; and Singapore Base, 203; state and reorganisation after First World War, 204, 205; War Department policy towards, 204; reorganised with Territorial Army, 205-7; called "heavy artillery", 205; organisation 1927-33, 205-9; increase for Far East, 211; leaves Eire, 213; separated from field branch, 213, 214; ready in 1939, 218; order of battle at home and abroad 1939, 219-24; situation at home 1939, 225; takes over Defence Electric Lights, 225; expansion to man Emergency Coast Batteries, 227-9; at last Coast Artillery, 229; during invasion threat 1940-1, 233; at Dover, 233-238; makes use of radar, 234, 235; after threat of invasion passed, 238; reduction of 240, 241; replaced by Home Guard, 240, 241; forms relief force for Channel Islands, 241; training establishments, 241, 242; in Crete, 243-6; in North Africa, 251, 252; takes over coast-defences in Syria, 251; mobile units formed, 252; in Sicily and Italy, 252-4; in East, 1941, 257, 274; at Hongkong, 258-263; at Singapore and Penang, 264-73; at its greatest strength during Second World War, 275; after Second World War, 275, 276; abolished, 276, 277; its epitaph, 277, 278.
Artillery, Director-General of, 175.
Artillery Districts, 150.

Artillery, Dutch Emigrant, 110, 119.
Artillery, Egyptian, 222.
Artillery, Field, 148, 152, 204, 213; Royal Field Artillery, 151, 154, 165, 175.
Artillery, Heavy, 148, 149, 151, 158, 165, 200, 201, 204, 205, 213.
Artillery, Indian, 208, 257, 266.
Artillery, Local, 160, 161.
Artillery, Medium, 205, 212, 213, 228.
Artillery, Mountain, 148, 149, 151, 165, 201, 208, 212, 213.
Artillery, Royal, Coast Artillery branch of 4; formation of, 24; includes Coast Artillery, 26; reinforcements of sent to Gibraltar, 27; command of at Gibraltar, 27; at Minorca, 34, 35, 36, 63-5, 123; in coast-garrisons 1759, 36; in 1763, 37; at St. John's Newfoundland, 37, 38; at Home 1758, 39, 40; abroad 1764, 41; terms of enlistment in, 41; formation of Invalids, 42; in West Indies, 56, 58, 59, 60, 108-9, 155, 156; at St. Lucia, 58, 156; at St. Kitts, 59; in North America, 60, 61; in Florida, 61, 62; in West Africa, 62; at Gibraltar, 65, 68, 69, 71, 72, 73, 74, 80, 126, 134, 251; role of Marching Companies 77, 131, 147; in Scotland, 79; abroad in 1789, 80, 131; on outbreak of war 1793, 82; at Toulon, 82, 83, 85; and provision of gun-detachments, 100, 102, 159; in coast-defences 1795-1804, 100, 101; in Ireland, 106, 131, 134; in Channel Islands, 107, 134, 155; in San Domingo, 108-12; at Guadeloupe, 112, 113; at Trinidad, 114; at Dominica, 115-8; in Corsica, 120; at Marcouf Islands, 121, 122; at Malta, 122, 134, 222, 246-50; at Madeira, 123; in Sicily, 123, 124; at Capri and Scylla, 123, 124; in Adriatic islands, 125; at Corfu, 125, 134 at Cadiz and Tarifa, 126-8; at Genoa, 120; at Capetown, 128, 134, 156; in Ceylon, 128, 134, 156; in Mauritius, 128,134, 156; at St. Helena, 128, 156; in defence of Canada 1812-14, 129-

31; during French Revolutionary and Napoleonic Wars 1793-1815, 131, 132; reduction and reorganisation 1820, 133; formed into battalions, 135, 138; role of companies, 139, 147; transferred to War Office, 142; situation after Crimea War, 147, 148; split into two branches, 148, 151-4, 165; reorganised 1860, 148, 149; goes to India, 149; Indian artilleries amalgamate with, 149; reorganised 1877, 1882, 1889, and 1891, 150, 151, 161; in India, 149, 156; formation of Territorial artillery, 165; in cost-defences, 1918, 201; re-united into one corps, 204, 205; in coast-defences, 1932-33, 206-9; leaves Eire, 213; split into two branches, 213, 214; Brigades called Regiments, 214; absorbs seardhlight personnel, R.E., 225; at Hongkong, 258, 259; at Singapore, 266, 267.
Artillery, Royal Alderney, 163, 188.
Artillery, Royal Canadian Garrison, 157.
Artillery, Royal Foreign, 118, 119.
Artillery, Royal Garrison, formation of, 151-4, 159; formed into Groups, 134; system of command organised, 154; Groups abolished, 154; garrisons 1900-14, 155-7, at Sierra Leone, 156, 157; leaves Canada, 157; and District Establishments, 158; inspected by King Edward VII at Portsmouth, 158; local artilleries abroad formed in, 160, 161; in India, 161; Volunteers in, 165; Territorials in, 165, 166; School of Gunnery for, 175; practice-camps, 176; ready for war in 1914, 181; duties in coast fortresses, 1914, 182-4; at outbreak of war, 1914, 185-91, 204; at Hartlepools, 197; employed in the field, 1914-18, 200, 201, 204; organisation after First World War, 204, 205; abolished, 205; leaves Cape of Good Hope, 208.
Artillery, Royal Guernsey, 163, 188.
Artillery, Royal Horse, 151, 165, 175, 213.
Artillery, Royal Irish, 44; coast-artillery functions of, 45, 46; forms Invalid Company 45; distribution in coast-defences, 46, 79, 131; in West Indies, 79, 112, 113, 114; abolished and incorporated in R.A., 79.
Artillery, Royal Jersey, 163, 188.
Artillery, Royal Malta, 122, 135, 160, 189, 208, 209, 217, 222, 246-250, 276.
Artillery, Siege, 139, 141, 147, 148, 149, 151, 200, 201, 204, 205, 239, 240.
Artillery Studies, Director of, 175.
Artillery Trains, 24, 138.
Arundel Haven (Littlehampton,) sea-battery at, 32, 33; state of in 1779, 50; garrison, 40, 43, 78, 136; armament, 97.
Ascension Island, 252.
Asia, 202, 210, 255.
Asiatic Companies, R.A. (R.G.A.), 160, 161.
Asiatic Gun Lascars, 135, 159, 160, 161.
Askifou (Crete), 245.
Atherfield Point (Isle of Wight), 179.
Atlantic Ocean, 32, 37, 38, 48, 62, 63, 86, 115, 234, 255.
Attack, Scales and Forms of, 2, 171, 172, 182.
Augusta (Sicily), 248, 252.
Austerlitz, (Battle of), 92.
Australia, 255.
Australian Forces, 8th Division, 266, 269, 271, 272.
Austria, 82, 92, 202.
Austrian Succession, War of, 27-31.
Automatic Sight, 174, 183, 216, 277.
Auxiliary Force, Burma, 257, 274, India, 223.
Avonmouth, 98, 229.
Axemouth, 98.
Axis, Berlin-Rome, 213, 247, 250.
Aynge, George, Master-Gunner, 136.
Ayr, Volunteers, 103.
Babali Hani (Crete), 245.
Badajoz, 127.
Bahamas, 134.
Balaclava, 142.
Balearic Islands, 34.
Ballinamuck (Ireland), 106.
Baltic Sea, 55.
Bamborough Castle, 95.
Band, R.G.A. (Portsmouth), 158.
Bantry Bay, French invasion attempt at, 86, 87, 104; defences of, 105.

Index

Barbados, 56, 59, 110, 112, 118, 135, 195; garrison of 58, 60, 108, 113, 114, 118, 134, 155.
Barbette, 172, 180.
Barfleur, Cape, 16, 120.
Barfleur, H.M.S., 169, 170.
Bari, 252, 253.
Barletta, 252.
Barmouth, Volunteers, 103.
Barnstaple, 232; Volunteers, 103.
Barr and Stroud Range Finder, 215, 217.
Barrington, Admiral, 56.
Barron, James, Master-Gunner, 136.
Barrosa (Battle of), 127.
Barrow, 166, 187, 207, 221, 229.
Barry Island, 98, 166, 187, 229.
Bases, Naval, defence of, 1, 24, 62; importance of, 2; in War of Spanish Succession, 20; in War of American Independence, 60, 62; in French Revolutionary and Napoleonic Wars, 75; in Mediterranean, 120, 122; around Straits of Gibraltar, 126; on Great Lakes, 129-131; landward defence of, 132, 172, 178, 182, 184, 257, 258, 259, 264, 265; in Crimean War, 142; in India, 149; manning of abroad, 160, rearming of 1865, 169; forms of attack on, 171, 172, 182; defence of, 1906, 172; Royal Commission on defence of, 1859, 177-80; method of defending, 1914, 182-4; defences of, 1914, 184-91; defences of 1915-8, 200; needed in Far East, 203; air threat to, 213; defences of, 1939, 220-24; defences of at Home, 1940-41, 229-30; in Crete, 244; in the East, 1941, 255; defence of after Second World War, 276; surely defended by Coast Artillery, 277, 278.
Basic Training Regiment (Coast Artillery), 242.
Basilisk (gun), 7, 13.
Basseterre (St. Kitts), 59.
Basseterre (Guadeloupe), 113.
Bastia (Corsica), 120.
Bastille (Fall of), 77.
Bastions, design of, 7, 20; at Gibraltar, 68, 71, 72; at Tilbury, 96.
Bastion-Trace, 13, 20.
Bathurst (Gambia), 251.

Battalions, Royal Artillery, 133, 135; 9th Battalion, 135.
Battalion, Invalid R.A., formation of, 42; disbanded, 135.
Batten, Admiral, 11.
Batteries, Armament, No. 245, Battery, 276.
Batteries, Beach: see Batteries, Emergency Coast.
Batteries, Coast, role of, 2; at times of invasion, 3; at time of Armada, 9; at Tynemouth, 33; new in 1779, 50, 51; municipal, 51, 55, 77; at Gibraltar, 66, 68; in 1789, 77; in 1797, 91; during 1797 to 1805, 94-9; manning of, 100-2; in Ireland, 105, 106; at Bere Island, 105; and warships, 132; equipped with shell, 140; equipment of in 1850, 141; bombarded by howitzers and mortars, 141; Capt. Parkin's views on, 142-6; issue of R.B.L. guns to, 171; superiority over warships, 172; recommendations of Royal Commission 1859 concerning 178-80; fortification of 1880-14, 180; organisation of 1914, 182-4; organisation of in 1920-1, 204, 205; called Heavy Batteries, 205; at Singapore, 210, 218, 267; Close Defence, 216; fortifications of 1919-39, 218; protection against aircraft, 218; Emergency, 227-32; at last called Coast Batteries, 229; use of Radar, 234, 235; reduction of and manning by Home Guard, 240, 241; returned to United Kingdom from Malta, 250; at Gibraltar, 1943, 251; mobile, 252-4; at Hongkong, 259; Batteries (Units):— 4th, 251; 7th, 267; 11th, 267; 12th, 259; 20th, 259; 22nd, 267; 24th, 259; 26th, 251; 27th, 251; 30th, 259; 31st, 267; 32nd, 267; 36th, 259; 40th, 251; 41st, 251; Rangoon, 257, 254.
Batteries, Defence, No. 965 Battery, 258.
Batteries, Emergency Coast, 32, 93, 101, 227-32, 240, 241, 275.
Batteries, Field, 133, 147, 151, 152.
Batteries, Floating, at Dublin Bay, 46; at Gibraltar, 70-4, 132.
Batteries, French, at Minorca, 35, 64,

65; at Toulon, 83; at Tarifa, 126.
Batteries, Garrison R.A., 148-53.
Batteries, German, 237-40.
Batteries, Heavy, 158, 200, 201, 205, 208, 209, 212, 22-4; Batteries Units:—1st, 208; 2nd, 209, 224; 4th, 209, 222; 6th, 209, 222; 7th, 212, 223; 8th, 208; 9th, 209, 223; 10th, 209, 222; 12th, 209, 224; 13th, 208, 223; 14th, 208, 222; 15th, 208, 223; 17th, 208, 222; 18th, 208, 223; 19th, 208, 213, 222; 20th, 209, 224; 21st, 205; 22nd, 209, 212, 223; 23rd, 222; 24th, 209, 224; 25th, 209, 223; 26th, 158, 222; 27th, 209, 222; 29th, 209; 30th, 212, 224; 31st, 212, 223; 32nd, 212, 223; 108th (R.G.A.), 158; Indian, 208. Sierra Leone, 213, 224.
Batteries, Instruction, 147.
Batteries, Italian, 252-4.
Batteries, Maintenance, 276, 277.
Batteries, Medium, 205, 212.
Batteries, *Royal Garrison Artillery, 155, 158.
Batteries, Russian, 140.
Batteries, Sea, 32, 33, 101.
Batteries, Spanish, at Minorca, 64, 65; at Gibraltar, 66, 69-72.
Battery Commanders, duties of in 1914, 182-4.
Battle Cruisers, 181, 196-200, 225, 235, 236, 266.
Battleships, 171, 172, 181, 182, 214, 215, 234, 243, 245, 255, 266, 274. Pocket, 225.
Batu Berlayer (Singapore), 267, 273.
Batumaung (Penang), 267, 273.
Batu Pahat (Malaya), 269.
Bayonne, 120.
Beach Batteries, see Batteries, Emergency Coast.
Beach Head, 93.
Beatty, Admiral Sir David, 199, 200.
Beer and Seaton Volunteers, 103.
Behala Reping (Singapore), 267, 273.
Belcher's Battery (Hongkong), 259, 261.
Belfast, 188, 208, 213, 221, 229.
Belford, Captain R.A., 31.
Belgium, 85, 86, 93.
Belize (Honduras), 114.
Belle Grove (Guerrnsey), 99.
Belleisle, Marshal (French General), 32.
Belleisle, 36, 39.

Bellow, Peter, Master-Gunner, 137.
Belvedere (Chatham), 95.
Bembridge Down (Isle of Wight), 178
Bengal, Bay of, 193, 273, 274.
Bengal, Army (East India Company), 149.
Benghaisa Battery (Malta), 246.
Benghazi, 247, 251.
Bentinck, General Lord William, 128.
Berbice (Guiana), 79, 114, 119.
Berehaven and Bere Island (Bantry Bay) garrison, 137, 188, 208; armament, 105, 188; general, 207.
Berlin-Rome Axis, 213, 247, 250, 251.
Bermuda, garrison, 82, 134, 155, 190, 208, 209, 224; armament, 170, 190, 224; Invalid Company at, 82; Militia, 161, 190, 208, 209, 224; general, 176, 200.
Berry Head, 97.
Berville (Guadeloupe), 112.
Berwick-on-Tweed, coast-fort at, 7; garrison, 11, 25, 40, 43, 78, 81, 101, 136; armament, 11, 19, 25, 95, 230; state of, 16; Volunteers, 103; Militia, 162; general, 91, 94.
Betang Kush (Singapore), 267.
Bethlehem (Kent), 231.
Bexhill, 232.
Bideford, 88.
Binjemma Battery (Malta), 246.
Bizothon, Fort (San Domingo), 110.
Black, William, Master-Gunner, 137.
Black Legion (Free Legion), 87-90.
Blackness Castle, 77; garrison, 78, 136; armament, 98.
Blacknor, Point (Portland Defences), 179.
Blakang Mati (Singapore), 210, 273.
Blakeney, General Sir William, 34, 36.
Blanc Nez (German Battery), 240.
Blatchington, coast-battery at, 51; garrison, 78, 136; armament, 97.
Blenheims (Aircraft), 267.
Blockhouse Battery (Gravesend), 96.
Blockhouse, Fort (Gosport), 78, 97, 136.
Blockships, 172, 174, 182, 215, 216.
Bloody Foreland, 107.
Blucher (German warship), 181, 196-200.
Blue Water School, 1.

Index

Bluff Head Battery (Hongkong), 259.
Blyth, 95, 165, 229.
Boag, Lieutenant R.A., 73, 74.
Board of Ordnance, see Ordnance, Board of.
Bogebba (Malta), 246.
Bognor, 232.
Bokhara (Hongkong), 259, 261, 262.
Bombay, 149, 223, 255; garrison, 156, 189, 208, 223; armament, 189, 223, 256; Volunteers, 189, Auxiliary Force, 223.
Bombay Army (East India Company) 135, 149.
Bomb Vessels, 141.
Bompart (French admiral), 106, 107.
Bonaparte, Joseph (King of Naples and Spain), 123, 124.
Bonaparte, Napoleon, see Napoleon.
Bonhomme Richard (Flagship of Paul Jones), 55.
Boomsmashers, 182, 216.
Bordeaux, 92.
Borneo, 257.
Boston (Massachusetts), 48.
Boston (Lincolnshire), 230.
Bouille, de (French General), 56, 59.
Boulogne, 101, 233, 235, 239.
Bourne, Lieutenant R.N., 121.
Bouset Point (Guernsey), 99.
Bovisand (Plymouth), 242.
Bowen, Ensign David (Fishguard Fencibles), 88, 89.
Boxer, Captain, R.A., 141.
Boxer Fuze, 141.
Boyd, General Sir Robert, 82, 83.
Bradford, Leiutenant, 23rd Fusiliers, 110.
Bremen, 85.
Brennan, Thomas, Master-Gunner, 137.
Brest, 37, 53, 82, 86, 87, 88, 106, 107, 115, 118, 235.
Bridgewater, 98; Volunteers, 103.
Bridlington Volunteers, 103.
Bridport, Admiral Lord, 87.
Bridport, 97; Volunteers, 103.
Brigades, Coast, 276.
Brigades, Garrison R.A., 148, 149, 150, 151, 162. Brigades (Formations):—1st (Northern), 148, 150, 162; 2nd (Lancashire), 148, 149, 150, 162 3rd (Eastern), 148, 149, 150, 162; 4th (Cinque Ports), 150, 162; 5th (London), 148, 149, 150; 6th (Southern), 149, 150, 162; 7th (Western), 148, 149, 150, 162; 8th (Scottish), 150, 162; 9th (Welsh), 150, 162; 10th (North Irish), 150, 162; 11th (South Irish), 150, 162; 12th, 148, 149; 15th, 148, 149; 17th, 149; 21st, 149; 22nd, 149.
Brigades (and Regiments), Heavy R.A., 208, 209, 212, 213, 217, 219-24, 246. Brigades and Regiments (Units):—1st, 208; 2nd, 208; 3rd, 209, 222; 4th, 209, 217, 222, 246; 5th, 208; 6th, 208; 7th, 208; 8th, 208, 224; 9th, 212; 12th, 224.
Brigades, Infantry, 258-63, 269, 270, 271. Brigades (Formations):— 12th Indian, 271; 22nd Indian, 270; 44th Indian, 271; 45th Indian, 269; 53rd, 269; Malayan, 271.
Brigadier, Coast Defences or Coast Artillery, 205, 228, 229.
Brigadier, Royal Artillery, 205.
Brightlingsea, 230.
Brighton, sea-battery at, 32, 33, 232; state of in 1779, 49; garrison, 40, 43, 78, 136; armament, 97, 232; Volunteers, 103; general, 164, 165, 226.
Brimstone Hill (St. Kitts), 59, 118.
Brindisi, 252.
Brisbane, Captain Thomas, 49th Foot, 110.
Bristol, 87, 88, 94; Volunteers, 103.
Bristol, Channel, 87, 88.
Britain, Battle of, 233.
Britain, Great, see Great Britain.
Britannia, H.M.S., 46.
British Empire, 159, 169, 184, 200, 273, 275.
British Expeditionary Force of 1793-94, 82, 85, 86; of 1939-40, 226.
British Government, see Government, British.
British Isles, see Great Britain.
Brittany, 31, 49, 52.
Brixham, 230, 242; Volunteers, 103.
Broadstairs, 51, 96.
Brook (Isle of Wight), 179.
Broken Curtain (Chatham), 95.
Broughty Ferry, 166.
Bruges, 9.
Bruntisland, 166.
Buchanan, Captain R.A., 42, 43.
Buckingham, W., Master-Gunner, 137.
Buffalo, Fort, 129, 131.

Buffaloes (Aircraft), 267, 270.
Bulwarks, construction of, 6, 7, 9.
Bunkers Hill (Battle of), 48.
Buona Vista Battery (Singapore), 267, 270, 272.
Burghesi, Marshal (French general), 70.
Burgoyne, General Sir John, 48.
Burma, 257, 273, 274; Auxiliary Force, 257, 274.
Burma Oil Company, 193.
Burnham, 230.
Burton, Fort (Guernsey), 99.
Bute, Militia, 162.
Byng, Admiral John, 26, 30, 33, 34, 36.
Byron, Admiral, 55.
C9 (Submarine), 198, 199.
Cabbo Bay (Guernsey), 99.
Cabinet (British Government), 76, 210.
Cabril, Fort (Dominica), 113, 115, 117, 118.
Cadiz, 66, 68, 86, 115, 126-8.
Cagigal (Spanish general), 70.
Cairo, 251.
Caister Heights, 50.
Calabria, 123.
Calais, 30, 138, 233, 239.
Calcutta, 149, 156, 189, 193, 256; Volunteers, 189.
Calder Harbour (Singapore), 267, 270.
Caledonia, H.M.S., 130.
Calshot Castle, coast-fort at, 6; garrison, 8, 25, 40, 43, 78, 136, 186; armament, 19, 25, 50, 97, 186; state of in, 1779, 50.
Calvi (Corsica), 120.
Camber Castle (Rye), coast-fort at, 6; garrison, 8, 78.
Camden, Fort (Cork Harbour), 105, 180.
Campanella Battery (Naples), 253.
Campbell, Brigadier, 62.
Campbell, Battery (Malta), 246, 249.
Campbelltown Volunteers, 103.
Camranh Bay (Indo-China), 267, 268.
Canada, 27, 36, 39, 48, 80; operations in defence of 1812-4, 129-31; coast-defences handed over to, 156, 157.
Canadian Forces, Fencibles, 130; Militia, 130; Royal Canadian Garrison Artillery, 157; 2nd Corps, 239, 240; at Hongkong (Royal Rifles of Canada and Winnipeg Grenadiers), 258-63.
Canea (Crete), 243-5.
Cannon(type of gun), 7, 12.
Canseau (Nova Scotia), 28.
Cantania (Sicily), 252.
Canton, 259.
Cape Breton Island, 27, 28, 29, 39.
Cape Garrison Artillery, 161, 190, 201.
Cape of Good Hope and Capetown, taken by British, 128: handed over to Union of South Africa, 207; sea-route via, 251; garrison, 128, 134, 156, 190; armament, 170, 190; general, 131, 134, 135, 149, 156, 207.
Cape Passaro (Battle of), 26.
Cap Gris Nez, 233, 235, 237-40.
Capri, Island of, 123, 124; Battery (Naples), 253.
Captain of the Fort, 8, 9, 11, 12.
Carbonera Island (Newfoundland), 38, 39.
Cardiff, 164, 166, 187, 221, 229.
Cardigan Bay, 88.
Cardiganshire Militia, 88, 90, 162.
Cardwell Reforms, 149.
Caribbean Sea, command of, 48, 55, 56, 58-60; British and French activities in 1778-82, 56-60; service and health in, 81; British operations in 1793-1815, 118-119; Royal Artillery in 113; at Peace of Amiens, 114; French raid into 1806, 118; German cruisers in, 192; general, 119.
Carisbrook Castle, armament, 19, 25; garrison, 25, 40, 43; state of in 1779, 50.
Carlisle, Fort (Cork Harbour), 46, 105, 180.
Carman, W. Y. (author), 47.
Carmarthenshire Militia, 162.
Carnarvon, 230.
Carnot, Lazare (French War Minister), 141.
Carnot System of Fortification, 141.
Carolina, South, 61, 87.
Caroline Islands, 192.
Carriages, Gun, 12, 20, 47, 69, 138, 141, 157, 172, 218.
Carrick Island, 137.
Carrickfergus, Paul Jones at, 55; garrison, 44, 45, 79, 104, 137, 188; armament, 106, 188; general, 150.

Index

Carron Company, 47.
Carronades, 47, 76, 95-9, 121, 122, 170.
Carter, Captain R.A., 41.
Carthagena, 115.
"Cascable" (Author), 168.
Casemates, 6, 141, 142, 145, 146, 180.
Castagnier (French naval officer), 88, 89.
Castlebar (Ireland), 106.
Castle Cornet (Guernsey), 6, 99, 136.
Castlemartin, 88, 89.
Castles, Henry VIII's coast-forts, 6, 30, 277.
Castries, Port (St. Lucia), 56, 58.
Casualties, at Minorca, 36, 64, 65; French at St. Lucia, 58; at Rhode Island, 60; at Gibraltar, 68, 69, 70, 73; at Toulon, 83, 85; in San Domingo, 108, 110; at Guadeloupe, 112; at Dominica, 118; at Marcouf Islands, 122; at Cuxhaven, 125; at Madras, 194; to Russian cruiser Zhemchug, 194; to H.M.S. Pegasus, 195; at Hartlepools, 199; at Dover, 238; at Crete, 245.
Catalinas (Aircraft), 267.
Cavalier Battery (Plymouth), 53.
Cavaliers Battery (Sheerness), 95.
Cawdor, Lord, 88:90.
Cawsand Bay (Plymouth), 52.
CD/CHL (Radar Early Warning Sets), 234, 235.
Centurion, H.M.S., 217.
Cephalonia (Ionian Islands), 125.
Cerigo (Ionian Islands), 125.
Ceuta, 126.
Ceylon, taken by British, 128; Japanese attacks on, 273, 274; garrison, 134, 156, 160, 191, 208, 223, 257, 274; armament, 170, 191, 211, 223, 274; general, 131, 135, 149, 159, 161, 211.
Ceylon Garrison Artillery, 161, 191, 208, 223, 257, 274.
Ceylon-Mauritius Royal Artillery (Royal Garrison Artillery), 135, 160.
Chagos Islands, 193, 256.
Changi (Singapore), 210, 212, 267, 270, 272.
Channel, English, and invasion of Britain, 3; Spanish Armada sails up, 9; command of, 30, 32, 48, 53, 87, 92, 115, 226, 233, 277; French concentrate troops at ports of, 32, 52; Franco-Spanish fleet in, 53; French concentrations in, 86; French preparations for invasion across, 91, 92, 101, 116, 118; German activities in 1914-8, 192; Germans occupy French shore, 226; British difficulties in, 226; Hitler plans to invade Britain across, 226; Germans fail to get command of the air over, 233; Germans use French ports on, 234; German warships sail up, 235, 236; Germans evacuate coast of, 236, 239; general German activities in, 238.
Channel Fleet (British), 30, 32, 49, 68, 87, 115.
Channel Islands, coast-defence of, 19; Invalid Companies R.A. in, 42, 43, 44, 77, 78, 107; French attacks on, 53-5; defences of 1805, 94, 98-100; during 1793 to 1815, 107; garrison of, 134, 136, 155, defences of 1914, 188; dismantling of defences, 1929, 207; occupied by Germans, 229, 232; relieved, 241; Militia, 54, 107, 163, 188, 207.
Chapel Bay (Pembroke Defences), 179.
Charges, 7, 141, 168, 171, 173.
Charles I, King, 11, 12.
Charles II, King, 11, 13, 16.
Charles Edward, Prince (Young Pretender), 30, 31.
Charles Fort (Kinsale), garrison, 44, 45, 46, 79, 104; armament, 45, 46, 106.
Charleston (South Carolina), 60.
Charleton, Captain R.A., 35, 39, 43, 54.
Charlotte, Fort (Mobile), 61, 62.
Charlotte, Fort (St. Vincent), 113.
Chatham, coast-fort at, 7; Dutch attack on 14; Commission on defences of 1859, 177-80; garrison, 8, 11, 24, 25, 30, 40, 43, 77, 78, 81, 82, 101, 136, 155; armament, 11, 19, 25, 95; general, 16, 32, 41; Volunteers, 103.
Chatham, Lord, 99.
Chauncey, Captain Isaac (U.S.N.), 129-30.
Chebacto Harbour (Nova Scotia), 29.

Cheque, Clerk of the, 76.
Cherbourg, 32, 54, 91. 120.
Cheshire, 91; Volunteers, 164; Territorials, 166, 187, 206, 207, 221.
Chester, 88, 164.
Chichester, 32.
Chief Gunners, 7, 11.
Chile, 193.
China, 202.
China, 62, 255, 257, 264, 278.
Chinese, 211, 212, 257, 258.
Choiseul (French Minister of War), 37.
Christmas Island, 256, 257, 274.
Chung-am-Kok (Hongkong), 259.
Churchill, Sir Winston, 233.
Cinque Ports, at time of Armada, 9; Volunteers, 51, 52, 103, 164; general, 16, 32, 41.
Citadel, Royal (Plymouth), 13, 52, 97, 137, 276.
Civilian Members and Subordinates (Board of Ordnance), 76, 77, 142.
Civil War (Great), 10, 11.
Civita Vecchia, 254.
Clacton, 93, 231.
Claparede (French General), 117.
Clare Militia, 162.
Clarence Barracks (Portsmouth), 158.
Clay (Kent), 9.
Clayton, Colonel Jasper, 26.
Clerk of the Cheque, 76.
Clerk of Deliveries, 76.
Clerk of the Ordnance, 76.
Cley Eye (Norfolk), 231.
Cliff End (Isle of Wight), 178.
Cliffords Fort (Tynemouth), 33.
Clock-Code, 216.
Close-Defence Guns, 216, 240, 250.
Clugheen (Ireland), 46.
Clyde, Firth of, defences of, 77, 187, 188, 207, 221, 229; Territorials, 166, 187, 206, 207, 221.
Coalhouse Battery (Tilbury), 96, 185, 231.
Coast Artillery see Artillery, Coast.
Coast Artillery Brigades, 276.
Coast Artillery, Radar Sets, 234, 235.
Coast Artillery School, new name for School of Gunnery, 203; depot battery for, 205; between 1919 and 1939, 217, and Emergency Coast Batteries, 228; during Second World War, 241, 242; moves to Llandudno, 242;
expansion of, 242; moves to Plymouth, 276.
Coast Artillery Training Centre, 227, 228, 242, 276.
Coast Batteries, see Batteries, Coast.
Coast Brigade R.A., 157, 158.
Coast Defences, role of 1; expenditure on, 2, 10; first forts for, 6, 277; in Civil War, 10; during Commonwealth, 11; state of, 1667, 16; of England, Wales, and Channel Islands, 1716, 19, 24, 26; alerted in '45 Rebellion, 30; during Seven Years War, 32-40; inspection of East Coast defences, 33; during last part of Seven Years War, 39; in India, 41; manned by R.A. Invalids, 42; Irish, 44, 45-7, 79, 104-6, 188, 205, 207, 209, 213; state of south and east coast, 1779, 49, 50; garrisons of in, 1779, 51; of Plymouth, 1779, 52, 53; of North America, 60; manning of by Volunteers, 75; in 1790, 76, 77, 131; garrisons of 1789, 77, 78; stations, 1789, 77, 78; of Scotland, 77, 79; garrisons abroad, 1789, 80; in 1793, 82; in 1794, 91; during 1797 to 1805, 93-9, system, 1804, 100; provision of detachments for 100, 102; Volunteers in, 101-4, 164, 165; of the Channel Islands, 107, 155; of San Domingo, 110; of Dominica, 115; in Corsica, 120; of Marcouf Islands, 121; of Adriatic islands, 125; of Straits of Gibraltar, 126; of Great Lakes 129-31; changes during 19th Century, 133; manned by Invalid Detachment, 135; Master-Gunners and Invalids in 1824, 136, 137; introduction of Carnot system of fortification, 141; transferred to War Office; Captain Parkin's views on, 142-6; of India, 149; to be manned by R.G.A., 151; R.G.A. in, 155, 156; on formation of Coast Brigade, 157; expansion of 159; Militia in, 162, 163; T.F. in 165, 166; armed with R.M.L. guns, 168; armament of 1856, 1867, and 1881, 169, 170; first issue of R.B.L. guns to, 171; situation in 1900, 171; re-arming of

Index

1904-14, 177-80; manning of and method of fighting, 1914, 180-4; garrisons of 1914, 185-91; of Scapa Flow and Cromarty, 1914, 195, 196; at Hartlepools, 197; of British Empire 1915-8, 200, 201; at Singapore Base, 203; state and organisation of after First World War, 204, 205; at Home handed over to T.A., 204-6; organisation of 1927-33, 205-9; of Channel Islands dismantled, 207; of Singapore Base, 210, 211, 218, 267; in Far East and Indian Ocean, 211; in 1937, 213; changes in during 1919-39, 215-7; moving guns of, 218; fortifications of, 1919-39, 218; protection of against aircraft, 218; at Home and abroad, 1939, 219-24; situation at Home in 1939, 225; to resist German invasion, 227-32; at Dover, 1941-5, 233-8; introduction of Radar in, 234; reduction of and manning by Home Guard, 240, 241; of Crete, 243; of Malta, 246; in Mediterranean, 250; of Syria and North Africa taken over by British Coast Artillery, 251, 252; mobile, 252-4; in Sicily and Italy, 252-4; strengthened in East, 255-7; and Japanese, 258; of Hongkong, 259; of Singapore and Penang, 267; of Ceylon, 274; after Second World War, 275, 276; abolished, 276, 277.
Coast Defence and Anti-Aircraft Branch, R.A., 213, 214.
Coast Forts and Fortresses, See Forts and Fortresses Coast.
Coast Observer Detachments, 234.
Coast Regiments, R.A., see Regiments, R.A., Coast.
Coast Watching Cordon, 234.
Cochin, 256.
Cockburn, Lieut-Colonel, 59.
Cocos Islands, 194, 256, 257.
Colaba (Bombay), 149.
Coleridge Volunteers, 103.
Collective Security, 210, 213.
Collier, Captain W., R.A., 83.
Collingwood, Admiral Lord, 125.
Collinson (Hongkong), 259, 262.
Colne, River, 95.
Colombo, 128, 134, 191, 208, 223, 255, 256, 257, 274.
Colonial Establishments, 211.
Colville, Admiral Lord, 38.
Command of the Air, 226, 258, 263, 268, 277.
Command of the Seas, 2, 3, 26, 32, 37, 39, 55, 56, 63, 74, 75, 82, 92, 126, 132, 142, 192, 203, 204, 243, 255, 257, 258, 263, 268, 277, 278,
Commander Corps Coast Artillery. 228, 229, 235, 239.
Commander Royal Artillery, duties of in Coast Fortress, 182; of Malta, 246.
Commissions, 1623, 10, 107; 1704, 16; Royal, of 1859, 177-80.
Committee of Public Safety (Jacobin), 83.
Committees, Engineer (Ordnance Board), 99, Parliamentary of 1854, 142; Lord Morley's, 150; General Arbuthnot's, 152, 153; Lord Harris', 153; General Marshall's, 154; General Stirling's, 154; General Owen's, 172.
Commons, House of, 276.
Commonwealth, 11.
Companies, Asiatic or Local R.A. (R.G.A.), 160, 161.
Companies, Fortress, R.E., 174, 182, 206, 225; 26th Company, 266.
Companies, Royal Artillery, 24, 25, 26, 27, 29, 31, 34, 39, 41, 56, 60, 61, 65, 77, 100, 101, 108, 114, 118, 119, 121-34, 136, 139.
Companies (Units):—Belford's, 31; Bentham's, 120; Buchanan's, 42; Carter's, 41; Charleton's, 39; Downman's, 60; Dover's, 37, 38; Eyre's, 65, 69; Fead's, 63; Flight's, 34-6; Gall's, 58, 60; Goodyear's, 31; Gregory's, 39; Groves', 65; Holcroft's, 130; Holman's, 26, 27; Hughes', 24; Innes', 60; Jamaica, 58, 60; James', 39; Johnston's, 62; Lambert's, 63; Lewis', 65; Lloyd's, 65; Martin's, 65, 71; Ord's, 24, 29; Pattison's, 24, 39; Parry's, 60; Rogers', 29; Standish's, 56, 58; Stephens', 83, 85, 120; Stepkin's, 24; Waller's, 115, 117; Walton's, 63; Williamson's, 56, 58.
Companies, Garrison, R.A., 133, 135, 151, 158; No. 1 Company,

9th Battalion, 135.
Companies, Invalid, R.A., formation of, 26, 41, 42; first moves, 42; formation of Invalid Battalion, 42; distribution of, 42, 43, 78; establishment, 42; description of, 44; in 1779, 51; in Channel Islands, 54, 77, 78, 107; in 1789, 77, 78; sent to Bermuda, 82; sent to Heligoland, 124; sent to Anholt, 125; strength in 1815, 133; garrisons of, 43, 78; disbanded, 135. Companies (Units):—Anderson's, 43, 78; Buchanan's, 43; Charleton's, 43, 54; Dover's, 43; Fairlamb's, 78; Godwin's, 78; Gostling's, 78; Hind's, 43; Macleod's, 43; Standish's, 78; Tiffin's, 78; Toriano's, 43, 78; Van's, 78; Webdell's, 43; Westman's, 78; Whitmore's, 43, 78; Winter's, 43.
Companies, Royal Garrison Artillery, 155, 160, 161, 185-91, 200, 204. Companies (Units):—
No. 1, 189; No. 2, 185; Nos. 3, 190; No. 4, 189; No. 5, 189; No. 6, 189; No. 7, 189; No. 8, 189; No. 9, 189; No. 10, 188; No. 11, 186; No. 12, 186; No. 13, 185; No. 14, 185; No. 15, 188; No. 16, 158, 186; No. 17, 188; No. 18, 185; No. 19, 185; No. 20, 188; No. 21, 187; No. 22, 185; No. 23, 158; No. 28, 186; No. 29, 158, 186; No. 30, 186; No. 32, 158, 186; No. 33, 186; No. 34, 158, 186; No. 36, 187; No. 37, 158, 186; No. 38, 187; No. 40, 185; No. 41, 187; No. 42, 158, 186; No. 43, 188; No. 45, 187; No. 46, 185; No. 47, 186; No. 49, 188; No. 50, 190; No. 52, 189; No. 54, 189; No. 55, 189; No. 56, 191; No. 61, 190; No. 62, 189; No. 63, 189; No. 64, 189; No. 65, 189; No. 66, 190; No. 67, 186; No. 69, 189; No. 70, 190; No. 75, 189; No. 76, 190; No. 77, 189; No. 78, 191; No. 80, 181; No. 83, 191; No. 84, 189; No. 85, 189; No. 87, 191; No. 88, 191; No. 93, 191; No. 95, 190; No. 96, 189; No. 97, 190; No. 99; 189; No. 100, 189; No. 102, 189.

Conception Bay (Newfoundland), 39.
Congreve, General Sir William, 139.
Connaught Battery (Singapore), 267, 270, 272, 273.
Constance (French Warship), 88.
Co-Ordinate Converter, 215, 217.
Copons (Spanish general), 126.
Corbet, Mayse (Lieut-Governor, Jersey), 54.
Cordova, de (Spanish admiral), 66, 70, 72.
Corfu, 125, 134, 156.
Cork (and Harbour), garrison, 44, 45, 46, 79, 101, 104, 134, 137, 155, 188, 208; armament, 45, 46, 170, 188; Royal Commission, 1859, on defences of, 177-80; defences of 1914, 188; general, 68, 150, 151, 157, 176; Militia, 162, 163, 188, 207.
Cornwall, 49, 52, 91, 102; Militia, 162; Volunteers, 52, 102, 164; Territorials, 166, 187, 206.
Cornwallis, Lord, 106.
Cornwallis Battery (Penang), 267, 269.
Cornwallis Batteries, Lower and Upper (Chatham), 95.
Cornwallis, River (Nova Scotia), 28.
Coronel (Battle of), 193.
Corps, 12th, 235; 2nd Canadian, 239, 240; 3rd Indian, 266, 269.
Corsica, 100, 120.
Corsican Rangers, Royal, 123, 124.
Counter-Bombardment Guns, 215, 216, 234, 235, 236, 238, 240, 250.
Courses, Gunnery, 175, 176, 217, 242.
Cove (Cork Harbour), 46, 105.
Covehithe, 231.
Cowes Castle, coast fort at, 6; garrison, 11, 25, 40, 43, 78, 136; armament, 11, 19, 25, 97; state of, 1779, 50.
Coxheath, 49.
Craddock, Admiial, 193.
Craigs Battery (Sheerness), 95.
Creoq Point, Le (Guernsey), 99.
Crete, 243-6.
Crillon, Duke of (French General), 64, 65, 70, 72, 73.
Crimea, 142.
Crimean War, 133, 140, 142, 147, 152, 156, 161,¶163, 166, 169, 146.
Crinnis Cliff (Cornwall), coast-battery at, 97; Volunteers, 103.

Cromarty, 166, 195, 196; Volunteers, 164.
Cromer, 95, 231.
Cromwell, Oliver (Lord Protector) 11, Crossands Lightship, 196.
Cruisers, 171, 172, 181, 182, 214, 215, 216, 225, 234, 235, 236, 243, 245, 255, 266, 274; German abroad, 1914, 192-5, 204; German in North Sea, 1914-5, 196-200.
Culverin (gun), 3, 7, 12, 13.
Cumberland, 91.
Cumberland, Duke of, 94.
Cumberland, Fort (Portsmouth defences), 97, 136.
Cuppage, Lieutenant, R.A., 74.
Cuxhaven, 125.
Cwm Wnda Heights (near Fishguard), 89, 90.
Czechoslovakia, 219.
D'Aguilar (Hongkong), 211, 259, 262.
Dakar, 36.
Dalmatian Islands, 125.
Daniel, Andrew, Master-Gunner, 137.
Danish West Indies, 118.
Darby, Admiral, 68.
Darell, Captain (Duke of York's Maritime Regiment), 15.
Dar-Es-Salaam, 256.
Dartmouth, coast-fort at, 6, 8; garrison, 25, 40, 43, 78, 136; armament, 25, 97, 229; state of in 1779, 50; general, 53; Volunteers, 103.
Dawlish, 232.
Day, 2/Lieut, R.A., 35.
Deal, coast-fort at, 6; during Civil Wat, 11; during Seven Years War, 32; garrison, 8, 11, 25, 40, 43, 77, 137; armament, 11, 19, 25, 96, 231; Volunteers, 103.
Deben, River, 93, 95.
Defence Battery, R.A., No. 965, 258.
Defences, Coast, see Coast Defences.
Defence Electric Lights (Searchlights), 174, 182, 183, 216, 225, 227, 228, 262, 269, 273, 275.
Defences, Fixed, see Coast Defences.
Defence Landwards, 132, 172, 178, 182, 184, 257, 258, 259, 264, 275.
Defence Regiment, R.A., 16th, 266.
Defence, Minister of, 276.
Defended Ports, 1, 20, 24, 30, 42, 44, 56, 62, 77, 81, 123, 129, 132, 142, 149, 154, 155, 160, 169, 171, 172, 178, 182, 200, 205-9, 213, 219-24, 229, 230, 240, 243, 250, 252-4, 255-7, 275, 276-8.
Delaney Battery (Guernsey), 99.
Delimara Battery (Malta), 246.
Deliveries, Clerk of, 76.
Delman, Colonel, 14, 15.
Demerara (Guiana), 79, 114, 119.
Demi-Cannon (gun), 7, 13.
Dem-Culverin (gun), 7, 13.
Denmark, 48, 85, 86.
Depots, R.A. and R.G.A., 150, 151, 158.
Destroyers, 167, 171, 174, 176, 181, 183, 195, 196, 197, 199, 200, 215, 216, 225, 235, 236, 237, 243, 245, 255, 258, 266, 274.
Derfflinger (German warship), 196, 197-200.
Derna, 247.
Depression Range-Finder, 173, 183, 215, 216, 227.
Detachment, Invalid, R.A., 135-7, 157.
Detachments, Emergent, 156.
Detroit, Fort, 129.
Detroit (U.S.S.), 129, 130.
Devil's Peak (Hongkong), 261.
Devon, 49, 52, 88, 91; Militia, 162; Volunteers, 52, 88, 164; Territorials, 165, 187, 206, 220.
Devonport, 134, 137, 148, 151, 156, 164, 165.
Diamond Harbour (Calcutta), 149.
D'Iberville (French Warship), 194.
Diego Garcia (Chagos Islands), 193, 256, 257.
Direction Island (Cocos Islands), 194.
Directors (Instruments), 215, 216.
Director of Artillery Studies, 175.
Director-General of Artillery, 175.
Disdale Point (Portland defences), 179.
Dispart Sight, 141.
Districts, Artillery, 150.
District Establishments, 8, 24, 135, 158, 197, 204-9, 275, 276.
District Gunners, 135, 138, 275.
Districts, Military, 49, 52, 91, 92, 94, 95, 97, 98.
District Officers, 158.
Divisions, Garrison, R.A., 151, 161, 162, 163; Eastern, 151, 162; Southern, 151, 162, 163; Western, 151, 163.
Divisions, Infantry, Formations:—
8th Australian, 266, 269, 271,

272; 9th Indian, 266; 11th Indian, 266, 269, 271; 18th, 269, 271.
Dixon, Lieut-Colonel, R.E., 52, 53.
Dixon, Robert, Master-Gunner, 136.
Dockyard-Workers, as Additional Gunners, 51, 159.
Dogger Bank (Battle of), 192, 199, 200.
Dogra Regiment, 266, 270.
Dolores, Los (Spanish Floating Battery), 72.
Dominica (West Indies), 39, 56, 60, 80, 108, 113, 114; French attack on, 115-9.
Donegal Militia, 162.
Donibristle (Scotland), 276.
Donne Battery (Guernsey), 98.
Dorset, 49, 52, 91; Volunteers, 164; Territorials, 166, 186, 206, 220.
Douglas, Captain Sir Charles, R.N., 47.
Douglas, General Robert (Lieut.-General of Ordnance), 137, 138.
Dover, Captain, R.A., 37, 38, 43.
Dover, coast-fort at 6; garrison 8, 11, 25, 30, 40, 43, 51, 77, 81, 101, 134, 137, 155, 185, 207, 220, 276; armament, 11, 19, 25, 96, 170, 185, 220, 229; new battery constructed at 51; Royal Commission, 1859, on defences of, 177-80; during Second World War, 233-8, 238, 240; action against German battle-cruisers Gneisnau and Scharnhorst, 235, 236; shelled by Germans, 237, 238, 240; after Second World War, 276; Volunteers, 102.
Dover, Straits of, 226; during Second World War, 233-8, 240.
Dowman, Captain, R.A., 60.
Downs, The, 8, 30.
Doyle, Fort (Guernsey), 99.
Drake, H. M. S., 55.
Drake, Admiral, 59.
Drake's Island (St. Nicholas Island, Plymouth), 9. 52, 97, 137, 179.
Dreadnought, H.M.S., 172.
Dresden (German warship), 192, 193, 195.
Drogheda, 79, 104.
Drop Redoubt (Dover), 96.
Drummond, General Gordon, 131.
Drysdale, Romeo, Master-Gunner, 136.
Dublin (and Bay), 46, 79, 104, 106, 134, 135, 137, 155, 157; Militia, 162.
Dublin Gazette, 45.
Duff, Admiral, 66.
Dumaresques, 227.
Dumbarton Castle, garrison, 40, 43, 77, 78, 236; armament, 98, 170; general, 166, Volunteers, 103, 164.
Dumpton Point (Kent), 231.
Dunbar, 94, 98, 151, 164; Volunteers, 103.
Duncan, Lieutenant J., R.A., 83.
Duncannon, Fort (Waterford), garrison, 44, 45, 46, 79, 104, 134, 137; armament, 45, 46, 106, 170; general, 135.
Dundalk, 79, 104.
Dundas, General, 83.
Dundee, 98, 164, 229; Volunteers, 103.
Dungeness, 96, 101, 231.
Dunkirk, 9, 30, 32, 226, 228, 233, 236, 239.
Dunn, Lieutenant W., R.A., 123.
Dunree, 137.
Dunwich, 231.
Dupleix, Marquis de, 29.
Durham, 91; Militia, 162; Volunteers 164; Territorials, 166, 186, 197-9; 206, 207, 220.
Dutch, England at war with, 10, 11, 13-5; and North Sea 3; attack Sheerness and Harwich, 13-5; marines, 17; in War of American Independence, 3, 48, 74; in West Indies, 79, 114; lose Cape of Good Hope and Ceylon, 128; Navy, see Navy, Dutch.
Dutch Emigrant Artillery, 110, 119.
Dutch Wars, First, 11; Second, 10, 13; Third, 16.
Duvivier, Captain (French Army Officer), 28, 29.
Dymchurch, 100, 131.
Dyson, Lieutenant-Fireworker, R.A., 27.

Earle, John, Master-Gunner, 137.
Early Warning Sets (Radar), 234, 235.
East and Far East, 202, 203, 210, 213, 215, 255.
East Africa, 192, 195, 212, 243, 255.
East Anglia, German naval raids on, 195-200.
Eastbourne, 32, 96, 100, 232; Volunteers, 103.

Index

Easter Island, 192.
Eastern Brigade, R.A., 150, 162.
Eastern Coast Defences, 185, 207.
Eastern Counter-Bombardment Fire Command (Dover), 236.
Eastern District (Military), 91, 95.
Eastern Group, R.G.A., 154.
Eastern King (Plymouth), 33, 179.
East Florida, 74.
East Gun Battery (Tilbury), 96.
East India Company, 29, 41, 128.
Eastleigh, 165.
Eastney Battery (Portsmouth defences), 78, 97, 136.
East Riding (Yorkshire) Volunteers, 164; Territorials, 166, 186, 206, 207, 220.
Eastware Bay, 93, 96.
"E" Boats (German M.T.B.), 226, 238, 239, 240, 250, 253.
Eddiston, William, Master-Gunner, 136.
Edinburgh, 166; Militia, 162; Volunteers, 164.
Edmonds, William, Master-Gunner, 137.
Edward VI, King, 6.
Edward VII, King, 158.
Elswick (Ordnance Factory), 169.
Egypt, 86, 92, 93, 246.
Egyptian Artillery, 222.
Eire (Irish Free State), 205, 207, 213, 226.
"Elaulic", 218.
Elba, 120, 122, 131.
Elbe, River, 236.
Eldridge, William, Master-Gunner, 137.
Electric Lights, School of (Gosport), 228.
Eliott, General George (Governor of Gibraltar), 65, 69, 71.
Elizabeth I, Queen, 6, 8, 277.
Elizabeth Castle (St. Helier, Jersey), 54, 55, 98, 136.
Elliott, George, Master-Gunner, 136.
Emden (German warship), 192-4.
Emergent Detachments, 156.
Emergency Coast Batteries, see Batteries, Emergency Coast.
Emergency Reserve, Officers, 228.
Empire, British, see British Empire.
Emplacements, construction of, 6, 12.
Ems, River, 85.
Encoder, 215.
Engineer Committee (Board of Ordnance), 99.
Engineers, Ordnance, reports of, 16.
Engineers, Royal, 69, 92, 120, 142, 174, 182, 201, 206, 225, 228, 258, 266.
England, strength in narrow seas, 5; attacked by Armada, 8, 9; at war with Spain, 5, 6, 8-10; with France, 5, 10; with Holland, 10, 11, 13; see also Great Britain.
England, French Army of, 91, 92, 115.
English Channel, see Channel, English.
Erie, Fort, 129, 130, 131.
Erie, Lake, 129, 130, 131.
Esquimalt (British Columbia), 155, 157.
Essex, 49, 91; Territorials, 166, 185, 206.
Establishments, Colonial for C.A., 211.
Establishments, District, see District Establishments.
Estaing, d' (French Admiral), 55, 56, 58, 60, 61.
Etranger, Royal (Foreign Corps), 119.
Eugen, Frinz (German Warship), 235, 236.
Europa Point (Gibraltar), 70.
Europe, 10, 48, 75, 85, 86, 118, 120-8, 203-4, 226, 250.
Evie (Orkney Islands), 166.
Ewe, Loch, 230.
Examination Service (Anchorage, Battery, Officer, and Vessel), 184, 225, 240.
Exchequer, 7, 12, 24.
Exford (Collier), 194.
Exmouth Volunteers, 103.
Expedition Force, British, 1793-4, 82, 85, 96 ; 1939-40, 226.
Eyre, Captain, J., R.A., 65.
Faber (Singapore), 267.
Fairlamb, Captain, R.A., 78.
Fairlight (Hastings), 235.
Falton (gun), 7, 11.
Falcon, H.M.S., 140.
Falconet (gun), 7, 11.
Falkland Islands, 224, 257; Defence Corps, 224.
Falsgrove (Scarborough), 197.
Falmouth, coast-forts at, 6; and Armada, 9; garrison, 8, 25, 30, 40, 43, 78, 81, 136, 137, 155, 178, 206, 220; armament, 19, 25, 97,

187, 220, 229; engages German "E" Boats, 239; general, 164, 166, 186; Volunteers, 52, 102, 103.
Famagusta (Cyprus), 250.
Fan Bay Battery (Dover), 234.
Fanning Islands, 257.
Far East, 202, 203, 210, 211, 213, 215, 225.
Faroe Islands, 232, 275.
Faron Heights (Toulon), 83.
Farringdon Volunteers, 103.
Fead, Captain, R.A., 63, 64, 65.
Federated Malay States Volunteers, 266.
Felixstowe, 15, 231.
Fencibles, 49, 88, 89, 91, 93, 130, 135, 159, 160. Units:—Canadian, 130; Fishguard, 88, 89; Glengarry (Canada), 130; Malta, 135, 159, 160; Newfoundland, 130; Pembrokeshire, 88, 89; Sea, 101.
Fermain Bay (Guernsey), 98, 107; Militia, 107.
Ferrajo, Porto (Elba), 120.
Ferrol, 49.
Fever, Yellow, 56, 81, 108, 110, 112, 114, 119, 157, 278.
Fidra, 230.
Field Artillery, see Artillery, Field.
Field Batteries, see Batteries, Field.
Fife: Militia, 162; Volunteers, 164.
Fighting Book, 184.
Fiji, 257.
Filey, 230.
Finginhoe, 95.
Fino, Porto (Italy), 254.
Fire Commander, duties of, 1914, 182-4; general, 204, 205, 239.
Fire Commands, 182, 204, 235, 236, 239.
Fire Direction Table, 216.
Firing Tubes, 47, 141, 168, 169, 173.
First of June (Battle of), 82.
First Rate (Ship-of-the-Line), 46.
First World War, see War, First World.
Fishguard, French raid on 87-90, 91; Fort, 88, 89, 98, 230; Fencibles, 88, 89; Volunteers, 103.
Five Power Naval Treaty, see Washington Treaty.
Fixed Defences, see Coast Defences. Units:—201st; 203rd, 252-4th.
Flagstaff Battery (Sheerness), 95.
Flamborough Head, 55.
Flanders, 5, 9, 82, 201, 226.
Fleets, British, Channel, 30, 32, 49, 68, 87, 115; Home, 30, 32, 74; Mediterranean, 18, 26, 27, 33, 36, 82, 120.
Fleetwood, 230.
Fleur D'Epee (Guadeloupe), 112.
Flight, Captain, R.A., 34.
Flint-Lock, 47.
Floating Batteries, in Dublin Bay, 46; at Gibraltar, 70-4.
Florida, 36, 39, 41, 61, 62, 74.
Flotilla, Naval Local Defence, 195-7.
Flushing, 92.
Folkestone, sea-battery at, 32, 33, 231; garrison, 40, 136, armament, 96, 231; shelled by Germans, 237, 240; general, 100, 157, 226; Volunteers, 103.
Food, shortage at Malta, 247, 248.
Foot, Regiments of, see Infantry, Regiments of.
Foreign Artillery, Royal, 118, 119.
Foreign Infantry Corps, 117, 119, 122, 123, 124, see also Infantry, Foreign Corps.
Forfar, Militia, 162; Volunteers, 164.
Forman, 2/Lieut, R.A., 35.
Forts and Fortresses, Coast, role of, 1; Henry VIII's, 6, 77, 277; command of, 8, 9, 12, 80, 150, 154, 155; organisation and administration of, 7, 12, 19, 24, 25, 76, 77, 80, 150; design of 6, 20, 47, 77, 99, 100; losses of, 62, 263, 273, 275, 278; landward defence of 132, 172, 178, 182, 257-9, 264, 275; changes in during 19th Century, 133; transferred to War Department, 142; Captain Parkin's views on 142-6; of India, 149; Militia and Volunteer Artilleries allotted to, 155, 162, 163; increase of between 1860 and 1880, 157; rearming of 1904-14, 171; very vulnerable on landside, 172, 178; practice seawards from, 176; Royal Commission, 1859, on, 177-80; objects of, 177; problem of defence of, 178; method of defending, 1914, 182-4; decision to build at Singapore, 203; Japanese attacks on, 257, 258; loss of 263, 273, 275, 278; effects on of loss command of the sea, 278.

Index

Individual Forts are shown against their names.
Fortescue (Historian), 58, 59, 110.
Forth, Firth of, defences of, 55, 77, 93, 98, 187, 188, 195, 207, 221, 229; Territorials, 166, 187, 206, 207, 221.
Fortifications, for Coast Artillery, 3; for Tudor forts, 6, 7; at time of Armada, 8, 9; at time of Civil War, 10; Vauban system, 20, 47; responsibility for, 94; teaching of, 139; changes in between 1815 and 1850, 141; Carnot system of, 141, 142; Captain Parkin's views on 142-6; between 1860 and 1914, 180; between 1919 and 1939, 218; protection against aircraft, 218; for Emergency Coast Batteries, 228; general, 12, 20, 77.
Fort Major, 80.
Fort Record Book, 184.
Fortress Companies, R.E., 174, 182, 206, 225, 266; 36th Company, 266.
Fortress Plotting Room and Plotter, 215, 234, 237.
Fortress System (Observation), 211, 215, 217, 218, 235.
Forward, H.M.S., 198, 199.
Four Days Fight (Battle of the), 13.
Four Power Agreement (1938), 219.
Fowey, coast-batteries at, 50; armament, 97, 230; Volunteers, 103.
Foyle, Lough, 137.
France, at war with England, 5, 10, 13; with Great Britain in War of Spanish Succession, 17-20; in War of American Independence, 48-74; in Revolutionary and Napoleonic Wars, 75, 77, 82-132; and '45 Rebellion, 30; losses in Seven Years War, 36, 37, 39; and Gibraltar, 70; gains in War of American Independence, 74; her situation in 1793 and 1795, 75, 82; and invasion of Ireland, 86, 87, 105-7; and invasion of Britain, 92; and rupture of Peace of Amiens, 92; Napoleon's last campaign in, 128; losses of during Revolutionary and Napoleonic Wars, 128; century of war with Great Britain, 132; introduction of Carnot System of fortification in, 141; her attitude in 1859, 163; experiments in armoured warships, 167; after First World War, 202; agrees to Washington Treaty, 214; goes to war with Germany, 219; surrenders to and occupied by Germans, 226; Germans defeated in, 239; Italy declares war on, 243; general 44, 201, 234.
Fraser, General, 59.
Fraser, Alexander, Master-Gunner, 136.
Frederick Der Grosse (German Warship), 181.
Free Legion (Black Legion), 224, 251.
Freetown (Sierra Leone), 224, 251.
French, command Western Mediterranean, 3; at first siege of Gibraltar, 17, 18; colonies in North America, 27; posts in West Africa, 36, 62; expedition against Annapolis, 28, 29; attack Minorca, 33-6; expedition against Newfoundland, 37-9; threats of invasion of Great Britain, 30, 32, 33, 34, 36, 39, 49, 52, 53, 75, 86, 91-9, 101, 114, 115, 163; attack Jersey, 53-55; activities in West Indies, 1778-82, 55-60; in North America, 60, 61; in Mediterranean, 63; at second attack on Minorca, 63-5; at third siege of Gibraltar, 66-75, 132; at Toulon, 82, 83, 85; victorious in Europe, 1794-6, 85, 86; invasions of Ireland, 86, 87, 104-7; raid on Fishguard, 87-90; in San Domingo, 108, 110; recapture Guadeloupe, 112, 113; attack Dominca, 115-8; in Corsica, 120; try to retake Macouf Islands, 121, 22; occupy Malta, 122; invade Naples, 123; retake Scylla and Capri, 124; besiege Cadiz and Tarifa, 126-8; in Italy, 128; attack Sierra Leone, 129; Channel ports and airfields occupiaed by Germans, 226; forces evacuated from Dunkirk, 226; German batteries on Channel coast of, 233, 237, 238; railways, 234; Germans evacuate Channel coast, 236; general, 27, 32. Navy, see Navy, French.
French Revolution, 75, 82, 104, 108.

French Revolutionary and Napoleon
French Revolutionary and Napoleonic Wars, see Wars, French Revolutionary and Napoleonic.
Freshwater (Isle of Wight), 97.
Freyberg, General, 245.
Frinton, 231.
Frisian Islands, 236.
Funchal Bay (Madeira), 123.
Fundy, Bay of, 28, 80.
Fuzes, 141, 167, 173, 217.
Gabaron Bay (Cape Breton Island), 29.
Gage, General, 48.
Gallissonniere, Marquis de (French admiral), 33, 34.
Gall, Captain, J. D., R.A., 58, 60.
Galle (Ceylon), 128.
Galleons (Spanish), 9.
Galway Bay, defence of 104-6; garrison, 44, 45, 78, 104, 137; Militia, 162.
Gambia, 251.
Gantry (Shoeburyness), 176.
Garnish Island (Bantry Bay), 105.
Garrick, Robert, Master-Gunner, 137.
Garrison Artillery Training (1911), 177.
Garrison Batteries, R.A., see Batteries, Garrison.
Garrison Brigades, R.A., see Brigades, Garrison.
Garrison Companies, R.A., see Companies, Garrison.
Garrison Divisions, R.A., see Divisions, Garrison.
Garrison Point (Sheerness), 13, 179.
Garrisons, Coast-Artillery, first, 7; at time of Armada, 8; during First Dutch War, 11; after War of Spanish Succession, 20, 25; after formation of Royal Artillery, 24; during '45 Rebellion, 30; during Seven Years War, 36, 40; of Newfoundland, 37, 39; in 1764, 41; formed from Invalid Companies, 43, 45; in Ireland, 44-6, 79; in 1779, 51; in West Indies, 58, 60, 108, 112, 113, 114, 118, 119; in North America and Florida, 60-2; in West Africa, 62; of Minorca and Gibraltar, 63, 65; in 1789, 77-60; in San Domingo, 110; of Dominica; in Corsica, 120; of Marcouf Islands, 121; of Scylla, 123; in Sicily, 123, 124; of Heligoland, 124, 125; of Anholt, 125; of Adriatic islands, 125; of Corfu, 125; around Straits of Gibraltar, 126; of Capetown, Ceylon, Mauritius, & St. Helena, 128; on Great Lakes, 129, 130; during 1820 to 1850, 134; formed from Invalid Detachment, 135-7; R.G.A., 154, 155; 1860-1910, 155, 156; withdrawn from Corfu, Ionian Islands, Newfoundland, and New Brunswick, 156; at St. Helena, and Sierra Leone, 156, 157; withdrawn from Canada, 157; Local Companies in, 160, 161; composition, 1914, 182; rules for calculating, 184; at outbreak of war, 1914, 185-91; 1932-3, 206-9; for Far East, 211; leave Eire, 213; at outbreak of war, 1939, 219-24; of Malta, 246, 250; in East, 257; of Hongkong, 258, 259, 262, 263; of Singapore, 266, 267.
Garrisons, Infantry, 8, 25, 30, 54, 56, 63, 65, 69, 80, 81, 94, 102, 108, 110, 115, 117, 123, 124-6, 129, 130, 159, 258, 262, 263, 266, 270.
Garrisons, Rules for Calculating, 184, 211.
Gascoign (Inventor), 47.
Gatt, Brigadier G. C. (Royal Malta Artillery), 246, 247.
Gatwick Sands (near Fishguard), 90.
Gauteaume (French Admiral), 118.
Genoa, 128.
George II, King, 30.
George III, King, 125.
George, Fort (Scotland), 42, 78.
George, Fort (Guernsey), 98, 136.
George, Fort (Lake Ontario), 129, 130, 131.
George, Fort (Mauritius), 209.
Georgeopolis (Crete), 243.
Georgia, 60.
Georgia, South, 257.
German Batteries, 233, 237-40.
Germans, failure to take Channel ports, 192; submarine campaign, 192, 200; raids on Grea; Britain, 1914-5, 192, 195-200t activities of raiding cruisers, 192-5; their plan at Hartlepools, 198; threatening attitude, 1938; successful campaigns, 1940, 226;

Index 299

plans and preparations to invade Britain, 226; occupy Channel Islands, 229, 232; fail to obtain air-command over Channel, 233; mount heavy guns at Cap Gris Nez, 233; attempt to pass ships and convoys through, Straits of Dover, 234, 236, 238; evacuate French Channel coast, 236, 240; shell Dover and English south-coast towns, 237, 238, 240; defeated in Normandy, 239; attack Crete, 243-6; join Italians in attacking Malta, 247, 248; in Italy, 252-4; lose Tsingtao to Japanese, 1914, 257, 258. Airforce (Luftwaffe), see Airforce, German Navy, see Navy German.

Germany, 86, 92, 124; her strength at sea, 1914, 181, 204; at war with Britain, 1914-8, 192-201; her naval policy, 200; naval situation after First World War, 202; joins in naval armaments race, 214; Hitler and Nazi Party gain complete control of, 219; naval situation of 1939; accepts surrender of France, 226; victorious, 1940, 226; return of warships to from Brest, 235, 236; invasion of by Allies 241; surrenders, 241.

Ghent, 9.

Ghent, van (Dutch admiral), 13.

Gibraltar, first garrison sent to, 16, 17; captured by British, 17; first siege, 17, 18; second siege, 26, 27; third siege, 3, 65-74, 278; final relief, 74, 132; during Peninsular War, 126; during First World War, 195; during Second World War, 243, 251; garrison, 17, 24, 26, 41, 65, 70, 71, 73, 80, 126, 134, 156, 189, 209, 222, 251, 276; armament, 17, 20, 26, 66, 74, 169, 170, 189, 222, 251; general, 33, 34, 36, 63, 66, 81, 83, 108, 122, 123, 148, 173, 176, 278.

Gibraltar Bay, 68, 71-4, 251.

Gibraltar, Straits of 63, 66, 74, 126.

Gilkicker Point (Gosport), 178.

Gillingham, coast-fort at 14; garrison, 25, 40, 43, 136; armament, 19, 25, 96; Volunteers, 103.

Gindrinkers Line (Inner Defences, HongKong), 259, 261, 263.

Glamorgan, Militia, 162; Volunteers, 164; Territorials, 166, 187.

Glasgow, 164, 166.

Glasgow, H.M.S., 193, 195.

Glengarry, Fencibles (Canada), 130.

Gloire, La (French Warship), 167.

Gloucestershire, 91.

Gneisenau (German warship), 192, 235, 236.

Godfrey, Lieutenant, R.A., 73, 74.

Godwin, Captain, R.A., 78.

Golden Hill (Isle of Wight), 176.

Gonaives (San Domingo), 110.

Good Hope, Cape of, see Cape of Good Hope.

Good Hope, H.M.S., 193.

Goodyear, Captain, R.A., 31.

Gordon Highlanders, 266.

Gordon, James, Master-Gunner, 137.

Goree Island (West Africa), 36, 62.

Gorleston, coast-battery at, 50; armament, 95; German cruiser raid on, 196.

Gosport, garrison, 25, 40, 43, 134, 136, 155, 158; armament, 19, 25, 97, 170, 186; general, 151, 228; Volunteers, 104.

Gostling, Captain, R.A., 78.

Government, British, on threat of French invasion, 32, 33; re Gibraltar, 66; and defence of Firth of Forth, 77, 79; and situation in 1797, 91; breaks Peace of Amiens, 92; and Volunteers, 102, 163; decides to occupy Madeira, 123; takes over Ceylon, 128; re R.M.L. and R.B.L. guns, 168; and armaments race 202; decides to build Singapore base, 203; Labour and Singapore base, 203, 210, 264; decides to re-arm Far East, 210; and Japan, 254; abolishes Coast Artillery, 277.

Governor of Fort, 12, 80.

Gozier (Guadeloupe), 112.

Graham, Brigadier, 112.

Graham, General Sir Thomas, 127.

Grain (Sheerness), 180.

Gramont, H.M.S., 37, 38.

Grandy Barracks (Devonpprt), 156.

Grand Alliance, War of, 16.

Grand Army (French), 92, 93.

Grand Battery (Gibraltar), 66, 71.

Grand Bay (Guadeloupe), 112.

Grand Bay (Dominica), 115, 117.
Grandeterre (Guadeloupe), 113.
Grand Fleet, 195, 196, 197.
Grand Harbour (Valetta), 248, 249.
Grand Hotel (Scarborough), 197.
Grand Rocq Battery (Guernsey), 99.
Grant, General, 56.
Grasse, Count de (French admiral), 55, 58, 59, 60, 74.
Gravelines, 9.
Graves, Admiral, 115.
Gravesend, coast-fort at, 6; Dutch attack on, 14; new coast-battery at, 50; and Coast Brigade, 157; garrison, 8, 25, 30, 40, 43, 78, 81, 136; armament, 19, 25, 96; general, 164, 166; Volunteers, 102, 104.
Great Britain, defence of coast-line of, 2 and Command of the Seas, 2, 3, 26, 37, 39, 55, 62, 63, 75, 87, 129, 142, 192, 203, 243, 255, 257, 258, 263, 268, 277, 278; threatened with invasion, 3, 30, 32, 33, 48, 49, 52, 58, 75, 86, 87, 91, 92, 93, 226-33; during '45 Rebellion, 30, 31; opposition to French Revolution and Napoleon, 75, 82, 85, 92, 131; expeditionary force evacuated 85; defence position, 1795, 86; and Ireland, 87; preparations to defend, 1797-1805, 91-4; defences of 1805, 94-9, 100; and West Indies, 1802, 114; gains during Revolutionary and Napoleonic Wars, 128; at war with U.S.A., 129-31; century of warfare with France, 132; supremacy on the seas during Crimean War, 142; expansion of defences of, 159; and armoured warships, 167; Royal Commission, 1859, on defences of, 177-80; defences of 1914, 185-9; German submarine threat to, 192, 200; German naval raids on 192, 195- 200; naval situation of after First World War, 202; interests of in Far East, 202, 203; attitude towards U.S.A. and Japan, 202, 203, 210; defences of, 205-7; her position in Narrow Seas and Mediterranean, 213; air-defence of, 213, 214, 225; agrees to Washington Treaty, 214, 512;

and Munich Crisis, 1938, 219; defences in 1939, 219-21; naval situation of 1939, 225; Germans plan to invade, 226; expeditionary force evacuated to, 226; defences to resist invasion, 227-32, 275; Radar Cordon established around 234; southern coast towns of shelled by Germans, 238; mobile C.A. units formed in 252; inferior at sea to Japan, 255; protected by Coast Defences and Artillery, 275, 277; her greatness founded on sea-power, 278.
Great Lakes (Canada), 80, 129-31.
Greatstone, 231.
Greece, 156, 243, 244.
Greeks, 244.
Greenock, 136, 162; Volunteers, 104.
Green's Battery (Gibraltar), 66.
Greenwich, 41, 42.
Gregory, Captain, R.A., 35, 39.
Grenada (West Indies), 39, 55, 56, 79, 80, 108, 113, 114, 119.
Grey, General, 164.
Grimsby, 231; Volunteers, 103.
Gris Nez, Cap, 233, 235, 237-40.
Grosser Kurfurst (German Battery), 233, 240.
Groups, R.G.A., 154.
Grove, Poncet la (French Historian), 35.
Groves, Captain G., R.A., 65, 73.
Grumley, Captain, R.A., 73.
Guadeloupe, 36, 39, 60, 79, 112; retaken by the French, 112-4, 118.
Guards and Garrisons, 24, 39, 41, 42.
Guard, Home, 51, 240, 241.
Guard, Town, 51, 159.
Guernsey, coast-fort at, 6; Invalid Company moves to, 42, 107; redoubts added, 51; garrison, 25, 40, 42, 43, 77, 78, 81, 134, 136, 155, 188; armament, 19, 25, 51, 98, 170, 189; defences, 91, 98, 99, 107; Militia, 107, 163, 188.
Guided Missiles, 275, 277.
Guildford Battery (Dover), 96.
Gunboats, 68, 69; manned by Sea Fencibles, 101; at Marcouf Islands, 120-122, at Capri, 124; on Great Lakes, 131; steam, 140; mine-sweeping, 196; Italian, 248, 249; at Hongkong, 258; at

Index

Singapore, 266.
Gun Drill, 12, 20, 71, 168, 169.
Gun Lascars (Asiatic), 135, 159, 160.
Gunloops, 6.
Gunner Magazine, 168.
Gunners, District, 135, 138, 275.
Gunners, Additional, see Additional Gunners.
Gunner's Quadrant, 7.
Gunnery, Coast-Defence, 7, 12, 20; with red-hot shot, 71; changes in methods, 133; effects on of steam warships, 140; in 1850, 141; new instruments for, 173, 174; effects on of Singapore base, 203; between 1919 and 1939, 215-7.
Gunnery Courses, 175, 176, 217, 242.
Gunnery, Instructors of, 175, 176, 217.
Gunnery, School of, founding of at Shoeburyness, 139, 174, 175; courses at, 175, 176; Instructors of Gunnery sent from, 176; practice-camps supervised by, 176; re-named Coast Artillery School, 203.
Guns, Coast-Defence, object of, 2; superiority over naval guns, 2; early types, 6, 7; smooth-bore, 7, 12, 20, 132, 140, 159, 167, 169, 170; rifled, 3, 4, 152, 159, 177; R.M.L. 168-72; R.B.L. 167-9, 172; standardised by weight, 12, 20; favourite types, 26; in 1770, 46; method of mounting, 47, 141, 228; in 1790, 76; changes in during 19th Century, 133; moving and shifting, 138, 139, 218; designated by calibre, 140; special for firing shell, 140; struggle with armour, 167; conversion of smooth bore to R.M.L. 169, 170; first issue of R.B.L. and Q.F., 171; situation in 1900, 171; re-arming of coast-defences, 1904-14, 171-3; progress, 1856-1914, 173; between 1919 and 1939, 215-7; counter-bombardment, 215, 216; close-defence, 216; for Emergency Coast Batteries, 227-32; after Second World War, 275; linked with naval guns, 277.
Types:—Those guns designated either by weight or calibre are indexed here: guns designated by names, etc. are indexed against those names. 80-pdrs., 169; 68-pdrs, 140, 167, 169; 64-pdrs., 169; 56-pdrs., 140; 42-pdrs., 17, 20, 25, 46, 76, 95-9; 40-pdrs., 167; 36-pdrs., 68, 95-9; 32-pdrs., 20, 25, 26, 52, 66, 73, 95-9, 115, 121, 169; 24-pdrs., 3, 20, 25, 26, 32, 35, 37, 38, 45, 51, 53, 54, 66, 76, 79, 95-9; 105, 106, 115, 121, 124; 20-pdrs., 98, 99, 18-pdrs., 20, 25, 26, 32, 33, 37, 45, 50-3, 66, 76, 95-9, 105, 115; 16-pdrs., 26; 15-pdrs., 79; 12-pdrs., 11, 20, 25, 26, 32, 46, 50, 51, 53, 58, 66, 95-9, 105, 106, 115; 12-pdr. Q.F., 171-3, 182, 185-91, 216, 217, 219-21, 229-31, 232, 243, 246, 250-3, 256, 267, 273; 9-pdrs., 20, 25, 26, 46, 50, 51, 66, 88, 89, 95-9; 8-pdrs., 11, 6-pdrs., 11, 17, 20, 25, 26, 32, 37, 46, 50, 51, 53, 54, 66, 85, 95-9, 105, 106, 117, 121, 167; 6-pdr. Q.F., 171; 6-pdr. Twin, 215, 216, 220-4, 229, 249, 250, 251, 267; 4-pdrs., 20, 25, 95-9, 121; 3-pdrs., 20, 25, 26, 50, 58, 98; 3-pdr. Q.F., 171, 195; 18-inch, 255; 16-inch, 233; 15-inch, 215, 216, 218, 224, 233-9, 240, 243, 256, 267, 270, 272; 14-inch, 229, 233, 239, 240; 13.5-inch, 167; 12.5-inch (38 ton), 168, 169; 12-inch, 171, 181, 197; 11-inch, 181, 196, 197, 225, 233; 10-inch (84 cwt.), 140; 10-inch (25 ton), 168, 169; 10-inch (naval), 169, 171; 9.2-inch, 170-3, 182, 185-91, 210, 211, 215, 217-24, 229, 234, 235, 238, 246, 250, 251, 256, 261, 262, 267, 270, 273, 274; 9-inch (12 ton), 168; 9-inch (naval), 171; 8-inch (65 cwt.), 140; 8-inch (naval), 181, 215, 225, 243; 7.5-inch, 223, 256; 7-inch, 167; 6-inch, 167, 169-73, 182, 185-91, 194, 197, 210, 213, 215, 216, 219-24, 227-32, 234, 238, 239, 243, 246, 250-2, 256, 257, 259, 261, 267, 269-74; 5.9 inch, 225; 5.5-inch, 227-32, 251, 252; 4.7-inch Q.F., 170-3, 182, 185-91, 216, 217, 219-24, 227-32, 246, 256, 257, 259, 261; 4.5-inch, 251, 256, 4-inch, 181, 195, 198, 227-32, 243, 246, 250-7, 259.

Gustavus Adolphus (King of Sweden), 12.
Haddington, Militia, 162; Volunteers, 164.
Haifa, 213, 222, 243, 250.
Haisland, General, 53.
Halcyon, H.M.S., 196.
Half Moom Battery (Sheerness), 95.
Halifax (Nova Scotia), 29, 38, 81, 149, 157; garrison, 36, 41, 80, 134, 155; armament, 170.
Hamburg, 92, 120.
Hamoaze (Plymouth), 179.
Hampshire, 49, 91; Militia, 162; Volunteers, 164; Territorials, 165, 186, 206, 219.
Hanover, 85.
Hanoverian Troops, 63, 65, 69.
Hanseatic Towns, 5.
Happisburgh, 231.
Harcourt, John, Master-Gunner, 136.
Hardy, Admiral, Sir Charles, 49.
Harfleur, 138.
Harris, Lord, 153.
Harris, Lieut-Colonel G. N., Royal Marines, 195.
Hartlepools, 95, 158, 166; German bombardment of, 196-9, 200, 207, 220, 229; Volunteers, 103.
Harwich, defence against Armada, 8, 9; building of Landguard Fort at, 10; Dutch attack on, 1667, 3, 14, 15; new coast-bty. at, 50; garrison, 185, 207, 220; armament, 95, 185, 220, 229; general, 16, 43, 93, 166, 195, 227, 236.
Hastings, sea-battery at, 32, 33; state of, 1779, 49; garrison, 40, 43, 78, 136, 232; armament, 96, 232; general, 93, 235; Volunteers, 52, 103.
Hatherwood Point (Isle of Wight), 178.
Haulbowline (Cork Harbour), 105, 180.
Haussonville, d' (French colonel), 27, 38.
Havana, 36, 39.
Haverfordwest, 89.
Havre, 92, 120, 121.
Hawaian Islands, 254.
Hawke, Admiral Lord, 39.
Hayle, 232.
Hayter Volunteers, 103.
Heaton Volunteers, 103.

Heavy Artillery, see Artillery, Heavy.
Heavy Batteries, see Batteries, Heavy.
Heavy Brigades and Regiments, see Brigades (and Regiments), Heavy.
Hebrides, 227, 232.
Hedauville (French general), 110.
Helensburgh, 166.
Heligoland, 124, 125, 196.
Hengisbury Head, 232.
Henry VIII, King, 5, 6, 77, 138, 277.
Henry, Fort (Jersey), 98.
Heraklion (Crete), 243-6.
Herbert, Sydney (Secretary of State for War), 177.
Hereward, H.M.S., 246.
Heron, Captain, 28.
Hesse-Darmstadt, Prince George of, 17, 18.
Heugh Battery (Hartlepools), 197-9.
Higham, 9.
High Cape (Norfolk), 231.
High Sea Fleet (German), 196, 197, 199, 200, 202, 225.
Hill Farm (Isle of Wight), 178.
Hind, Captain, R.A., 43.
Hipper, von (German admiral), 196, 197, 199.
Hitler, Adolf, invasion of Britain, 3; his rise to power, 219; assumes supreme power, 219; his intention to make war, 219; goes to war with Great Britain and France, 219; denounces Versailles Treaty, 225; plans to invade Britain, 226; abandons all intention of invading Britain, 233; invades Russia, 233; orders naval squadron at Brest to return to Germany, 235.
Hitler Youth Corps, 254.
Hoche, Lazare (French General), 86-8, 104.
Hoche (French warship), 106.
Hockings, Captain, R.E., 121.
Hogue, La, 16, 121, 122, 235.
Holcroft, Captain, R.A., 130.
Holland, at war with England, 10, 11, 13-5; in War of American Independence, 48, 74; ally of Great Britain, 82; invaded and occupied by French, 85, 86; losses during Revolutionary and Napoleonic Wars, 128; German naval squadron off shores of, 236.

Index

Hollesley Bay, 93, 95.
Holm (Orkney Islands), 166.
Holman, Captain, R.A., 26, 27.
Holyhead, 230.
Holy Island, 19, 40, 78, 136.
Home Fleet (British), 30, 32, 74.
Home Forces, 228.
Home Guard, 51, 240, 241.
Home Office, 102.
Honduras, 114, 119.
Hongkong, decision to re-arm, 1932, 210; report on coast-defences of, 1935, 211; Japanese attack on and capture of, 3, 63, 147, 257-63, 275, 278; garrison, 134, 135, 156, 160, 191, 208, 209, 212, 224, 258, 259; armament, 170, 191, 211, 224, 256, 259; general, 135, 149, 159, 200, 255, 278; Volunteers, 224, 258, 259.
Hongkong and Singapore Royal Artillery (Royal Garrison Artillery), 135, 160, 181, 191, 201, 208, 209, 212, 223, 224, 257, 258, 266, 267, 269, 272, 273, 276.
Hongkong Targets, 176.
Hood, Admiral Lord, 59, 82, 120.
Hooe Lake Point (Plymouth defences), 179.
Hope Point (Gravesend), 96.
Horizontal Screw, 47.
Hornsey, 231.
Horse Sand Fort (Spithead), 178.
Houge a La Pierre (Guernsey), 99.
Houmette Battery (Guernsey), 99.
House of Commons, 276.
Howe, Admiral Lord, 60, 74, 82.
Howitzers, 26, 66, 121, 139, 141, 205, 261.
Hoyland, A., Master-Gunner, 136.
Hoyland, John, Master-Gunner, 136.
Hubbard, James, Master-Gunner, 23.
Hudson, River, 48.
Hudson Lowe, General, 124.
Hudsons (Aircraft), 267.
Hughes, Captain, R.A.
Hull, state of defences, 16; Humber defences, 1914, 186; garrison, 11, 25, 30, 40, 43, 78, 81, 101, 134, 136, 186; armament, 11, 19, 25, 95, 186; general, 6, 55, 166; Volunteers, 104.
Hull Southend, coast-battery at, 95.
Human Torpedoes, 248, 249, 251, 253.
Humber, River, defences of, 186, 195, 207, 220, 229.
Humbert (French general), 106, 107.
Hungary, 202.
Hunstanton, 231.
Huron, Lake, 129.
Hurricanes (Aircraft), 269, 270.
Hurst Castle, coast-fort at, 6; state of in, 1779, 50; garrison, 8, 25, 40, 43, 78, 136, 186; armament, 19, 25, 97, 178, 186.
Hussars, 2nd, King's German Legion, 126.
Hyeres Bay, 120.
Hythe, sea-battery at, 32, 33, 36, 231; garrison, 40, 43, 78; general, 32, 93; Volunteers, 103.
Iceland, 232, 275.
Ilfracombe, 88, 97, 232.
"Illuminated Area", 174, 183, 216, 249.
Immortalite, H.M.S., 170.
Incendiary Ammunition (Shot and Shell), 69, 71-8, 110.
Inchgarvie (Firth of Forth), 98.
Inchgoln (Firth of Forth), 98.
India, 36, 41, 128, 149, 155, 161, 189, 208, 212, 257, 273, 274, 276.
Indian Army, 128, 149, 266. Formations:—3rd Corps, 266, 269; 9th Division, 266; 11th Division, 266, 269, 271; 12th Brigade, 271; 22nd Brigade, 270; 44th Brigade, 271; 45th Brigade, 269; State Forces, 270. Regiments:—1st Heavy A.A., 266; Dogra, 266, 270; Punjab, 258, 259, 261, 262, 266.
Indian Artillery, see Artillery, Indian.
Indian Ocean, 62, 192, 193, 211, 255, 257, 278.
Indian Mutiny, 149.
Indians (American), 28, 29.
Indies, West, see West Indies.
Indo-China, 257, 267, 268.
Indonesia, 273.
Industrial Revolution, 132.
Infantry, 14, 15, 26, 27, 28, 34, 37, 54, 56, 58, 59, 61, 65, 81, 94, 108, 110, 112, 114, 123, 127, 130, 149, 150, 258, 259, 261, 262, 266.
Infantry, Additional Gunners, 34, 36, 51, 66, 81, 100, 117, 130, 159.
Infantry Brigades, see Brigades, Infantry.
Infantry Divisions, see, Divisions, Infantry.
Infantry, Foreign Corps. Units:—

Hardenberg's Hanoverians, 69; Lowenstein's Fusiliers, 119; Royal Corsican Rangers, 123, 124; Maltese Light Infantry, 122; Royal Malta Regiment, 124; Royal Etranger, 119; Sicilian Rangers, 125; York Light Infantry, 177.
Infantry, French Marine, 28, 29, 37, 38.
Infantry Garrisons, see Garrisons, Infantry.
Infantry Regiments. Units:—Anstruther's, 27; Disney's, 27; Lord Douglas', 14; Egerton's, 26; Hayes', 27; Lord Mark Kerr's, 26; Newton's, 27; Pearce's, 26; Philip's, 27, 28; Dule of York's Maritime, 14, 15; 1st Foot, 59; 4th Foot, 34; 5th Foot, 58; 8th Foot, 130; 12th Foot, 69; 15th Foot, 59; 23rd Fusiliers, 34; 24th Foot, 34; 27th Foot, 123; 34th Foot, 34; 39th Foot, 65, 114; 40th Foot, 37; 41st Foot, 130; 43rd Foot, 112; 46th Foot, 115, 117; 48th Foot, 56; 51st Foot, 63; 56th Foot, 65; 58th Foot, 65, 123; 60th Foot, 61, 119; 61st Foot, 63; 62nd Foot, 123; 69th Foot, 59; 72nd Foot, 65; 73rd Foot, 65; 78th Highlanders, 54; 86th Foot, 59; 95th Foot, 54; 97th Foot, 65; Dogra, 266, 270; Gordon Highlanders, 266; Loyal 266; Middlesex, 258, 259, 262; Malay, 266; Manchester, 266; Punjab, 258, 259, 261, 262, 266; Royal Rifles of Canada, 258, 259, 262; Royal Scots, 258, 259, 261, 262; Rajputana Rifles, 258, 259, 261, 262; Veteran (Royal), 125; West India, 115, 117; Winnipeg Grenadiers, 258, 259, 262.
Inglis, Lieutenant, R.A., 35.
Innes, Captain John, R.A., 60.
Ionian Islands, 125, 134, 156.
Inspector of Royal Artillery, Reports of, 76, 77.
Inspectors, Ordnance, Reports of, 49, 50.
Instruction Batteries, 147.
Instructors of Gunnery, 175, 176, 217.
Instruments, Coast Artillery, 147, 152, 157, 159, 173, 174, 176, 180, 183, 206, 215, 216, 217, 227, 234, 275, 277.
Invalids, Infantry, 30, 42, 80, 81, 102, 121, 122.
Invalids, Royal Artillery, formation of, 26, 41, 42; first moves, 42; formation of battalion, 42; distribution of, 42, 43, 77, 78; establishment of 42; description of, 44; in 1779, 51; garrisons of, 43, 78, 135; in Channel Islands, 42, 54, 77, 78, 107; responsibilities of, 77; sent to Bermuda, 82; at Fishguard, 88, 89; at lesser forts, 100; at Pendennis Castle, 102; at Heligoland, 124, 125; sent to Anholt, 125; battalion disbanded but Invalids continue as Detachment 135; its organisation, 135; its distribution, 136, 137; abolished, 157.
Invalids, Royal Irish Artillery, 45, 79.
Invasions of Great Britain, threatened by the Spanish, 8, 9, 10; by the French, 30, 32, 33, 34, 36, 39, 49, 52, 53, 75, 86, 91-9, 101, 114, 115, 163; by the Germans, 226-33; general, 3; of Ireland, attempted by the French, 86, 90, 105, 106, 107; of Malaya, carried out successfully by the Japanese, 263-74.
Invergodron, 229.
Inverness, 230.
Ipswich, 164, 166.
Ireland, Coast Artillery in and Coast Defences of, 44, 45, 46, 79, 104-6, 188, 205, 207, 208, 213; defence of, 91, 93, 100, 104, 105, 106; garrison of, 104; French invasions of, 86, 90, 105, 106, 107; situation of in 1796-1811, 86, 104-7; government of, 105; rebellion in, 106; field-batteries in 133; distribution of Militia in, 162, 163; coast-defences in 1914, 188; formation of Irish Free State, 205, 207; Coast Artillery leave, 213.
Ireton, General, 11.
Irish, inhabitants of Newfoundland, 37; in Fishguard expedition, 87; rebellion of, 106; and French invasion, 106; Militia, 86, 104, 162; Volunteers, 104.

Index

Irish Brigades, R.A., see Brigades, Garrison, R.A.
Irish Coast Defences, North and South, 163, 188, 208, 213.
Irish Free State (Eire), 205, 207, 213, 226.
Irish Sea, 55.
Ironclads (Warships) 140, 167.
Iron Duke, H.M.S., 167.
Ischia Battery (Naples defences), 253.
Isla de Leon (Cadiz), 128.
Island Bridge (Dublin), 135.
Isle of Wight, see Wight, Isle of.
Isola Battery (Malta), 246.
Italians, in Mediterranean, 243; on way to Crete, 245; attack Malta, 247-50; efforts against Gibraltar, 251; surrender, 252; embodied in British Coast Artillery, 252-4; final efforts of Fascists, 254. Airforce, see Airforce, Italian. Batteries, see Batteries, Italian. Navy, see Navy, Italian.
Italy, after First World War, 202; agrees to Washington Treaty, 214; joins in naval armaments race, 214; declares war on Great Britain, 243, 247; her communications to North Africa, 247; invaded by Allies, 250, 252; armistice with 251, 252; Allied advance up 252-4; Coast Artillery in 252-4, 275; general, 91, 123, 125, 128, 163.
Ithica (Ionian Islands), 125.
Ivybridge, 242.
Jacobites, 30.
Jamaica: first Coast Artillery garrison sent to, 58; garrison, 58, 60, 80, 108, 113, 114, 118, 134, 155, 160, 190, 208, 209, 224; armament, 170, 191, 224; Jamaica Company, R.A. (R.G.A Jamaica Company, R.A. (R.G.A.), 58, 60, 161; Jamaica Militia, 119, 161, 190, 208, 209, 224; general, 59, 108, 110, 113, 114, 118, 161.
James I, King, 10, 107.
James, Duke of York, 13.
James, Captain, R.A., 39.
James, Alexander, Master-Gunner, 137.
Japan, 62; situation after First World War, 202; aggressive policy in Far East, 210; attains first-class naval power, 1919, 214; agrees to Washington Treaty, 214; denounces Washington Treaty, 214; enters Second World War, 255; her warplan, 257.
Japanese, gain command of Pacific, 3; attack on Port Arthur, 171; 172, 257-9; ambitions, 203, naval supremacy, 255; threat to British dominions and possessions, 255; gain command of Pacific and Indian Oceans, 257; attack on and capture of Hongkong, 257-63; force for attack on Hongkong, 259; invasion of Malaya and capture of Singapore, 264-73, 275; forces for invasion of Malaya, 268; move into Bay of Bengal, 273; invade Burma, 274; attack Ceylon from the air, 274; advance comes to an end, 274. Airforce, see Airforce, Japanese. Navy, see Navy, Japanese.
Jats, 212.
Java, 257.
Jerbourg Battery (Guernsey), 98.
Jeremie (San Domingo), 110, 112.
Jersey, Invalid Company moves to, 42, 107; French attacks on, 53-5; defence of, 91; garrison, 25, 40, 43, 77, 78, 81, 101, 134, 136, 155, 188; armament, 19, 25, 98, 170, 189; general, 149; Militia, 107, 163, 188.
Johnston, Captaink R.A., 62.
Johnston, James, Master-Gunner, 137.
Johore Battery (Singapore), 267, 270, 272.
Johore (State of), 269.
Johore Strait, 210, 264, 271, 272.
Jones, Major-General Harry, 177.
Jones, Paul (Captain, U.S. Navy), 55, 77.
Joss Bay (Kent), 231.
Joyeuse, Villaret (French admiral), 82.
Juan Fernandez, 195.
Jubilee Battery (Hongkong), 259, 261.
Jury's Gut (Kent), 231.
Jutland (Battle of), 200.
Kai Tak (Hongkong), 258, 259.
Kallang (Singapore), 271.
Kampar (Malaya), 269.

v

Kando (Japanese admiral), 268.
Kahe's Redoubt (Fort St. Philip, Minorca), 35.
Kano (Nigeria), 251.
Karachi, 149, 156, 189, 208, 223, 255, 256.
Karlsruhe (German warship), 181, 192, 195.
Kattegat, 125.
Keeling Islands, 194, 256, 257.
Keith, John, Master-Gunner, 136.
Kent, 9, 32, 33, 49, 91; Trained Bands, 14; Militia, 162; Volunteers, 164; Territorials, 166, 185, 206, 207, 220.
Kessinghand, 231.
Khartoum, 251.
Kilindini (East Africa), 211, 212, 223, 243, 256.
Killala Bay (Ireland), 106.
Kimberley, H.M.S., 246.
Kincardine, Militia, 162; Volunteers, 164.
King, Major, 82nd Foot, 126.
King's African Rifles, 212, 223.
King's Bastion (Chatham), 95.
King's Bastion (Gibraltar), 68, 71, 72.
King's Cup (National Artillery Association), 165.
Kingsdown, 231.
Kingsgate, 231.
King's German Legion, 126.
King's Lynn, 230.
Kingston (Jamaica), 190, 209, 224.
Kingston (Ontario), 129, 134, 155, 156.
Kinniburgh, Robert, Master-Gunner, 136.
Kinsale, 86, 105; garrison, 44, 45, 46, 79, 104, 137, 155; armament, 45, 46.
Kirkaldy, 164, 166.
Kirkwall, 164, 166.
Kluang (Malaya), 269.
Knattenbury Hill (Plymouth defences) 179.
Knowles, Michael, Master-Gunner, 136.
Knox, Lieutenant-Colonel Thomas, Fishguard Fencibles, 88, 89.
Koehler, Lieutenant G. F., R.A., 69, 85.
Konigsberg (German warship), 192, 194, 195.
Kota Bharu (Malaya), 268.
Kota Tinggi (Malaya), 264.
Kowloon (Hongkong), 259, 261.

Labrador Battery (Singapore), 267, 270, 273.
Ladang (Singapore), 267, 270.
Lagos, 252.
Lagrange (French general), 115, 177.
Lakes, Great, see Great Lakes.
Lamargue (French general), 124.
Lambert, Captain, R.A., 63, 64.
Lamma Island (Hongkong), 261.
Lamy, Fort (French Sahara), 251.
Lanarkshire Volunteers, 164.
Lancashire, 91; Militia, 162; Volunteers, 164; Territorials, 166, 187, 206, 207, 221.
Lancashire Brigade, R.A., 150, 162.
Lancaster Volunteers, 103.
Lancresse Bay (Guernsey), 99.
Landguard Fort (Harwich), construction of, 10; plan of, 12; Dutch attack on, 14, 15; state of, 16; rebuilt, 16; master-gunner for, 23; Pembrokeshire Militia at, 101; garrison, 15, 25, 30, 40, 43, 78, 81, 82, 101, 134, 136, 155, 185; armament, 13, 15, 19, 25, 170, 185.
Landings, Allied in North Africa, 1942, 252.
Landport (Gibraltar), 18, 66, 71.
Lands End, 226.
Landward Defence, see Defence Landwards.
Langleys, East and West (Eastbourns), 96.
Larne, 230.
Lascars, Asiatic Gun, 135, 159, 160.
Layang (Malaya), 270.
League of Nations, 210, 213.
Leake, Admiral Sir John, 18.
Leatham, Robert, Master-Gunner, 137.
Leeward Islands, 60.
Lefroy, Colonel J. H., R.A., 177.
Leipzig (German warship), 192.
Leith, 55, 77, 150, 207; Volunteers, 104.
Leith Fort, construction of, 79 manning of, 100; general, 133, 157, 164; garrison, 79, 82, 101, 134, 155; armament, 79, 98, 170.
Leogane (San Domingo), 110.
Leonardo Battery (Malta), 246, 250.
Leon, Isla de (Cadiz), 128.
Legge, Colonel, Duke of York's Maritime Regiment, 14, 15.
Leghorn, 120, 128, 254.

Legion, Free or Black, 87-90.
Lemoine, Lieutenant J., R.A., 83.
Levant, 92.
Lewes, 166.
Lewis, Captain G., R.A., 65, 68, 71, 73, 74.
Lexington, 48.
Libya, 247, 250, 251, 275.
Lieutenant-General of Ordnance, see Ordnance, Lieutenant-General.
Lieutenant-Governor of Fort, 12, 80.
Lighthouse Battery (Hartlepools), 197, 198.
Lighthouse Battery (Ancona), 254.
Lights, Defence Electric, see Defence Electric Lights (Searchlights).
Lillie, John, Master-Gunner, 137.
Limerick, 44; Militia, 162.
Lincoln (U.S. general), 61.
Lincolnshire, 49, 91.
Lindemann (German battery), 233, 239, 240.
Lindsay, General Sir David, 52, 53.
Lion, H.M.S., 200.
Lisbon, 123.
Lisle, Major de, Guernsey Militia, 107.
Lissa (Vis), 125.
Littlehampton, 230; Volunteers, 103.
Liverpool, 88, 93, 94, 133, 150, 151, 157, 164, 166, 207; garrison, 78, 134, 187; armament, 98, 187; Volunteers, 104.
Lizard, The, 53, 238.
Llandudno, 242, 276.
Llanelly, 230.
Lloyd, Captain V., R.A., 65.
Local Defence Flotillas (Naval), 195-7.
Local Companies, R.A. (R.G.A.), 160, 161.
Loch Ewe, 230.
Lockhart, William, Master-Gunner, 136.
Lombardy, 128.
London, 9, 30, 76, 93.
Londonderry, 229; Militia, 162.
Looe, coast-battery at, 51; armament, 97, 230; general, 166, Volunteers, 166.
Lords-Lieutenant, 30.
Lord Warren (of the Cinque Ports), 77.
Lossiemouth, 230.
Lothian, Mid-, Volunteers, 103, 164.
Lothian, Thomas, Master Gunner, 137.

Lough Foyle, 137.
Lough Swilly (defences of), 107, 137, 155, 188, 207, 208.
Louis XIV (King of France), 3, 18.
Louis XV (King of France), 37.
Louis XVI (King of France), 75.
Louisburg, 27-9; British garrison of, 29, 36, 39.
Louis, Fort (West Africa), 36.
Louis, Port (Mauritius), 128, 191, 209, 223.
Lowe, General Hudson, 124.
Lowenstein's Fusiliers, 119.
Lowestoft, battle of 13; coast-batteries at, 50; garrison, 78; armament, 95, 229; Volunteers, 103.
Loyal Regiment, 266.
Luftwaffe, see Airforce, German.
Lumps Fort (Portsmouth defences), 78, 97, 136.
Lundy Island, 88.
Lutong (Sarawak), 256, 257, 274.
Luxemburg Volunteers (French), 54.
Lydd Volunteers, 104.
Lydden Spout (Dover), 235.
Lyemun (Hongkong), 262.
Lyme Regis, coast-battery at, 51, 232; Volunteers, 103.
Lytham, 232.
Mablethorpe, 231.
Mace, Major, R.A., 34.
Maclean, Brigadier Fitzroy, 125.
Macleod, Captain, R.A., 43.
Macleod, Brigadier T., R.A., 259.
Madalena Battery (Malta), 246, 250.
Madeira, 123.
Madras, 29, 37, 149; bombarded by Emden, 193; garrison, 156, 189; armament, 189, 256; Volunteers, 189, 193.
Madras Army (East India Company), 149.
Mahon (U.S. admiral), 1.
Mahon, George, Master-Gunner, 137.
Mahon Battery (Spanish at Gibraltar), 71.
Mahon, Port (Minorca), 17, 18, 33, 34, 63, 64, 70.
Maida (Battle of), 123.
Maintenance (of guns, stores, ets.), 21-5, 30, 76, 77, 135, 138, 157, 205, 206, 240, 275, 276.
Maintenance Batteries, 276, 277.
Maitland, Colonel, 61.
Majestueux (French warship), 115.

Maker Heights (Plymouth defences), 97; Volunteers, 103.
Malaya, Japanese invasion of, 257, 264-74; Infantry Brigades, 271.
Malay Regiment, 266.
Malay States, Federated, Volunteers, 266.
Maldive Islands, 256.
Maleme (Crete), 243, 244.
Malta, seized by French, 122; occupied by British, 122; in First World War, 195; in Second World War, 243, 246-50; garrison, 122, 134, 135, 156, 160, 189, 209, 222, 246, 276; armament, 170, 190, 222, 246; general, 126, 131, 149, 176, 203, 217, 235, 246. Artillery, see Artillery, Royal Malta. Regiment, 124, Fencibles, 135, 159, 160.
Maltese Light Infantry, 122.
Manchester, 133.
Manchester Regiment, 266.
Manchu Dynasty (of China), 264.
Manfredonia, 252.
Manning Tables, 166.
Manoeldone Battery (Malta), 246.
Manual of Organization and Fighting of a Coast Fortress, 1, 5.
Marazion, 166.
Marchant, Fort Le (Guernsey), 99.
Marching Companies, R.A., see Companies Royal Artillery.
Marcouf Islands, 120-2.
Margate, coast-battery at, 51, 231; armament 96, 231; Volunteers, 104.
Maria-Theresa (Austrian Empress), 27.
Marine, French Colonial (Infantry), 28, 29, 37, 38.
Marines, Royal, 17; at Minorca, 34; at St. Kitts, 59; at Savannah, 60; at Gibraltar, 73; at Marcouf Islands, 120-2; at Capri, 123; at Anholt, 125; at Balaclava, 142; at Scapa and Cromarty, 195, 196; and Emergency Coast Batteries, 227; at Dover, 233, 239, 240; in Crete, 243-6.
Maritime Regiment, Duke of York's 14, 15.
Martello Towers, 99, 100, 105, 107.
Martin, Admiral, 30.
Martin, Captain P., R.A., 65.
Martinique, 36, 39, 56, 59, 79, 108, 112, 113, 115, 118.
Marshall, Major-General, 154.
Mary, Queen, 6.
Maryport, 232; Volunteers, 104.
Mascarene, Lieut-Colonel (Governor of Annapolis), 28, 29.
Master-General of Ordnance, see Ordnance, Master General.
Master Gunner of England, 7.
Master-Gunners, 7; at time of Armada, 8; at time of First Dutch War, 11; after War of Spanish Succession, 20; warrant of appointment, 23; in garrisons, 8, 24, 25, 30, 40, 44, 77, 136, 137, 205; in sea-batteries, 32, 33, 101; in Guards and Garrisons, 41; and Invalid Companies, 44; in Ireland, 45; at Dartmouth, 50, 53; at the Royal Citadel, Plymouth, 53; status and responsibilities of, 77, 135, 138, 159; list of in 1824, 136, 137; duties, history, and importance of 138; part of Coast Brigade, 157; present at review by Edward VII at Portsmouth, 158; at Hartlepools, 197; in reorganization, 1927, 205, 206; after Second World War, 275.
Matheson, Alexander, Master-Gunner, 136.
Matilda, Fort (Guadeloupe), 113.
Mauritius, 128, 131, 148, 149, 256; garrison, 134, 156, 160, 191, 208, 209, 223, 257; armament, 170, 191, 223, 256; Territorial Coast Regiment, 257.
Mayo (County, Ireland), 106.
McCormick, Mr. 53.
McKenzie, William, Master-Gunner, 137.
Mediterranean Fleet (British), 18, 26, 27, 33, 34, 82, 120, 243.
Mediterranean Sea, command of, 3, 26, 32, 33, 63, 278; Byng's operations in, 33, 34; evacuation of, 48, 63; British bases in, 122; Great Britain's ppsition in 1937, 213; war extended to, 242; Malta's position in, 246, 247; coast-defences around, 1941, 250; Axis Powers activities in, 251; operations end in 254; effects of loss of command of, 278; general, 74, 99, 120, 122, 125, 255.

Index 309

Medium Battery (and Batteries), see Artillery, Medium and Batteries, Medium.
Medows, General, 58.
Medway Battery (Sheerness), 95.
Medway, River Dutch attack on, 13, 14; Commission on defences of 1859, 177-80; defences of, 185, 207, 220, 229; general, 170; Territorials, 207, 220.
Melville Battery (Dominica), 115, 117.
Melville, General, 47.
Menage (French general), 106.
Merchant Marine, volunteers from, 102; additional gunners from, 117.
Mersa Matruh, 250.
Mersey Island (Essex), 231; Volunteers, 103.
Mersey, River, defence of, 94, 187, 207, 221, 229.
Mersing (Malaya), 264.
Messina, 123, 124, 252.
Methods, Coast Artillery, 3, 4, 7, 12, 20, 47, 132, 141, 152; with red-hot shot, 71; with R.B.L. guns, 173, 174; in 1914, 181-4; from 1919 to 1939, 215-7; linked with naval, 277.
Mevagissy, 51, 97.
Micmac Indians, 28.
Middle East, 201.
Middlesex Regiment, 258, 259, 262.
Mid-Lothian Volunteers, 103, 164.
Mid-Ulster Militia, 162.
Millazzo (Sicily), 123, 124.
Milford, Lord, 89, 90.
Milford Haven, 6, 88, 94, 155, 166, 170, 179, 187, 207, 221, 229, 242; Volunteers, 103.
Military Districts, 49, 52, 91, 94-8, 103, 158.
Military Repository, Royal, 139, 176.
Militia, as additional gunners, 51, 81, 100, 104, 117, 119, 159; connection with first volunteers, 51; at Plymouth, 52; strength in 1797, 91; in Ireland, 104; revitalised, 1854, 161; artillery units, 161-3. Units and Corps:— Antrim, 162, 163, 188, 207; Argyll & Bute, 162; Bermuda, 161, 190, 224; Berwick & Haddington, 162; Canadian, 130; Cardiganshire, 88, 90, 162; Carmarthenshire, 162; Channel Islands (Jersey, Guernsey, & Alderney), 54, 107, 163, 188, 207; Clare, 162; Cork, 162, 163, 188, 207; Cornwall & Devon Miners, 162; Devon, 162; Donegal, 162; Dublin, 162; Durham, 162; Edinburgh, 162; Fife, 162; Forfar & Kincardine, 162; Galway, 162, 163; Hampshire, 162; Jamaica, 119, 161, 190, 224; Lancashire, 162; Limerick, 162; Londonderry, 162; Mid-Ulster, 162; New England, 28, 29; Norfolk, 162; Northumberland, 162; Pembrokeshire, 101, 162; Sligo, 162; Suffolk, 162; Sussex, 162; Tipperary, 162; Waterford, 162; West Cork, 162; West Indies, 56, 58, 117, 119, 161; Wicklow, 162; Wight, Isle of, 162; Yorkshire, 162.
Miller, George, Bandmaster, R.G.A., 158.
Mill Point (Kent), 231.
Milton (Kent), 9, 102.
Minehead, 232.
Miners, as additional gunners, 53, 159; Devon and Cornwall Militia, 162.
Minesweepers, 266.
Minion (gun), 7.
Minister (Sheerness), 95.
Minorca, first garrison sent to, 16; first attack on and capture of, 33-6; second attack on and capture of, 63-5, 278; retained by Spain, 74; retaken by Britain, 122, 123; returned finally to Spain, 123; garrison, 18, 24, 34, 36, 41, 63, 64, 123; armament, 18, 20, 34, 63; general, 3, 39, 63, 65, 70, 99, 126, 132, 278.
Minsinere, 231.
Missiessy (French admiral), 115, 117, 118.
Missiles, Ballistic and Guided, 275, 277.
Mitchell, Colonel J. W., R.A., 175.
Moat Bulwark (Dover), 96.
Mobile (West Florida), 61, 62.
Moffat, James, Master-Gunner, 137.
Mole St. Nicholas (San Domingo), 108, 110, 112.
Moltke (German warship), 196-9.
Moncrieff Battery (Coast of Kent), 96.

Monitors, 195.
Monkton, Fort (Gosport), 97, 136.
Monmouth, H.M.S., 14, 193.
Mont Arrive (Guernsey), 99.
Mont Crevet (Guernsey), 99.
Mont Cuette (Guernsey), 99.
Monte di Procida (Naples), 253.
Montgomery, Major-General, 158.
Montrose, 164, 166, 230.
Moreno, Buenoventura (Spanish admiral), 72.
Morley, Lord, 150.
Mortars, at Gibraltar, 17, 26, 66, 71, 73; at Minorca, 34, 35, 64, 65; at St. John's, 37; fired on Woolwich Common, 139; used against coast-defences, 141.
Mortella Bay (Corsica), 100.
Motor Boat Carriers, 248, 249.
Motor Explosive Boats, 248, 249, 254.
Motor Launches, 248, 249, 254.
Motor Torpedo Boats (and Gunboats), 215, 216, 226, 236-9, 248-50, 254, 258.
Mouat, Captain, R.N., 37, 38.
Moulin, La Huette (Guernsey), 98.
Mountain Artillery (and Batteries), see Artillery, Mountain.
Mount Davis (Hongkong), 259, 261, 262.
Mount Edgcumbe (Plymouth defences), 179.
Mountings, 47, 141, 215, 216, 218.
Mount Orgeuil (Jersey), 98.
Mounts Bay (Cornwall), 97; Volunteers, 104.
Mount Wise (Plymouth), 52, 53.
Mousquet (French warship), 194.
Moyse Corbet (Lieut-Governor of Jersey), 54.
Muar (Malay), 269.
Mulbousquet, Fort (Toulon), 83.
Mulgrave, Fort (Toulon), 83.
Mulgrave, General Lord, 82, 83.
Muller, von (General naval captain), 193, 194.
Munich Crisis, 1938, 219.
Municipal Coast Batteries, 51, 55, 77.
Murat, Joachim (King of Naples), 124.
Murray, General James (Governor of Minorca), 63, 64, 65.
Murray, John, Master-Gunner, 137.
Musashi (Japanese warship), 255.
Mussalmans, Punjabi, 160, 212.
Musselborough, 94; Volunteers, 104.

Mussolini, Benito, 213.
Mutiny, Indian, 149.
Myers, General, 118.
Naples, 123, 124, 253.
Napoleon Bonaparte (Emperor), at Toulon, 83; plans to invade Britain, 91-4, 101, 107, 114, 115, 118, 226, 227; goes to Egypt, 93, 122; and South Italy, 123; and Adriatic, 125; invades Spain and Portugal, 126; empire collapses, 128; at Elba, 131; defeated at Waterloo, 131; at St. Helena, 131, 135, 159, 163; general, 3, 44, 124, 233.
Napoleon, Louis (Emperor), 163.
Narrow Port (Guernsey), 98.
Narrow Seas, 8, 92, 93, 95, 213, 225, 255.
National Artillery Association, 165.
Nations, League of, 210, 213.
Navy, American (U.S.A. fleet), 55, 129-31, 202, 274.
Navy, Dutch, 13-5, 86.
Navy, French, 17, 18, 30, 33-5, 39, 49, 52, 54, 55, 60, 61, 63, 66, 71, 72, 74, 75, 82, 86, 87-9, 106, 107, 112, 115, 118, 124, 140, 194, 220.
Navy, German, 181, 192-5, 196-200, 202, 225, 253, 254.
Navy, Italian, 202, 243, 245, 253.
Navy, Japanese, 202, 203, 255, 259, 261, 268, 274.
Navy, Royal, role of, 1; expenditure on, 2; and Armada, 9; in Civil War, 10, 11; under Stuarts, 13; in Second Dutch War, 13, 14; in War of Grand Alliance, 16; in '45 Rebellion, 30; in Seven Years War, 32-40; at Minorca, 34, 36, 63; situation in 1779, 48; in West Indies 1778-82, 55-60; in North American waters, 60, 61; in Home waters, 63; at Third Siege of Gibraltar, 66, 68, 69; at Toulon, 85; raises Sea Fencibles, 101; and invasions of Ireland, 86, 87, 106, 107; and blockade of France, 115, 120; and Napoleon's invasion plans, 115, 118; at Marcouf Islands, 120-2; at Malta, 122; at Madeira, 123; at Capri and Scylla, 123, 124; captures Anholt, 125; at Tarifa, 126; on Great Lakes, 129-31; in the Crimean War, 142; from

1856 to 1914, 167, 169-71; in defence of coast-forts, 182, 184; at battles of Coronel and Falkland Islands, 193; operations against German cruisers, 193-5; in North Sea, 1914-5, 195-200; and its bases at Scapa and Cromarty, 195, 196; at raid on Hartlepools, 198, 199; at battle of Dogger Bank, 199, 200; need in Far East after First World War, 203; its difficulties in the Channel, 1940, 226; and Emergency Coast Batteries, 227, 228; in engagement with Scharnhorst and Gneisenau, 235, 236; in co-operation with Dover coast-defences, 237, 238; in Mediterranean, 1940, 243; at Crete, 244, 245; on coast of Italy, 254; at Hongkong, 258, 263; in defence of Malaya and Singapore, 266, 268; in the East based on Addu Atoll, 274; replaces Coast Artillery, 277; links with Coast Artillery, 278.
Navy, Russian, 140, 142, 202.
Navy, Spanish, 8-10, 26, 49, 53, 61-3, 66, 71-4, 86, 115, 118.
Navy, Turkish, 140.
Nazi Party, 219.
Neapolitans (troops), 82.
Needles, The, 178.
Negroes, in San Domingo, 108-12; revolt, in West Indies, 113; in West India Regiment, 116, 117; in Local Companies, R.A., 161; in East African coast-defences, 212.
Nelson, Admiral Lord, 86, 118, 120.
Nettleston Point (Portsmouth defences), 178.
New Brunswick, 80, 155, 156.
Newcastle, 150.
New England, 27-9, 48.
Newfoundland, first garrison sent to 16, 17; attacked by French, 37-9; Invalid Company at, 42; Coast Artillery withdrawn from 156; garrison, 24, 26, 27, 29, 36, 37, 39, 41, 80, 155; Fencibles, 130.
New Guinea, 257.
Newhaven, sea-battery at, 32, 33; state of in, 1779, 49; garrison, 40, 43, 78, 136, 185, 207, 220; armament, 97, 185, 220, 229;

Volunteers, 52, 103.
New Mole (Gibraltar), 66.
New Orleans, 61.
Newport, 229.
Newquay (Cornwall), 232.
New South Wales, 156.
New Tavern (Gravesend), 50, 96.
New Work (Dover), 50.
New York, 38, 41, 48, 60.
New Zealand, 255.
Niagra, Fort (Lake Ontario), 129, 130, 131.
Niagra, River, 130, 131.
Nicobar Islands, 193.
Nieuport, 9.
Nisida Battery (Naples), 253.
No Man's Land Fort (Spithead), 178.
Nore Mutiny, 102.
Norfolk, 49, 91; Militia, 162.
Normandy, 31, 49, 52, 239.
Normans Bay (Sussex), 232.
North Africa, 247, 250, 251, 252, 275.
North America, see America, North.
North Batteries (Liverpool), 98.
North Cape, 86, 226.
North Eastern Coast Defences, 186.
Northern Brigade, R.A., 150, 162.
Northern Coast Defences, 186.
Northern District, 91, 95, 103.
Northern Ireland (Ulster), 205, 213, 219.
North Foreland (battle of), 13.
North Irish, Brigade, R.A., 150, 162; Coast Defences, 163, 188, 208, 213.
North Keeling Island, 194.
North Scottish Territorials, 166, 187, 206.
North Sea, 3, 30, 55, 86, 192, 195-200, 204, 225, 235, 236.
North Shields, 165.
Northumberland, 91; Militia, 162; Volunteers, 164.
North Western Coast Defences, 187, 207.
North Western District, 91, 98.
North (or Great) Yarmouth, see Yarmouth, North or Great.
Nothe, The (Portland defences), 179.
Nottinghamshire Yeomanry, 251.
Noumea, 257.
Nourmont (Jersey), 98.
Nova Scotia, first garrison sent to, 16, 17; French expedition

312 History of Coast Artillery

against, 27-9; garrison, 24, 26-9, 36; handed over to Canadian government, 157.
Numbers 1 and 2, Forts (Coast of Kent), 96.
Nurnburg (German warship), 192.
Obeid, El (Sudan), 251.
Observation Posts, Air, 239; Battery, 239; Fortress, 215, 235.
Officers Cadet Training Unit (Coast Artillery), 242.
Officers Emergency Reserve, 228.
O'Hara, General, 83.
Okeham Ness (Sheerness), 180.
Okehampton, 175.
Old Mole (Gibraltar), 66, 68, 70, 72.
Oneida (U.S.S.), 129, 130.
Ontario, Lake, 129-31.
Operation Sea Lion (German Invasion Plan), 226, 233.
Orange Bastion (Gibraltar), 68, 71, 72.
Ord, Captain, R.A., 24, 29.
Ordnance, Board of, engineers' reports to, 16, 18; reduction of armaments by, 19; rules of, for care and preservation, 1716, 21-3; orders inspection of coast-defences, 33; and Minorca, 36; forms Invalid Companies, 41; in Ireland, 44; action of 1779, 49-52; sends reinforcements to West Indies, 58; constitution and policy of, 76; reports to 1790, 76, 77; sends Invalids to Bermuda, 82; responsibility for defence works, 93, 94; and Martello Towers, 99, 100; and provision of gun-detachments, 100; and Volunteers, 102; and Marcouf Islands, 120, 121; and Heligoland garrison, 125; authorises raising of Asiatic Gun Lascars, 135; and Master-Gunners, 138; abolished by Government, 142; general, 7, 12, 23, 24, 32, 77, 81, 88, 89, 92, 175; Civil Members of, 76, 142.
Ordnance, Clerk of the, 76.
Ordnance College (Woolwich), 175.
Ordnance Department, 227.
Ordnance, Lieut-General of, 23, 76, 137, 142.
Ordnance, Master-General of, 7, 12, 23, 24, 44, 49, 74, 76, 93, 99, 137, 142.
Orkney Islands, 98, 187, 188, 195,

227, 229, 232; Volunteers, 103, 164; Territorials, 166, 187, 195, 206, 221.
Orlando Battery (Naples), 253.
Orvillers, d' (French admiral), 53.
Orwell, River, 93.
Ottawa, River, 129.
Outer Hebrides, 227, 232.
Ouverture, Toussaint L' (Negro leader in San Domingo), 108, 110.
Owen, General Sir John, 172.
Pacific Ocean, 3, 192, 202, 255, 257, 274.
Padstow, 230.
Paignton, 242.
Pakshawan (Hongkong), 259, 261, 262.
Palliser, Major Sir William, 169.
Panther, H.M.S., 66.
Paoli (Corsican leader), 120
Par, 232.
Paris, 104, 108, 163.
Paris, Peace of, (1763), 37, 39.
Parkefield Battery (coast-battery), 50.
Parkin, Captain J. B., R.A., 142, 146.
Parliament, and the Stuarts, 10; Navy sides with in Civil War, 10; and Cromwell, 11; authorises raising of Volunteers, 49, 91, 101; and Crimean War, 142; and shelling of Dover, 238; abolishes Coast Artillery, 276, 277.
Parma, Alexander Farnese, Duke of, 9.
Parry, Captain, R.A., 60.
Pasirlaba (Singapore), 267, 270, 272.
Passage (Waterford), 44, 46.
Passage Point Battery (Plymouth), 53.
Passive Air Defence, 225.
Pastora (Spanish floating-battery), 72, 73.
Patani (Malaya), 268.
Patrol, H.M.S., 198, 199.
Patrol Forces, Naval, 195-9, 266.
Patterson, John, Master-Gunner, 137.
Pattison, Captain, R.A., 24, 39.
Paula, Primera and Segunda (Spanish floating-batteries), 72.
Pauls Cliff (Humber defences), 95, 231.
Pay, extra for Coast Artillery, 153.
Peace, of Amiens (1802), 86, 92, 93, 102, 114, 119, 123, 128; of Paris

Index 313

(1763), 37, 39; of Utrecht (1713), 16, 19.
Pearl Harbour (Hawai), 255, 257.
Pearson, Major Francis, 95th Foot, 54, 55.
Pegasus, H.M.S., 195, 277.
Pegwell Bay, 96.
Pembroke (and Dock), 88, 150, 155, 170 177-80, 187, 207.
Pembrokeshire, 88; Fencibles, 88, 89; Militia, 162; Yeomanry, 88, 89; Volunteers, 164; Territorials, 166, 187, 206, 207, 221.
Penang, Emden's attack on 193, 194, 277; arming of, 211; Japanese capture of, 269; garrison, 212, 224, 266; armament, 224, 256, 267; general, 255.
Penarth, 166.
Pendennis Castle (Falmouth), 3, 6, 11, 16; garrison, 8, 25, 30, 40, 43, 78, 102, 136; armament, 19, 25, 97; Volunteers, 102.
Pengerang (Singapore), 267, 270.
Peninsular War, 125-8.
Pensacola (Florida), garrison, 36, 41, 61; Spanish attack on, 62, 132.
Penzance, 230; Volunteers, 103.
Pepperell, William, 29.
Perak, River (Malaya), 269.
Percival, Lieut-General A. E., 266.
Perelle Bay (Guernsey), 99.
Perrin, Fort (Guernsey), 99.
Perry, Lieutenant (U.S.N.), 131.
Pessery Battery (Guernsey), 99.
Peterhead, 98, 230; Volunteers, 104.
Petit Bourg (Gaudeloupe), 112.
Pett Level (Sussex), 232.
Pevensey Bay, 32, 93, 232.
Philip II (King of Spain), 3.
Phillipine Islands, 257.
Philips, Admiral Sir Tom, 266, 268.
Picardy, 31.
Pickerie Battery (Guernsey), 99.
Picklecombe (Plymouth defences), 179.
Picton, General Sir Thomas, 114.
Piedmont, 128; Piedmontese troops, 82.
Pigeon House Fort (Dublin Bay), 105, 106, 135.
Pigott, General, 60.
Pinnock, M., Master-Gunner, 137.
Piombino, 254.
Pistcataqua (New Hampshire), 48.
Pitt, Fort (Chatham), 95.
Pitt, William (Lord Chatham), 36.

Placentia (Newfoundland), 17, 18, 37, 38; garrison, 29, 36; armament, 26.
Platforms, gun, 6, 12, 20, 141.
Plein Mont (Guernsey), 99.
Plotting Room, Fortress (and Plotter), 215, 234, 237.
Plumstead Marshes, 175.
Plymouth, coast-fort at 7; at time of Armada, 9; in Civil War, 11; Invalid Company arrives at, 42; additional-gunners for, 51; volunteers at, 52; defence of in 1779, 52, 53; admiral commanding at 53; manning of guns at, 100; Commission on defences of 1859, 177-80; Coast Artillery Training Centre at, 228, 242; Coast Artillery School moves to, 276; garrison, 8, 11, 24, 25, 30, 39-43, 53, 77, 78, 81, 82, 100, 101, 134, 137, 155, 187, 206, 220, 276; armament, 11, 19, 25, 52, 53, 97, 170, 187, 220, 229; general 13, 148, 149, 150, 151, 157, 165, 176, 179, 186, 206, 217; Volunteers, 104.
Pocket Battleships, 225.
Pohl, von (German admiral), 200.
Pointe-a-Pitre (Guadeloupe), 112.
Poland, 219.
Pooh (14-inch gun, Dover), 233.
Poole, 166, 230.
Pope, The, 5.
Popton Point (Pembroke Dock defences), 179.
Port Arthur (North China), 171, 257, 258, 259, 264.
Port Au Prince (San Domingo), 110, 112.
Port Castries (St. Lucia), 56, 58.
Portishead Point, 98, 232.
Portland, coast-fort at, 6; state of 1779; Commission on defences of 1859, 177-80; garrison 8, 25, 40, 43, 78, 137, 155, 206, 220; armament, 19, 25, 97, 186, 220, 229; general, 166.
Portland, Lord, 88.
Port Louis (Mauritius), 128, 191, 209, 223.
Port Mahon (Minorca), 17, 18, 33, 34, 63, 64, 70.
Porto Ferrajo (Elba), 120.
Porto Fino (Italy), 254.
Port o' Prince (Trinidad), 224.
Portreath, 97; Volunteers, 103.

314 History of Coast Artillery

Port Royal (Jamaica), 58, 108, 190, 208, 209, 224.
Port Said, 222, 243, 250.
Ports, Defended, see Defended Ports.
Portsea (Portsmouth), 97.
Portsmouth, early coast-defences of, 6; in Civil War, 11; state of, 16, 32; in Seven Years War, 32; Invalid Company sent to, 42; additional-gunners for, 51; manning of guns of, 100; and Coast Brigade, 157, 158; Edward VII reviews R.G.A. at, 158; Commission on defences of 1859, 177-80; garrison, 8, 11, 24, 25, 30, 39, 40, 42, 43, 51, 77, 78, 81, 100, 101, 134, 137, 155, 186, 206, 219; armament, 7, 11, 19, 25, 97, 170, 186, 219, 229; general, 121, 148, 149, 150, 151, 164, 165, 176, 206, 217, 227.
Port Stanley (Falkland Islands), 193, 224.
Port Sudan, 222, 243.
Port Talbot, 232.
Portugal, 86, 123; Portuguese, 5.
Port War Signal Station, 184, 197, 198.
Posillipo Battery (Naples), 253.
Position Range Finders, 173, 183, 215, 239.
Practice Camps, 176.
Practice Seawards, 176, 217.
Precautionary Period, 184, 219.
Prescott, General, 59.
Prescott, Brigadier, 113.
Press, The, 238.
Preston, 230.
Prevost, General, 61, 115, 117.
Price, Lieutenant, R.N., 121.
Priddy's Hard (Gosport), 97.
Prince Charles Edward (Young Pretender), 30, 31.
Prince Edward's Battery (Chatham), 95.
Prince Henry's Battery (Chatham), 95.
Prince of Wales Battery (Chatham), 95.
Prince of Wales, H.M.S., 266, 268.
Prince Rupert's Battery (Dominica), 115.
Prince William's Redoubt (Jersey), 98.
Principe Carlos (Spanish floating-battery), 72.
Prinz Eugen (German warship), 235, 236.
Privateers, 110.
Privy Council, 100.
Projectiles, 4, 167, 168, 172, 173, 198, 199, 217, 277.
Prome (Burma), 274.
Province Wellesley Volunteers, 266.
Pulau Brani (Singapore), 210, 273.
Pulau Hantu (Singapore), 267, 273.
Pulau Sajahat (Singapore), 267.
Pulau Ubin (Singapore), 267, 270.
Pulford, Air Vice-Marshal C. W., 267.
Punjabi Mussalmans, 160, 212.
Punjab Regiment, 258, 259, 261, 262, 266.
Purcell, Robert, Master-Gunner, 137.
Pyrenees, 85.
Quadrant, Gunner's, 7.
Quebec, 81, 148, 156; garrison, 36, 41, 80, 129, 134, 155.
Queen Anne's Battery (Gibraltar), 66.
Queensborough (Sheerness), 95.
Queensferry (Firth of Forth), 98; Volunteers, 104.
Queen's Redoubt (Fort St. Philip, Minorca), 35.
Queenstown (Canada), 130.
Queenstown (Cork Harbour), 180, 188, 207, 208.
Quiberon Bay (battle of), 39.
Radar, 234, 235-9, 242, 249, 253, 254, 275.
Raids, German naval, 1914-5, 195-200.
Railway Batteries (German batteries), 233, 240.
Rajputana Rifles, 258, 259, 262.
Ramhead (Cork Harbour), 46.
Ramparts, construction of, 6, 7; design of, 13, 20; batteries on, 77, 141, 180.
Ramsgate, 96, 230, 237; Volunteers, 104.
Range-Finders, 173, 183, 215, 216, 227, 234, 239.
Ranger (U.S.S.), 55.
Rangoon, 149 156, 186, 193, 256, 257, 274; Volunteers, 189; Coast Battery, 257, 274.
Rate-Clock, 227.
Rations, at Minorca, 64; at Malta, 248.
Ravelins, 20.
Raw, Brigadier C. W., R.A., 235, 239.
Rebellion, Irish (1898), 106.
Rebellion, The '45, 30, 31.

Index

Red-Hot Shot, 71-3, 110.
Red Noses Battery (Liverpool), 98.
Red Sea, 243.
Ree Point, La (Guernsey), 99.
Reeves, Captain, R.A., 73.
Regent, Fort (Jersey), 98, 107, 136.
Reggio, 123.
Regiments, Anti-Aircraft, R.A., see Anti-Aircraft Regiments, R.A.
Regiments, Coast, R.A., 228, 229, 242, 250, 251, 258, 259, 266, 267, 275, 276. Units:—3rd, 251; 4th, 250; 7th, 266, 267; 8th, 258, 259; 9th, 266, 267; 12th, 258, 259; 15th, 243-6; 19th, 251; 47th, 276; 574th, 253; Mauritius, 257; Training, 242.
Regiment, Defence, R.A., see Defence Regiment, R.A.
Regiments (Brigades) Heavy, R.A., see Brigades (and Regiments,) Heavy, R.A.
Regiments, Infantry, see Infantry Regiments.
Regiments, Searchlight, R.A., see Searchlight Regiment, R.A.
Renfrew Volunteers, 164.
Repository Exercises, 139, 176, 177, 217, 218.
Repository, Royal Military, 139, 176.
Repulse Bay (Hongkong), 266, 268.
Repulse, H.M.S., 266, 268.
Reserve, Officers Emergency, 228.
Resistance (French warship), 88.
Retimo (Crete), 244.
Retiro (Tarifa), 126.
Revenue (Coast-Guards), look-outs, 88; cutters, 88, 90.
Revolution, French, 75, 82, 104, 108.
Revolution, Industrial, 132
Reynier (French general), 123, 124.
Rhine, River 85.
Rhode Island, 48, 60.
Ricasoli Battery (Malta), 246, 249.
Richelieu, Duke of (French general), 33, 34.
Richelieu, River (Canada), 129.
Richmond, Duke of (Master-General of Ordnance), 74.
Richmond, Fort (Guernsey), 99.
Riguad (Negro Leader in San Domingo), 123, 124.
Ringsand (Dublin Bay), 105, 106.
Ringwould Volunteers, 104.
Riviera, 83.
Robertson, James, Master-Gunner, 136.

Robertson, Major, 35th Foot, 123, 124.
Robertson, Samuel, Master-Gunner, 137.
Robson, Lieut-Colonel L., R.G.A. (T.F.), 197-9.
Rocca or Roque, La (Jersey), 55, 98.
Rocco Battery (Malta), 246, 249, 250.
Rochefort, 106, 112, 115, 118.
Rochester, 14, 100, 166.
Roches Tower (Cork Harbour), 46.
Rock, The, see Gibraltar.
Rock Channel (Liverpool), 98.
Rockets, 275, 277.
Rock Gun (Gibraltar), 66.
Rocquain Bay (Guernsey), 99.
Rodney, Admiral Lord, 55, 59, 60, 63, 65, 66, 74.
Rogers, Captain, R.A., 29, 37, 38.
Rolt (Historian), 28, 30, 31.
Rommel (German general), 247.
Romney Volunteers, 103.
Rooke, Admiral Sir George, 17.
Rosario, El (Spanish floating-battery) 72.
Roseau (Dominca), 115-9.
Ross, Captain, 40th Foot, 37, 38.
Ross, Brigadier-General, 69.
Ross, Thomas, Masteer-General, 136.
Round Shot, 3, 4, 6, 12, 20, 21; fired at Minorca, 65; fired at Gibraltar, 68, 74; inferiority to shell, 140.
Round Tower (Gibraltar), 18.
Rouse Battery (Guernsey), 99.
Royal Air Force, see Air Force, Royal.
Royal Alderney Artillery, 163, 188.
Royal Artillery, see Artillery, Royal.
Royal Battery (Fort St. Philip, Minorca), 35.
Royal Canadian Garrison Artillery, 157.
Royal Charles, H.M.S., 14.
Royal Citadel (Plymouth), 13, 62, 97, 276.
Royal Commission, 1859, 177-80.
Royal Engineers, see Engineers, Royal.
Royal Foreign Artillery, 118, 119.
Royal, Fort (Martinique), 115.
Royal Garrison Artillery, see Artillery, Royal Garrison.
Royal Guernsey Artillery, 163, 188.
Royal Horse & Royal Field Artillery, 151, 154, 165, 175, 213.

Royalists (French), 82, 83, 85, 87, 108, 110, 119.
Royal Irish Artillery, see Artillery, Royal Irish.
Royal Jersey Artillery, 163, 188.
Royal Malta Artillery, see Artillery, Royal Malta.
Royal Malta Regiment, Royal Malta Fencibles, see against Malta.
Royal Marines, see Marines, Royal.
Royal Navy, see Navy, Royal.
Royal Oak Inn (Fishguard), 90.
Royal Rifles of Canada, 258, 259, 262.
Royal Scots, 258, 259, 261, 262.
Royal Trinity House Volunteers, 102.
Rufiji, River, 195.
Rules for calculating R.G.A. garrisons, 184, 211.
Rullecourt, Baron (French general), 54, 55.
Russell, James, Master-Gunner, 136.
Russia, 48, 202, 233, 259; Russians, 257, 258, Coast-Batteries, 140; Campaign of 1812, 127, 233; Navy, see Navy, Russian.
Ruyter, de (Dutch admiral), 13, 14.
Ryder Flares, 227.
Rye, coast-fort (Camber Castle) at, 6; sea-battery at, 32, 33; garrison, 40, 43, 78, 137, amament, 96; general, 32, 100; Volunteers, 103.

S24 (German destroyer), 181.
Saba (West Indies), 114.
Sacketts Harbour (Lake Ontario), 129.
St. Alban's Fort (Jersey), 98.
St. Angelo (Malta), 246.
St. Anne's Head (Pembrokeshire), 88.
St. Anthony's Castle (Falmouth), 97.
St. Augustine (Florida), 62.
St. Barbara, Fort (Gibraltar), 66.
St. Bartholomew (West Indies), 114
St. Carlos Battery (Spanish at Gibraltar), 71.
St. Croix (West Indies), 114.
St. David's Head (Pembrokeshire), 88
St. Elmo (Malta), 246, 248, 249.
St. Eustatius (West Indies), 59, 114, 118.
St. George (Bermuda), 209.
St. George, Fort (Madras), 37, 149.
St. George, Fort (Grenada), 113.

St. George, Fort (Lissa), 125.
St. George's Channel, 87.
St. George's Militia, 117.
St. Helena, 128, 131, 134, 135, 156, 163, 252.
St. Helen's Point (Isle of Wight), 179.
St. Helier (Jersey), 54.
St. Ives, 97, 166; Volunteers, 103.
St. John (New Brunswick), 80, 134, 155, 156.
St. John of Jerusalem, Order of, 122.
St. John (West Indies), 118, 119.
St. John's (Newfoundland), French attack on 37-9; garrison, 24, 29, 36, 37, 134, 155; armament, 37; general, 24, 132, 156.
St. Joseph Island (Lake Huron), 129.
St. Kitts (West Indies), 56, 58, 59, 113, 114, 118, 119.
St. Lawrence, River, 27, 80.
St. Louis (Guadeloupe), 112.
St. Lucia, 39, 56, 58, 60, 74, 79, 112, 113, 114, 119, 156, 157, 161.
St. Luke's Militia, 117.
St. Marc (San Domingo), 110, 112.
St. Margaret's Bay, 96, 231.
St. Martin (West Indies), 59, 114, 118.
St. Martin's Battery (Spanish at Gibraltar), 71.
St. Mary's Battery (Chatham), 95.
St. Maws Castle (Falmouth), 6, 78, 97, 137
St. Nicholas (Drake's) Island (Plymouth), 9, 52, 97, 137, 179.
St. Osyth, 100; Volunteers, 104.
St. Philip, Fort (Minorca), 34, 35, 36, 63, 64.
St. Philip, Fort (Gibraltar), 66.
St. Thomas (West Indies), 118.
St. Vincent (West Indies), 39, 58, 79, 80, 108, 113, 114, 118.
Saints (Battle of), 55, 60, 74.
Saints Bay (Guernsey), 98.
Sakai (Japanese general), 259.
Saker (gun), 7, 12, 13.
Salamanca (Battle of), 127, 128.
Salcoats Volunteers, 104.
Salcome, 230.
Sallerie (Guernsey), 99.
Saltwood Heights (Kent), 96.
Salutes, 22, 23, 138.
San Christoval (Spanish floating-battery), 72.
Sanday (Orkney Islands), 166.
Sandgate Castle, coast-fort at, 6; in Civil War, 11; in Seven Years

Index 317

War, 32; garrison, 8, 25, 40, 43, 77, 137; armament, 19, 23, 96.
San Domingo, 108-12, 118.
Sandown, coast-fort at, 6; garrison, 8, 25, 40, 43, 78 137; armament, 19, 25, 178; state of 1779, 50.
Sandwich, 7, 32, 96, 231; Volunteers, 103.
San Fiorenzo (Corsica), 120.
San Juan (Spanish floating-battery), 72.
Santa Anna (Spanish floating- batter
Santa Anna (Spanish floating-battery), 72.
Sarawak, 256, 257, 274.
Sark, 99, 136.
Saumarez, Fort (Guernsey), 99.
Saundersfoot, 166.
Saunderson, John, Master-Gunner, 136.
Savannah, 60, 61.
Savory (French naval officer), 106.
Sayer (Historian), 71.
Scapa Flow, 187, 188, 195, 200, 202, 221.
Scarborough (Castle), German naval raid on, 196, 197, 199; garrison, 11, 25, 39, 40, 43, 78, 137; armament, 11, 19, 25, 95, 230; general, 151, 164; Volunteers, 104.
Scattery Island (Shannon mouth), 105, 106, 137.
Scharnhorst (German warship), 192, 235, 236.
Scheer (German admiral), 196, 199, 200.
School of Electric Lights (Gosport), 228.
School of Gunnery, see Gunnery, School of.
Scilly Isles, garrison 40, 43, 78, 81, 137; armament, 19, 97; general, 49, 91.
"Scotch Up", 139, 176, 177, 217, 218.
Scotland, British fleet off coast of, 30; Invalids, R.A., in, 42, 43; coast-defences of, 77-9, 91-4, 98, 187, 188, 207, 221, 229, 230; general, 47.
Scott, W., Master-Gunner, 136.
Scottish Brigade, R.A., 150, 162; Coast-Defences, 187, 188, 207.
Scottish, North, Territorials, 166, 187, 206.
Scott's Head (Dominica), 115, 117.
Screw, Horinzontal, 47.

Scylla (Calabria), 123, 124.
Sea Batteries, see Batteries, Sea.
Seacombe, 166.
Sea Fenciebles, 166.
Seaford, sea-batteries at, 32, 232; state of 1779, 49; garrison, 40, 43, 78, 137; armament, 96, 232; general, 100; Volunteers, 103.
Seaham Harbour, 230.
Sea Lion, Operation (German invasion plan), 226, 233.
Seamen, R.N., as Additional Gunners, 14, 17, 25, 30, 34, 36, 38, 63, 73, 100, 121, 159.
Searchlights (Defence Electric Lights) 174, 182, 183, 216, 225, 227, 228, 242, 249, 250, 252, 253, 262, 269, 273, 275.
Searchlight Regiment, R.A., 5th Regiment, 266.
Seas, Command of the, see Command of the Seas.
Seaton (Devon), 232.
Seaton (Durham), 95, 97, 165, 230; Volunteers, 103.
Sebag-Montefiore (Historian), 103.
Second World War, see War, Second World.
Secretary of State for War, 99, 143, 175, 177.
Security, Collective, 210, 213.
Seine, River, 121.
Selected Military Officer, 184.
Selsey (Bill), 93, 94; Volunteers, 103.
Senegal, 36, 62, 74.
Serapis, H.M.S., 55.
Serapong (Singapore), 267, 270, 273.
Serpentine (gun), 7.
Service Overseas, Conditions of, 56, 81.
Sevastopol, 140, 142, 264.
Seven Years War, 32-40.
Severn Military District, 91, 94, 98.
Severn, River, defences of, 88, 207.
Seward, Captain, R.A., 73.
Seychelles, 256, 257.
Seydlitz (German warship), 181, 196-200.
Seymour Battery (Jersey), 98.
Shannon, River, 105, 155.
Shapensay (Orkneys), 166.
Sharp, William, Master-Gunner, 136.
Sheerness, Dutch attack on, 3, 13, 14; manning of guns, 100; Royal Commission, 1859, on defences of, 177-80; garrison, 14, 25, 30, 39, 40, 43, 78, 81, 137, 155, 185,

207; armament, 14, 19, 25, 95, 170, 185; general, 140, 150, 157, 166, 207.
Shell (Spherical), 65, 69, 73, 74, 140, 141, 167.
Shell (Cylindrical), see Projectiles.
Shellness, 231.
Sheppey, Isle of, 8, 9.
Sheringham, 231.
Shetland Islands, 98, 137, 230; Volunteers, 103.
Shingmun Redoubt (Hongkong), 261, 263.
Shirley (Governor of New England) 28, 29.
Shoeburyness, founding of School of Gunnery at, 139, 175; name changed to Coast Artillery School, 203; depot battery at, 205; Coast Artillery School between 1919 and 1939, 217; Emergency Battery at, 231; School leaves, 242; general, 155, 166.
Shoreham, 230.
Shoremead, 231.
Shorncliff, 96.
Shot, see Round Shot.
Shotley, 100.
Shrive, Cpatain, R.A., 218.
Siam, 257, 268.
Sicily, 123-8, 247, 248, 250; Sicilian Regiment, 125.
Sidmouth, 232; Volunteers, 103.
Siege Artillery, see Artillery, Siege.
Sierra Leone, 129, 208, 213, 224; garrison, 156, 157, 161, 190, 224; armament, 170, 190, 224, 251.
Sierra Leone Royal Artillery (R.G.A.) 161, 181, 190, 208, 224.
Sights (Gun), 7, 141, 174, 183.
Signal Staff Point (Pembroke defences), 179.
Signal Station, Port War, 184, 197, 198.
Sikhs, 160, 212.
Silingsing (Singapore), 267, 273.
Siloso (Singapore), 267, 273.
Simonstown, 128, 190, 200, 207.
Simple, John, Master-Gunner, 137.
Singapore, government decides to construct base at, 203; development and construction of, 203, 204, 207, 210, 218; additional-gunners for, 212; Japanese attack on and capture of, 3, 63, 85, 147, 257, 264- 273, 275, 278;

garrison, 156, 160, 191, 208, 209, 211, 212, 223, 266, 267; armament, 191, 215, 223, 256, 267; general, 149, 255, 258, 277, 278.
Singara (Malaya), 268.
Sinope (Black Sea), 140.
Six Gun Battery (Tilbury), 96.
Skegness, 231.
Sligo Militia, 162.
Slim, River (Malaya), 269.
Slough Fort (Thames & Medway defences), 185.
Smith, Admiral Sir Sidney, 120, 123.
Somerset, 91.
Somerville, William, Master-Gunner, 137.
Sotomayer, Alvarez de (Spanish general), 66.
Soult, Marshal (French general), 127, 128.
Soumilleuse Battery (Guernsey), 98.
South Africa, Union of, 207, 208.
South African War, 148, 154, 163.
South America, 192.
Southampton, 8, 165, 229.
South Bastion (Gibraltar), 68, 71, 72.
South Carolina, 61, 87.
South Eastern Coast Defences, 185.
Southend, 166.
Southern Brigade, R.A., 150, 162; Coast Defences, 185, 206.
Southern District, 91, 95, 158.
Southern Division, R.A., 151, 162, 163.
South Foreland Battery (Dover defences), 234-8.
South Georgia, 257.
South Gore Battery (Tees defences), 197, 198.
South Hook Point (Pembroke defences), 179.
South Irish Brigade, R.A., 150, 162; Coast Defences, 163, 188, 208, 213.
South Lowestoft, 50.
Southsea Castle, coast-fort, 6; capture of during Civil War, 11; garrison, 25, 40, 137, 158, 277; armament, 19, 25, 97, 178.
South Shields, 95, 230.
South Western Coast Defences, 186, 206.
South Western District, 91, 97.
Southwold, 95, 231; Volunteers, 104.
Spain, at war with England, 5, 6; and Armada campaign, 8-10;

Index

at war with Britain, 17-20, 26, 27-31, 48-74; losses in Seven Years War, 36, 39; enters War of American Independence, 48, 58, 61; Minorca returned to, 65; and Gibraltar, 70; gains in War of American Independence, 74; deserts her allies, 85; allies herself with revolutionary France, 114, 125, 126; and the Peninsular War, 126-8; losses during the French Revolutionary and Napoleonic Wars, 128.
Spanish, as discoverers, 5; in War of American Independence, 3, 61, 62; galleons, 9; at First Siege of Gibraltar, 26, 27; in Mediterranean, 63; attack Minorca, 64, 65; at Third Siege of Gibraltar, 66-75, 132; strength at Gibraltar, 70; at Toulon, 82, 83, 85; West Indies, 114; regain Minorca, 123; at Cadiz and Tarifa, 126-8; Germans reach frontier, 226; assist Italians, 251. Spanish Batteries, see Batteries, Spanish. Spanish Navy, see Navy, Spanish.
Spanish Succession, War of, see War of Spanish Succession.
Specialists, 151, 205, 206.
Spee, Count von (German admiral), 192, 193.
Sphakia (Crete), 245.
Sphinx Battery (Singapore), 210, 267, 270.
Spike Island (Cork Harbour) defences of, 135, 180; garrison, 46, 134; armament, 46.
Spithead, 33, 68, 74, 178; Forts, 158, 177, 178, 186.
Spit Sand Fort (Spithead), 178.
Spragge, Sir Edward (Governor of sheerness), 14.
Springhall Redoubt (Savannah), 61.
Spur Battery (Chatham), 95.
Stack Rock (Pembroke defences), 179.
Staddon Heights (Plymouth), 97, 179, 242.
Stair, Field-Marshal Lord, 30.
Stallingborough, 95, 231.
Standish, Captain, R.A., 56, 58, 78.
Stanley (Hongkong), 211, 259, 261, 262.
Stanley, Port (Falkland Islands), 193, 224.

Stannary Volunteers, 104.
Station-Board, 76.
Steam-Engine, 139, 140, 166.
Steam Warships, 140, 141, 152, 166, 177.
Stephens, Captain E., R.A., 83, 120.
Stepkin, Captain, R.A., 24.
Stewart, John, Master-Gunner, 137.
Stirling, Lieut-General, 154.
Stonecutters Island (Hongkong), 259, 261.
Stonehouse (Plymouth), Bridge Battery, 53; Volunteers, 103.
Storekeepers, 44, 45, 76; Principal Storekeeper, Board of Ordnance, 76, 138.
Stores, responsibility for and care and preservation of, 21-5, 30, 76, 77, 135, 138, 157, 275, 276.
Stornoway, 230.
Story, George, Master-Gunner, 136.
Straits, of Dover, 226, 233-8, 240; of Gibraltar, 63, 66, 74, 126.
Straits Settlements, 211, 212; Volunteers, 266, 271.
Stranrar, 230.
Stratford, 166.
Stromness, 166.
Stroud, 95.
Stuart, General Charles, 123.
Stuart, General Sir John, 123.
Stuarts, 10, 13, 30.
Sturdee, Admiral, 193.
Submarines, 167, 198, 199, 240, 247, 268; German submarine-campaign, 192, 200.
Subordinates, Civilian (Board of Ordnance), 76, 77.
Suda Bay (Crete), 243-5.
Sudan, Port, 222, 243; Detachment, R.A., 222.
Suez, 222, 243, 250.
Suffolk, 49, 91; Trained Bands, 14, 15; Militia, 162; Volunteers, 164; Territorials, 166, 185, 206, 207, 220.
Suffolk, Lord, 14, 15.
Sullivan (U.S. general), 60.
Sumatra, 257.
Sunderland, 101, 164, 166, 229; Volunteers, 103.
Superintendant, Royal Military Repository, 139.
Superior, Lake, 129.
Surinam (Guiana), 79, 114, 119.
Survey, 213.
Surveyor-General, 76.

Sussex, 32, 33, 49, 91; Militia, 162; Volunteers, 52, 164; Territorials, 165, 185, 206, 207, 220.
Sutherland Battery (Coast of Kent), 96.
Sutherland, Robert, Master-Gunner, 137.
Swanage, coast-battery at, 51, 232; garrison, 137; armament, 97, 232; general, 166.
Swansea, 94, 98, 187, 221, 229; Volunteers, 103.
Sweden, 48.
Swilly, Lough (defences of), 107, 137, 155, 188.
Swivels (guns), 105.
Swordfish (aircraft), 236.
Sydney, 149, 156.
Sydney, H.M.A.S., 194.
Syracuse, 123, 124, 252.
Syria, 251, 274.
Takoradi (Gold Coast), 251.
Talla Piedra (Spanish floating-battery), 72, 73.
Tamils, 211, 212.
Tanga, 256.
Tanjong Tereh (Singapore), 267, 273.
Taranto, 252.
Tarbert Island (Shannon mouth), 46, 79, 104, 105, 106, 137.
Tarifa, 126, 127.
Tarn, von der (German warship), 196-8.
Tate, Charles, Master-Gunner, 137.
Tate, William (French commander at Fishguard), 87-90.
Tayler, Silas, 15.
Tay, River, defences of, 187, 188.
Taxbiex Battery (Malta), 246.
Tees, River, defences of, 186, 197, 207, 229.
Teignmouth, 232.
Tekong Besar (Singapore), 210, 267, 270.
Tenaille, 20.
Tenby, 164.
Ternay, d'Arsac (French naval captain), 37-9.
Territorial Force and Army, Force formed, 163, 165; in R.A., 165; in Coast Artillery, 165, 166; conditions of service, 165; in coast-defences, 1914, 181, 185-91; employed in the field, 1915-8, 200; at end of First World War, 201; takes over coast-defence at Home, 204-6; becomes Territorial Army 206; Coast Artillery units in 1920, 206; Coast Artillery units in increased, 231; provides AA. artillery, 214; courses for at Coast Artillery School, 217; ready in 1939, 218; mans coast-defences at Munich Crisis, 219; mobilises September, 1939, 219; in coast-defences at Home on outbreak of war, 1939, 219-221; and Emergency Coast Batteries, 227, 228; reformed after Second World War, 276; takes over coast-defences again at Home, 276; Coast Artillery units of disbanded, 277. Units and Corps:—Antrim, 213, 221; Clyde, 166, 187, 206, 207, 221; Cornwall, 166, 187, 206; Devon, 165, 187, 206; Devon & Cornwall, 206; Dorset, 166, 186, 206, 220; Durham, 166, 186, 197-9, 206, 207, 220; East Riding (Yorkshire), 166, 186, 206, 207, 220; Essex & Suffolk, 166, 185, 206; Forth, 166, 187, 206, 207, 221; Glamorgan, 166, 187, 206, 207, 221; Hampshire, 165, 186, 206, 219; Kent, 166, 185, 206; Kent & Sussex, 207, 220; Lancashire & Cheshire, 166, 187, 206, 207, 221; Mauritius, 257; North Scottish, 166, 187, 206; Nottinghamshire, 152; Orkney, 166, 187, 195, 206, 221; Pembrokeshire, 166, 187, 206, 207, 221; Suffolk, 207, 220; Sussex, 165, 185, 206; Thames & Medway, 207, 220; Tynemouth, 165, 186, 206, 207, 221; Wallasey, 221; Wight, Isle of, 219.
Tesso, Marshal (French general), 18.
Tewson, Jonathan, Master-Gunner, 137.
Texel, River, 13, 55.
Thames, River, defence of, 6, 8, 9; Dutch attack on, 13; defences of, 185, 207, 220, 229; general, 55, 195, 226; Territorials, 207, 220.
Thirty Years War, 10.
Thornhead Battery, 96.
Thorpness, 231.
Tiburon, Fort (San Domingo), 110.
Tigne (Malta), 209, 246, 250.
Tiffin, Captain, Y.A., 78.

Index

Tilbury, coast-fort at, 6; Dutch attack on, 14; garrison, 8, 11, 25, 30, 40, 43, 78, 81, 137, 155; armament, 11, 19, 25, 96; Volunteers, 102, 104.
Tippery Militia, 162.
Tito, Marshal (Jugoslav leader), 125.
Tobago (West Indies), 59, 74, 108, 114.
Tobruk, 247, 251.
Todt Battery (German battery), 233.
Tomlin, Colonel S. C., R.A., 252.
Torbay, 51, 230.
Toriano, Captain, R.A., 43, 78.
Toronto, 130, 134.
Torpedo-Boats, 167, 170, 171, 176, 182, 183, 215, 216, 226, 235, 236, 237, 239, 254, 258.
Torpedo-Bombers, 258.
Torpedoes, 167, 170, 171, 174, 237, 248, 251, 253.
Torquay, 230.
Torre Battery (Naples), 253.
Torres, Count de las (Spanish general), 27.
Totland Point (Isle of Wight), 178.
Toulon, 33, 82, 83, 85, 93, 115, 120, 122, 126, 132.
Touquet, Le, 235.
Tovey, Lieut-Colonel A., R.A., 65, 68, 73.
Tower of London, 50, 138.
Towers, Martello, 99, 100, 105, 107.
Town Guard, 51, 159.
Townshend Battery (Dover), 96.
Trade-Routes, German attacks on, 192-5, 200, 204.
Trafalgar (Battle of), 86, 92, 135.
Trained-Bands, 8; West Kent, 14; Suffolk, 14, 15.
Training Centre, Coast Artillery, 227, 228, 242, 276.
Trains, Artillery, 24, 138.
Traverse, construction of, 20.
Treasury (Government), 153, 159.
Treaty, of Versailles 202, 225; of Washington, 211, 214, 215, 258.
Trengganu (Malaya), 269.
Trieste (Italian warship), 243.
Trincomalee, garrison, 128, 134, 223, 257, 274; armament, 211, 223, 256, 274; general, 131, 210, 211, 255, 274.
Trinidad, 79, 114, 118, 224; Volunteers, 224.
Trinity House Volunteers, Royal, 102.

Tripoli (Libya), 247, 251.
Tristan da Cunha, 135.
Triumph-Days, 22, 23.
Trotta Battery (Leghorn), 254.
Truro, 166.
Tsingtao (North China), 257.
Tubes, Firing, 47, 141 168, 169, 173.
Tudors and Tudor Coast Forts, 6-10.
Tuelle, La, 98.
Turkish Navy, 140.
Twales, James, Master-Gunner, 137.
Twiss Battery (Coast of Kent), 96.
Tynemouth, coast-fort at, 7; Armada and, 9; report on, 33; and Coast Brigade, 157; garrison, 8, 11, 25, 30, 40, 43, 78, 100, 101, 137, 155, 186, 221; armament, 11, 19, 25, 33, 95, 186, 221, 229; general, 164, 207; Volunteers, 104; Territorials, 165, 186, 206, 207, 221.
Tyne, River, defences of, 186, 195, 221, 229.
U-Boats (German submarines), 240.
Ulm, 92.
Ulster Militia, 162.
United Kingdom, see Great Britain.
United States of America, see America, United States of.
Upper Canda, 129.
Upnor Castle (Chatham), coast-fort at, 7; Dutch attack on, 14; garrison, 8, 11, 78, 137; armament, 11, 96.
Ushant, 49.
Utrecht, Peace of, 16, 19.
Vale Castle (Guernsey), 99.
Valetta (Malta), 122, 209, 247, 248, 249.
Van, Captain, R.A., 78.
Vantour (French warship), 88.
Vasto, 252.
Vauban, Marshal (French engineer), 20, 34, 141.
Vaubois (French general), 122.
Vaughan, Colonel Gwynne, 89.
Vazon Bay (Guernsey), 99.
Vengeance (French warship), 88.
Verne Hill (Portland defences), 179.
Vernon, Admiral, 30.
Versailles, Treaty of, 202, 225.
Veteran Regiment, Royal, 125.
Victor, Marshal (French general), 126, 127.
Vieuxport (St. Lucia), 113.
Vigie (St. Lucia), 58.

W

Vigo, 49.
Vilatte (French general), 127.
Vildebeestes (Aircraft), 258, 267.
Villadarias, Marquis of (Spanish general), 17.
Villaret Joyeuse (French admiral), 82.
Villeneuve (French admiral), 115, 118.
Vincent, General, 130, 131.
Virgin Islands (West Indies), 79.
Vis (Lissa), 125.
Vittorio Veneto (Italian warship), 243.
Vizagapatam, 256.
Volunteers, first raising of, 49; of 1779, 51, 52; conditions of service of, 51, 52, 102; in French Revolutionary and Napoleonic Wars, 75, 91, 93, 101-4, 131, 163; from 1859 to 1908, 163-5; formed into Territorial Force, 163, 165. Units and Corps:— Aberdeen, 103, 164; Aldeborough, 103; Appledore, 103; Ayr, 103; Barmouth, 103; Barnstaple, 103; Berwick, 103; Bombay, 189, 223; Bridgewater, 103; Bridlington, 103; Brighton, 103; Bristol, 103; Calcutta, 189; Campbelltown, 103; Caylon, 161, 191, 208, 223, 257, 274; Chatham & Gillingham, 103; Cheshire, 164; Cinque Ports, 51, 52, 103, 164; Coleridge, 103; Cornwall, 52, 164; Crinnis Cliff, 103; Cromarty, 164; Dartmouth, 103; Deal, 103; Devon, 52, 88, 164; Dorset, 164; Dover, 103; Dumbarton, 103, 164; Dunbar, 103; Dundee, 103; Durham, 164; Eastbourne, 103; East Riding (Yorkshire), 164; Edinburgh, 164; Exmouth & Sidmouth, 103; Falmouth, 103; Farringdon, 103; Fife, 164; Fishguard, 103; Folkestone, 103; Forfar, 164; Fowey, 103; Fraserburgh, 104; Glamorgan, 164; Gosport, 104; Gravesend & Tilbury, 104; Greenock, 104; Grimsby, 103; Haddington, 164; Hampshire, 164; Hartlepools, 103; Hastings, 103; Haytor, 103; Heaton, 103; Herne Bay, 103; Hongkong, 224, 258, 259; Hull, 104; Hythe, 103; Irish, 104; Kent, 164; Kincardine, 164; Lanarkshire, 164; Lancashire, 164; Lancaster, 103; Leith, 104; Littlehampton, 103; Liverpool, 104; Looe, 103; Lowestoft, 103; Lydd, 104; Lyme Regis, 103; Madras, 189, 193; Maker, 103; Malay States, 266; Margate, 104; Mersey Island, 103; Mid-Lothian, 103, 164; Milford, 103; Mountsbay, 104; Musselborough, 104, 104; Newhaven, 103; Northumberland, 164; Orkney, 103, 164; Pembrokeshire, 164; Pendennis, 102; Penzance, 103; Peterhead, 104; Plymouth, 104; Portreath, 104; Queensferry, 104; Ramsgate, 104; Rangoon, 189, 257, 274; Renfrew, 164; Ringwould, 104; Romney, 103; Rye, 103; St. Ives, 103; St. Osyth, 104; Salcoats, 104; Sandwich, 103; Scarborough, 104; Seaford, 103; Sheerness, 103; Shetland, 103; South Shields, 104; Southwold, 104; Stannary, 104; Stonehouse, 103; Straits Settlements, 266, 271; Suffolk, 164; Sunderland, 103; Sussex, 52, 164; Swansea, 103; Trinidad, 224; Trinity House, 102; Tynemouth, 104; Wellesley, Province of, 266; Weymouth, 104; Whitehaven, 103; Wight, Isle of, 104; Workington & Maryport, 104; Yarmluth, 104; Yorkshire (East Riding), 164.

Wales, 88, 91, 93.
Wales, Prince of, H.M.S., 266, 268.
Wallesey Territorials, 221.
Waller, General, 11.
Waller, Captain, R.A., 115, 117.
Walmer Castle, coast-fort at, 6; in Civil War, 11; in Seven Years War, 32; garrison, 8, 25, 40, 43, 77, 137; armament, 19, 25, 96.
Walton, Captain, R.A., 63.
Walton-on-the-Naze, 14, 100.
Wanstone Battery (Dover), 234, 236-240.
Warden, Lord (of the Cinque Ports), 77.
Warley, 49.
War Office (and Department), 1, 41, 53, 76, 91, 93, 94, 99, 102, 119, 120, 142, 152, 204, 205, 207, 210, 211, 213, 264.

Index 323

Warren, Commodore, R.N., 29.
Warren, Admiral Sir John, 107.
Warrior, H.M.S., 167.
Wars, American Independence, 42, 44, 45, 48-74, 75, 87, 278; American of 1812, 129-31; Austrian Succession, 27-31; Civil (Great), 10, 11; Crimean, 133, 140, 142, 147, 152, 161, 163, 166, 169, 246; Dutch, 10, 11, 13, 16; First World, 181-201, 202, 217, 218, 227, 246; French Revolutionary and Napoleonic, 44, 45, 47, 75, 82-132, 159, 163, 246; Grand Alliance, 16; Peninsular, 125-8; Second World, 125, 139, 178, 218; at Home, 219-42; in Mediterranean, Middle East, & Italy, 243-54; in Far East, 255-74; ends 275, 278; Seven Years, 32-40; South African, 148, 154, 163; Spanish Succession, 16, 17-20, 26, 27, 30, 45; Thirty Years, 10.
War, Secretary of State for 99, 143, 175, 177.
Warships, wooden, 3, 20; in 1770, 46; in 1790, 75, 76, 132, 138; in 1854, 140; in 1856, 167; armoured, 3, 4, 140, 152, 167, 168; steam, 140, 141, 152, 166, 177; ironclad, 140, 167; armament, 46, 75, 167-70; general, 147, 167, 169, 171-4, 181, 183, 184, 215, 216, 225, 243, 247; individual warships are shown against their names.
Warwick, Lord, 11.
Wash, The, 226, 238.
Washington Treaty, 211, 214, 215, 258.
Waterford, 44, 45, 46, 101, 104, 106; Militia, 162.
Waterloo (Battle of), 131, 133.
Waterport Battery (Gibraltar), 66, 71.
Watkin, Captain H. S., R.A., 173.
Watson, Lieut-Colonel Jonas, R.A., 27.
Watson, Arthur, Master-Gunner, 136.
Webdell, Lieutenant, R.A., 35, 43.
Webster, John, Master-Gunner, 137.
Wedge (elevating), 7, 47.
Weedon, 133.
Wellesley, Province of (Straits Settlements), Volunteers, 266.

Wellington, Duke of, 127, 128.
Wellington, Fort (Lissa), 125.
Welsh, 88; women at Fishguard, 90; Brigade, R.A., 150, 162; ports, defences of, 207.
West Africa, 36, 39, 129, 160.
West Bay (Dorset), 232.
Westbrooke, 96.
West Cork Militia, 162.
Western Approaches, 48, 52.
Western Brigade, R.A., 150, 162.
Western District, 52, 91, 94, 97.
Western Division, R.A., 151, 162, 163.
Western Heights (Dover), 180.
Western King (Plymouth), 52, 97, 179.
Western Redoubt (Bere Island), 105.
West Florida, 61, 62, 74.
West Fort (Stonecutters Island, Hongkong), 261.
West Gun Line (Tilbury), 96.
West Hoe Battery (Plymouth), 53.
West Indies, during War of American Independence, 55-60; Royal Irish Artillery in, 79, 112-4; service and health in, 81; during French Revolutionary and Napoleonic Wars, 108-19; Franco-Spanish fleets in, 115, 118; R. Foreign Artillery in, 119; raising of local Coast Artillery in, 160, 161; Coast Artillery continually in action in, 278; general, 19, 29, 36, 39, 61, 62, 63, 74; West Indian Battalion, R.A., 161; West India Regiment, 115, 117; West Indies Militia, 56, 58, 117, 119, 161.
West Kent, Trained Bands, 14.
West Lunette (Fort St. Philip, Minorca), 35.
Westman, Captain, R.A., 78.
Westmorland, Fort (Cork Harbour), 105, 180.
Wexford, 106.
Whiddy Island (Bantry Bay), 105, 137.
Whitby, 78, 95, 198, 230.
White, Brigadier C. J., R.A., 247.
White Booth, 95.
Whitehaven, attacked by Paul Jones, 55; batteries at, 94, 98, 230; Volunteers, 103.
Whitehead Torpedo, 170.
Whitepoint (Cork Harbour), 180.
Whitesand Bay (Plymouth defences), 179, 242.

Whitmore, Captain, R.A., 43, 78.
Whyte, Charles, Master-Gunner, 136.
Wick, 230.
Wicklow Militia, 162.
Wight, Isle of, 8, 19, 25, 40, 155, 158, 170, 176, 177, 178, 186, 217, 219, 242, 276; Militia, 162; Volunteers, 104; Territorials, 219.
Wilks, Captain, R.A., 83.
Willaumez (French admiral), 118.
William I, King (The Conqueror), 75.
William, Fort (St. John's, Newfoundland), 37, 38.
Williamson, Captain, R.A., 56, 58.
Willington, Lieutenant, R.A., 74.
Willis Battery (Gibraltar), 66.
Winchelsea, 232; Volunteers, 103.
Windward Islands, 58, 60.
Winnie (14-inch gun, Dover), 233.
Winnipeg Grenadiers, 258, 259, 262.
Winter, Captain, R.A., 43.
Winterton, 231.
Wisdom, Ezekiel, Master-Gunner, 137.
Witham, Lieutenant A., R.A., 69.
Woolston, 165.
Woolwich, 24, 29, 30, 31, 41, 42, 77, 78, 124, 133, 135, 139, 141, 147, 148, 149, 150, 151, 157, 175; Arsenal, 141; Common, 139, 174.
Workington, 230; Volunteers, 104.
World War, see against Wars, First and Second World.
Worthing, 232.
Yamashita (Japanese general), 268.

Yarmouth, North (or Great), German naval raid on, 196; general, 93, 150, 151; garrison, 25, 40, 43, 78, 137; armament, 19, 25, 95, 229; Volunteers, 104.
Yarmouth (Isle of Wight), coast-fort at, 6; garrison, 25, 40, 43, 78, 137; armament, 7, 19, 25, 97.
Yaverland (Isle of Wight), 178.
Yellow Fever, 56, 81, 108, 110, 112, 114, 119, 157, 278.
Yeo, Captain James, R.N., 130, 131.
Yeomanry, 88, 89, 102, 251; Nottinghamshire, 251; Pembrokeshire, 88, 89.
Yorck (German warship), 196.
York, James, Duke of, 13.
York, Frederick Duke of, 85.
York Light Infantry, 117.
Yorkshire, German naval raid on coast of, 192, 196-200; Militia, 162; East Riding of, Volunteers, 164; Territorials, 166, 186, 206, 207, 220.
Yorkshire District, 91, 95.
Youghal, 180.
Young, Fort (Dominica), 115, 117.
Young Pretender (Prince Charles Edward), 30, 31.
Yumato (Japanese warship), 255.
"Z" Force (Royal Navy), 266, 268.
Zante (Ionian Islands), 125.
Zanzibar, 195, 256, 277.
Zhemchug (Russian warship), 194, 277.

ERRATA

"However careful we may be the hands of Briareos and the eyes of Argos will not prevent them."

Cotton Mather.

ERRATA

Page 13, line 23.   *For* "1767" *read* "1667".
Page 17, line 24.   *For* "24th July (1707)" *read* "24th July, (1704)".
Page 51, line 5.   *For* "Blackington" *read* "Blatchington".
Page 55, line 21.   *For* "with" *read* "which".
Page 89, line 25 ⎫
Page 90, line 8 ⎬ *For* "Carn Wnda" *read* "Cwm Wnda".
         line 23 ⎭
Page 123, line 42. *For* "Calabrain" *read* "Calabrian".
Page 155, line 2.   *For* "time" *read* "true".
Page 166, line 3.   *For* "Bruntisland" *read* "Burntisland".
Page 181, line 5.   *Insert* "Royal Malta Artillery".
Page 208, line 28. *For* "22nd Heavy Brigade" *read* "2nd Heavy Brigade".
Page 209, line 35. *For* "22nd Heavy Battery" *read* "2nd Heavy Battery".
Page 255, line 18. *Delete* "18 Light Cruisers".
Page 258, line 32. *Delete* "1st Bn Winnipeg Grenadiers".
Page 267, line 30. *For* "Silingsby" *read* "Silingsing".

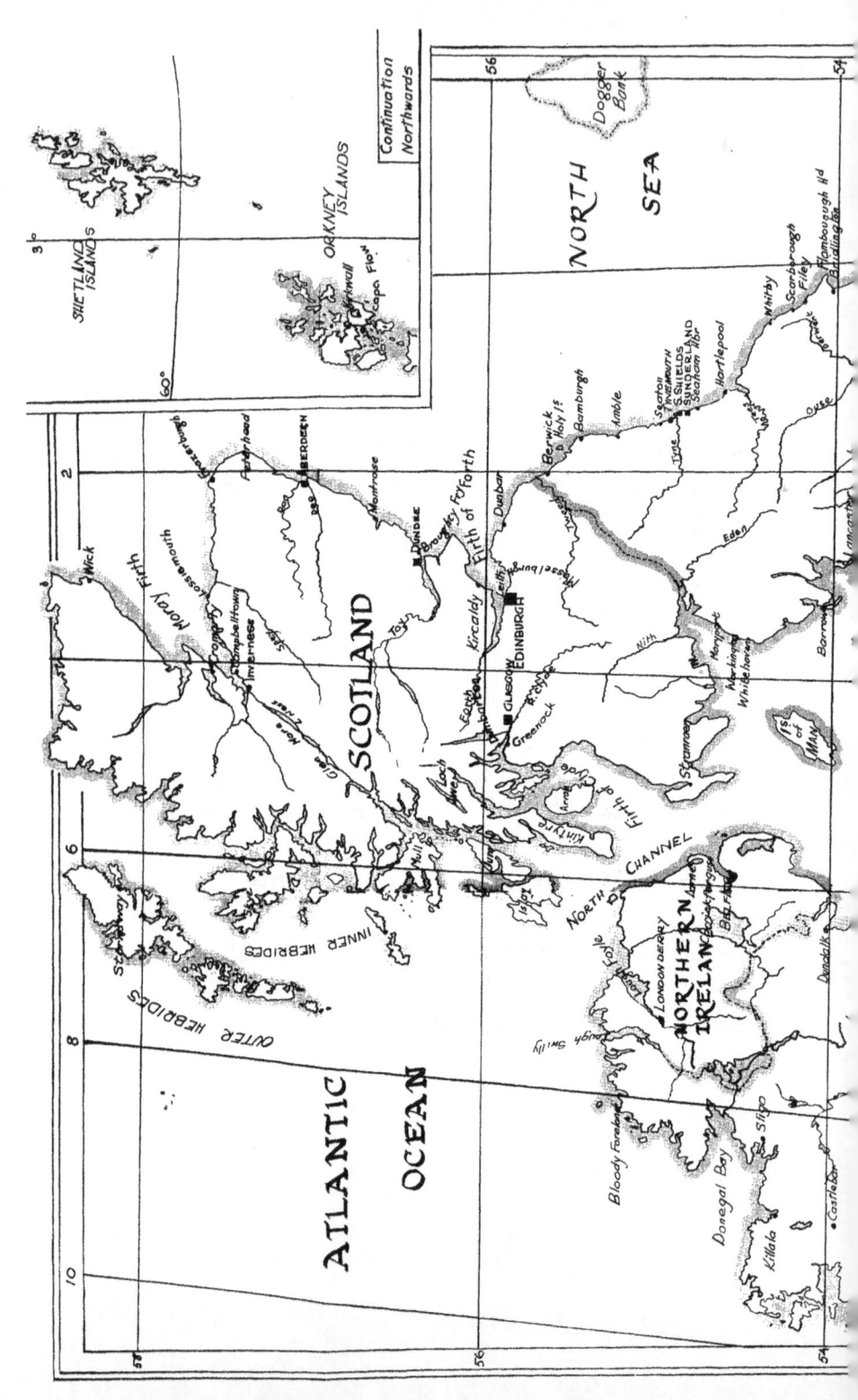

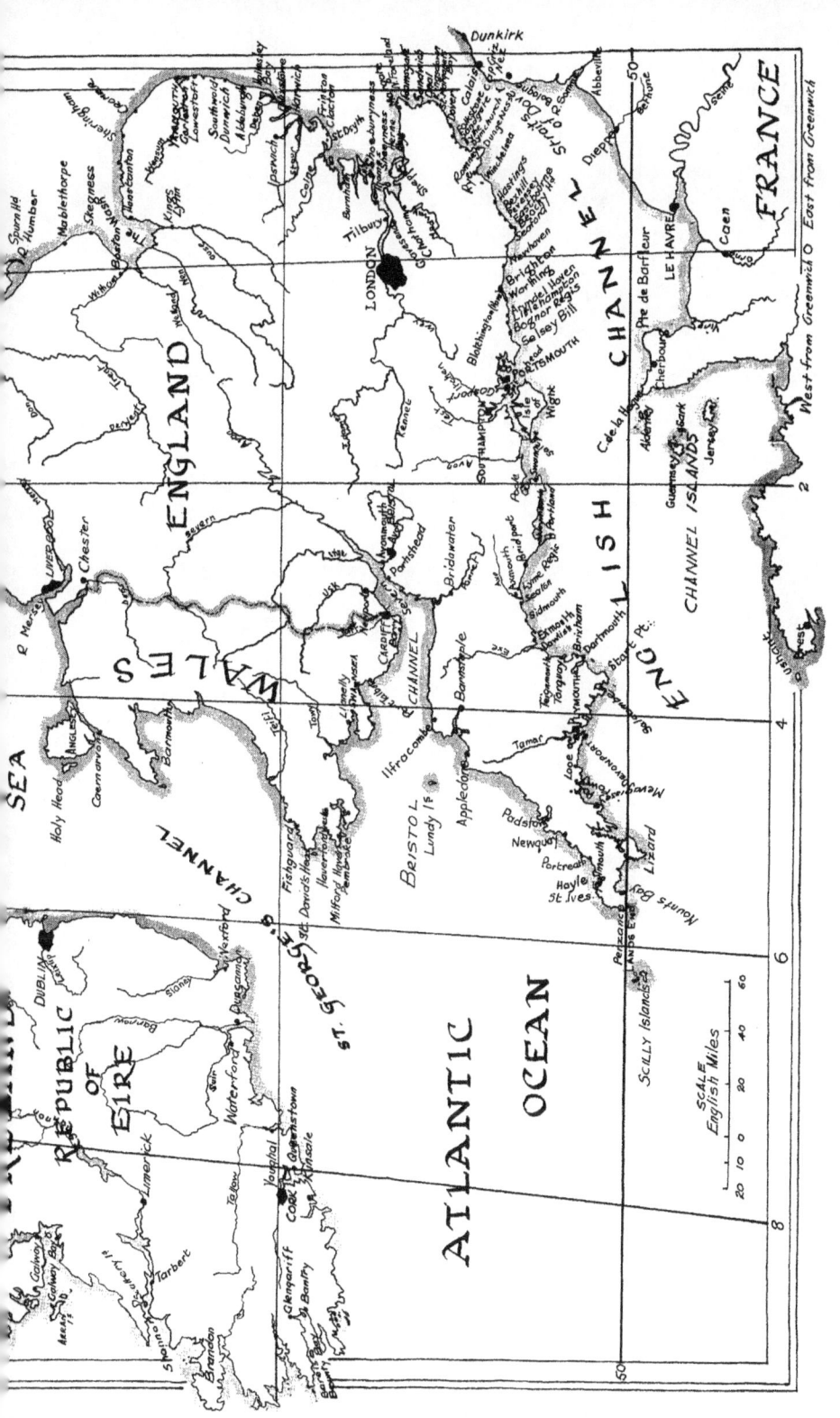

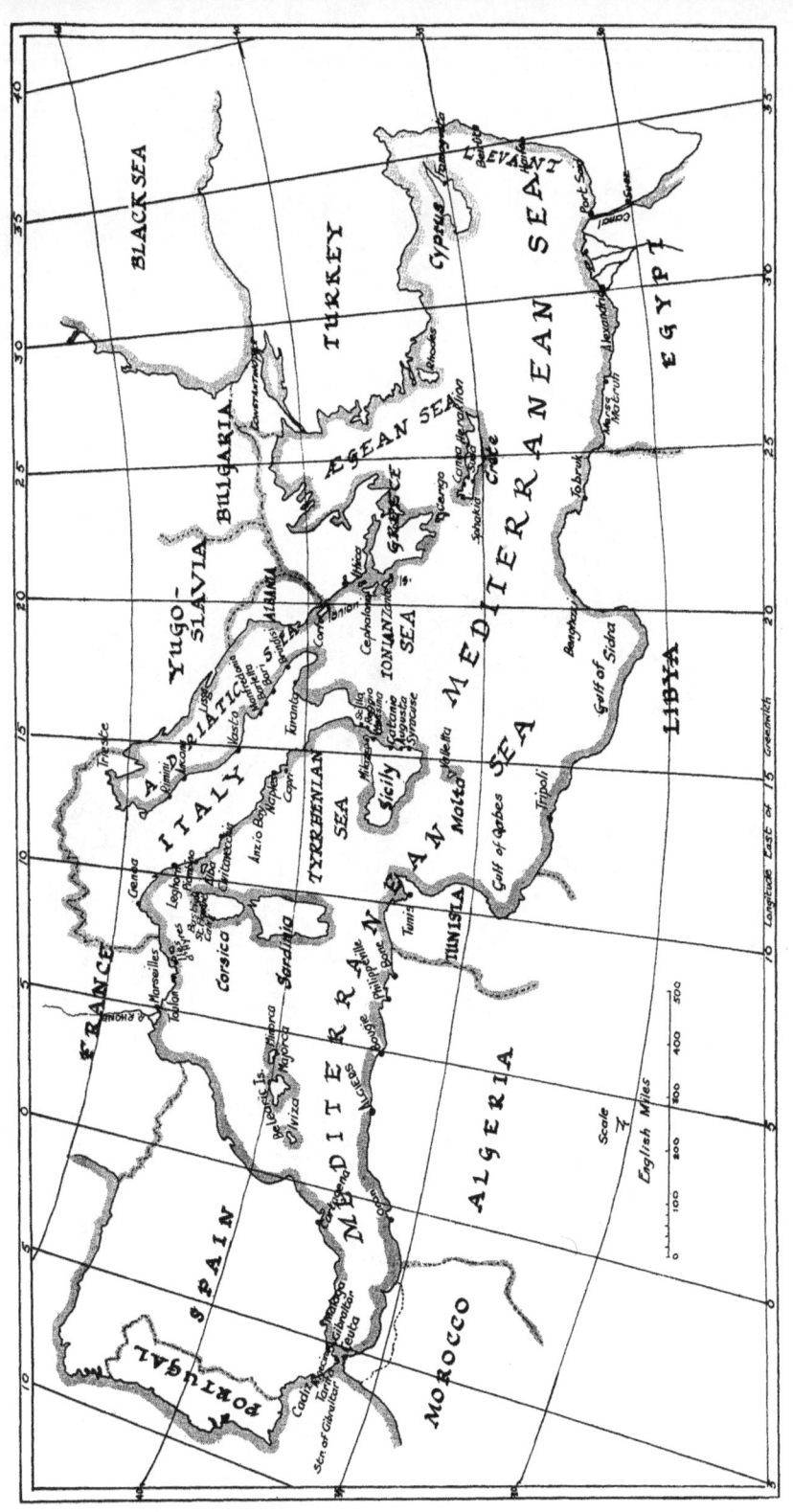

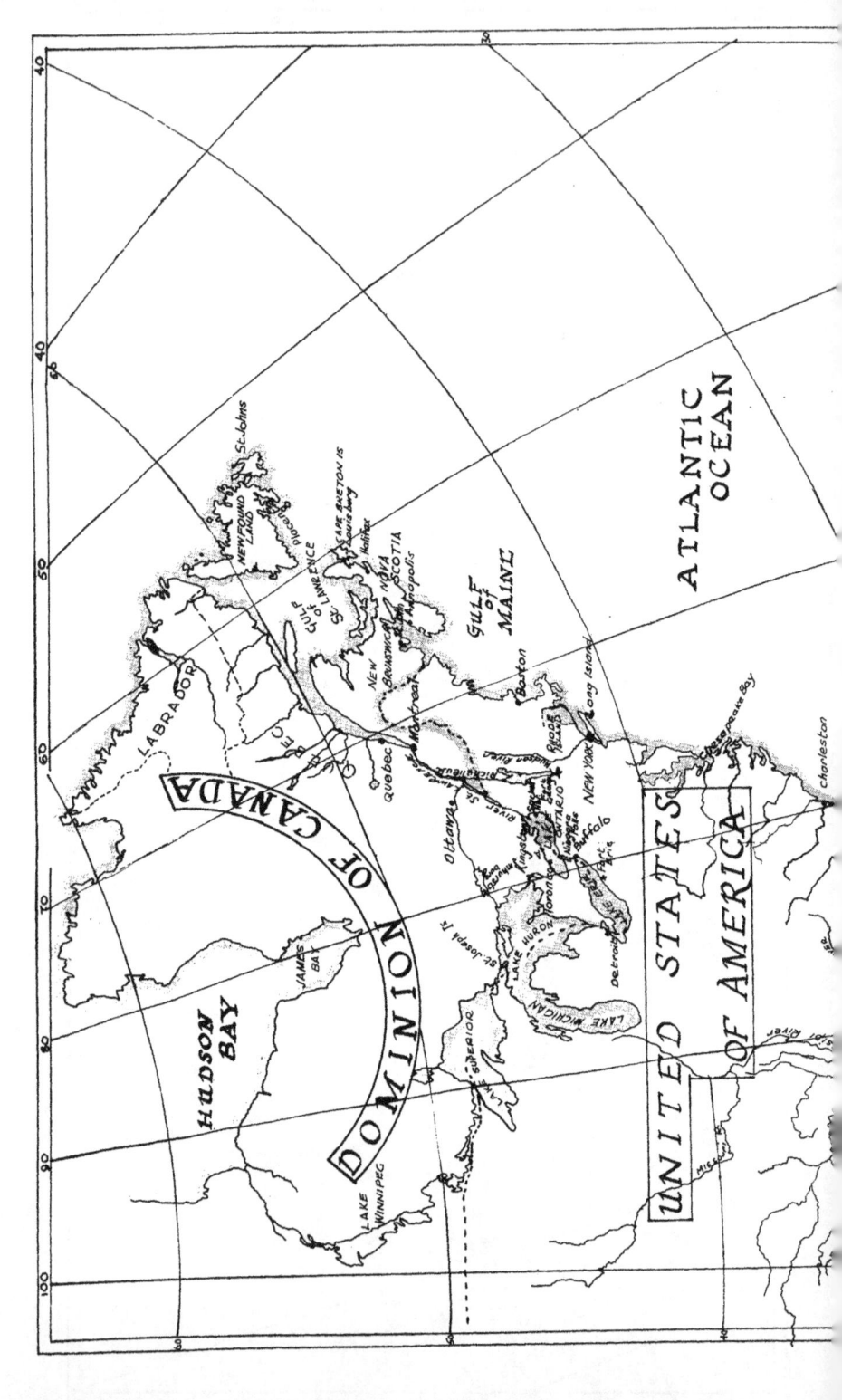

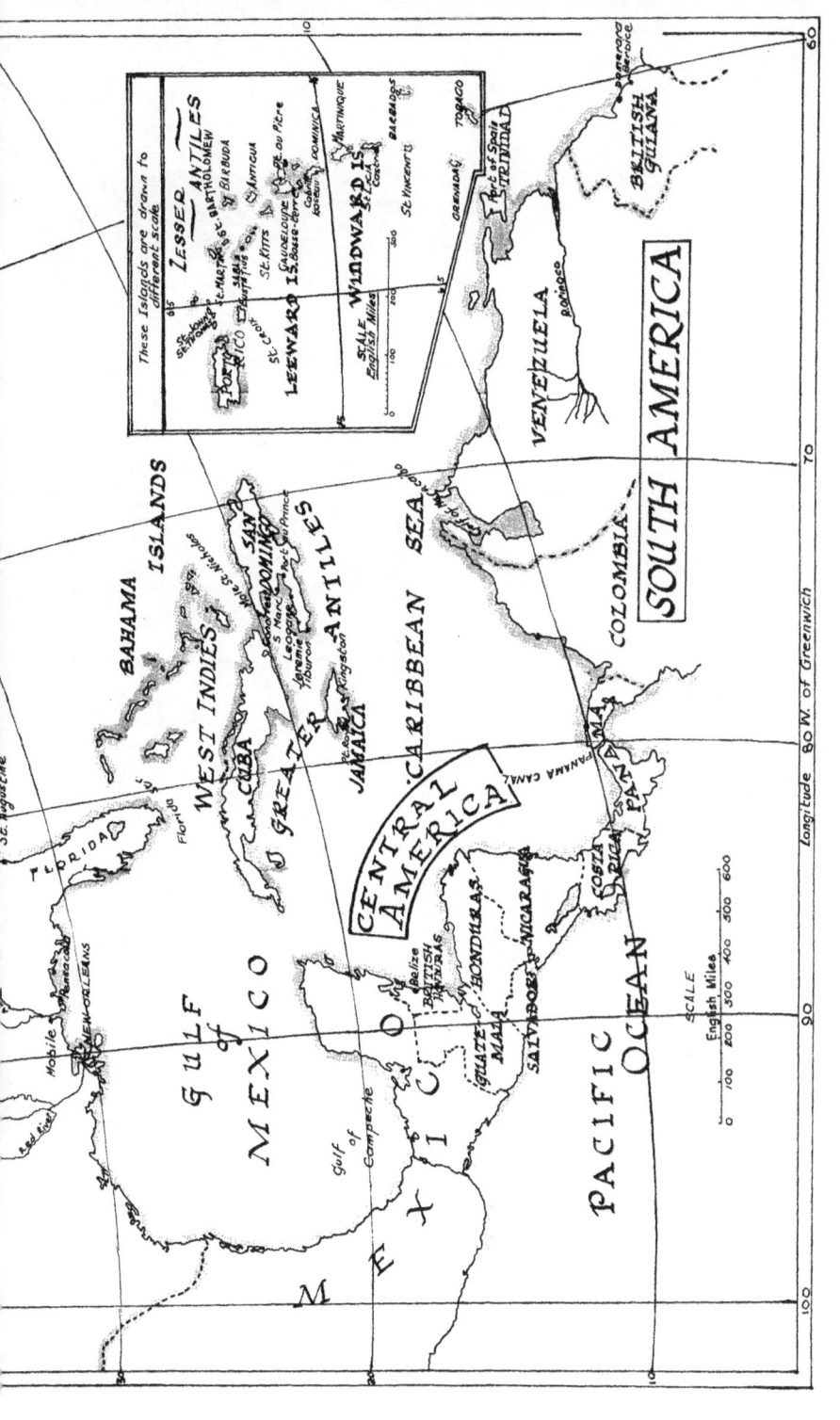

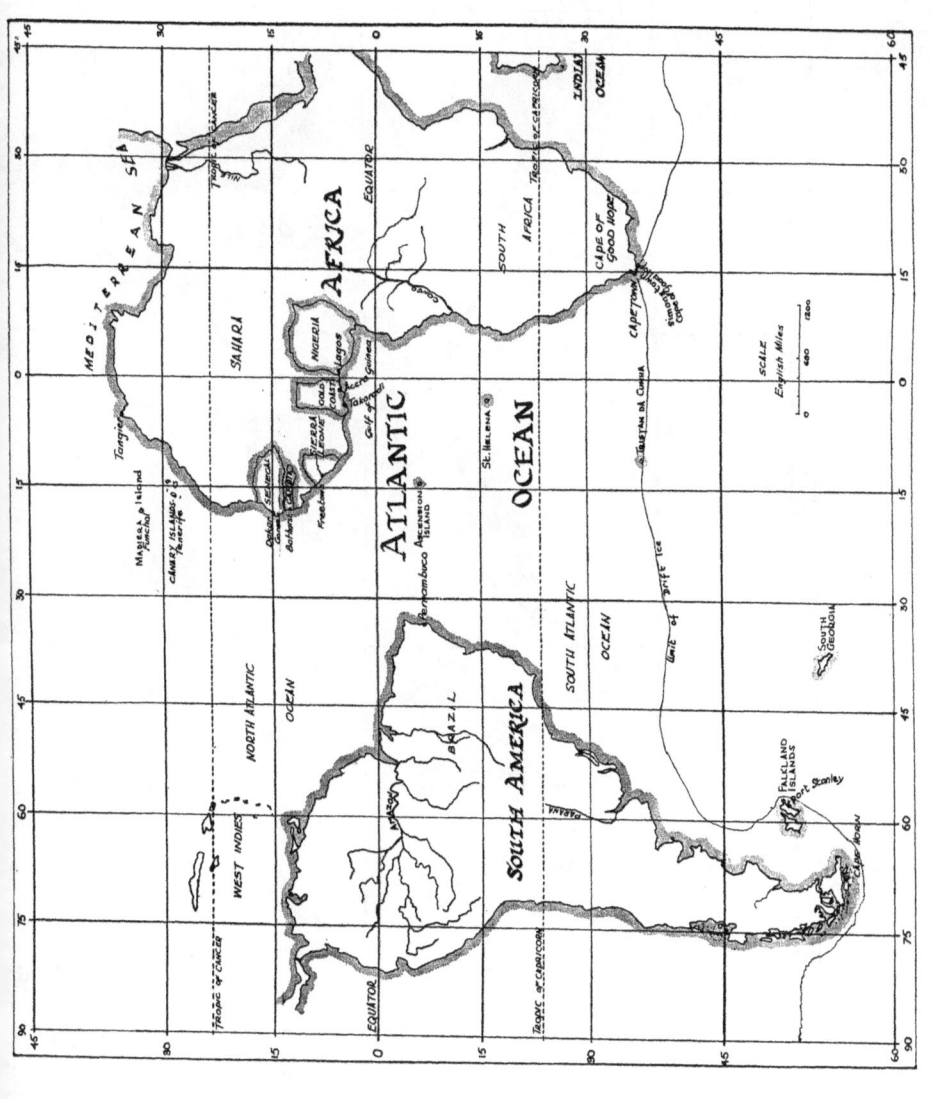

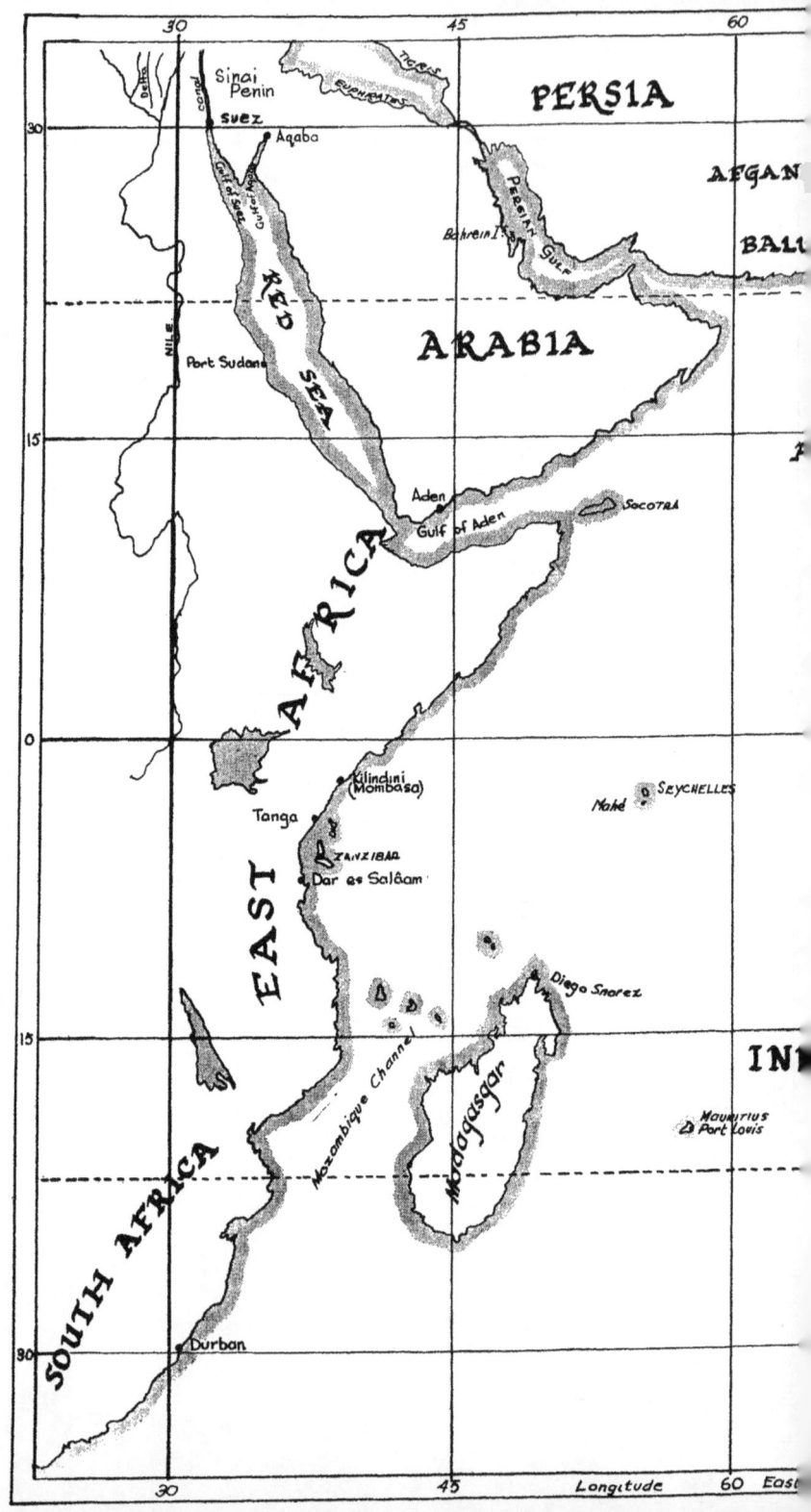

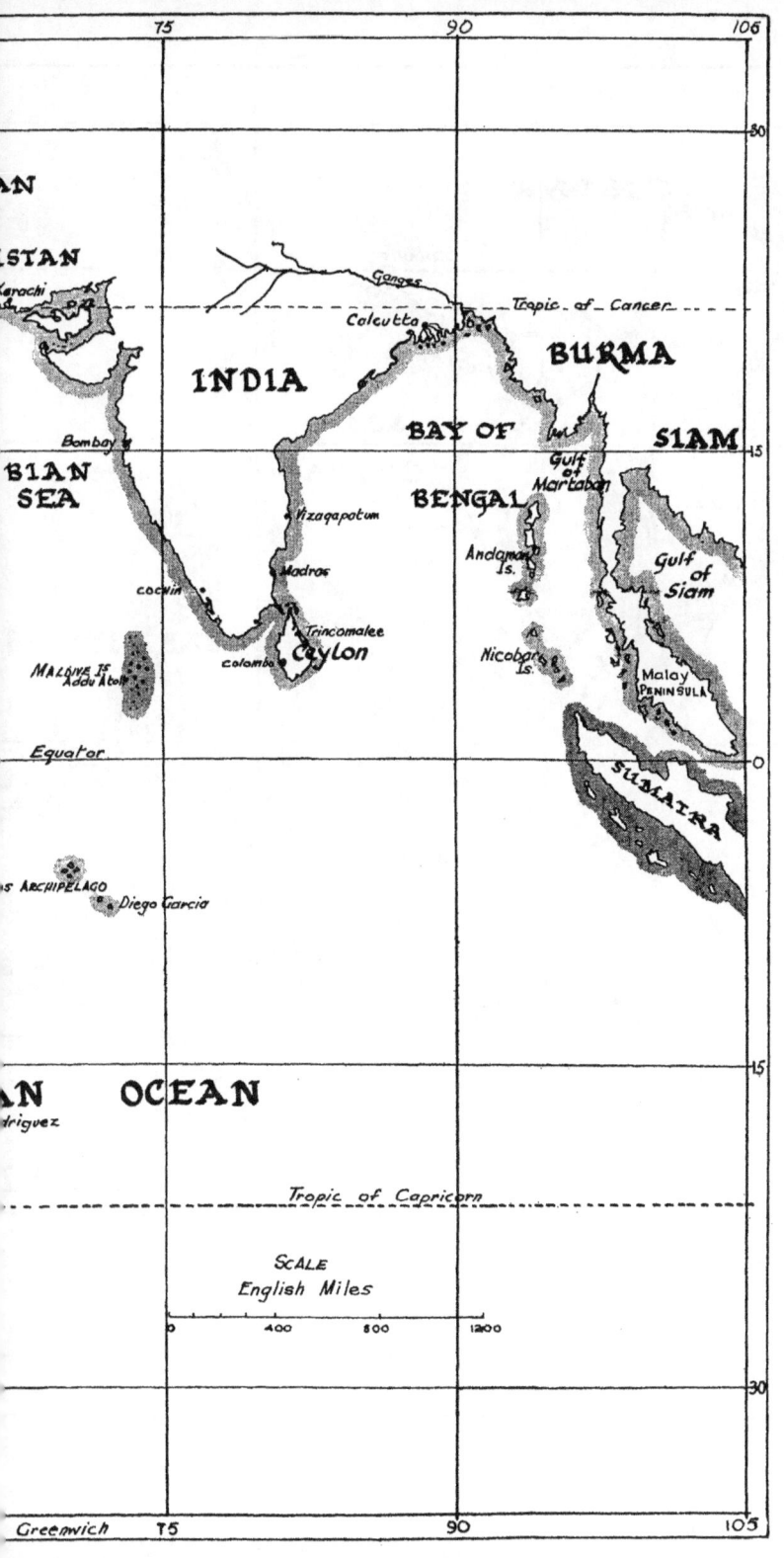

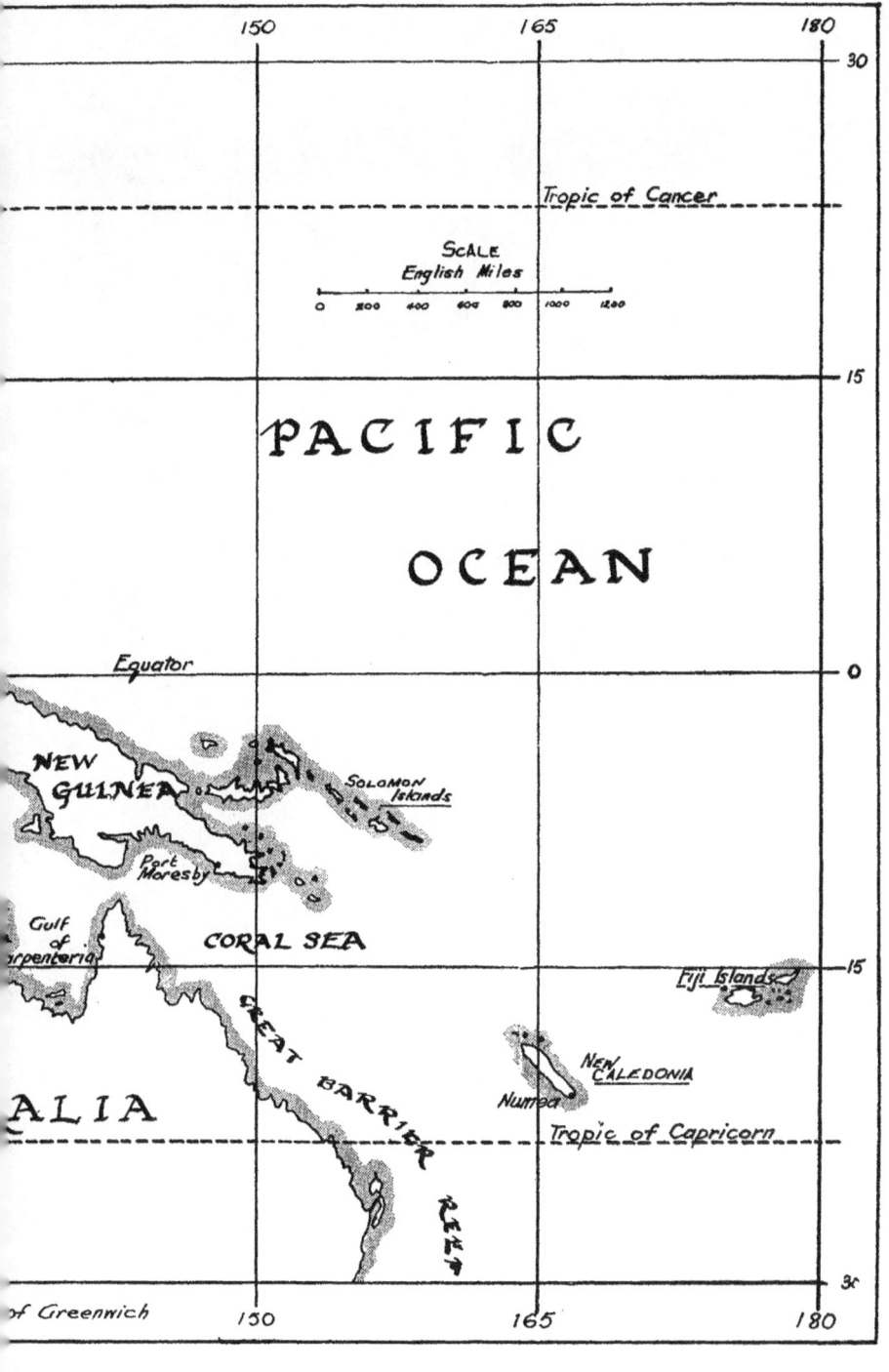

www.ingramcontent.com/pod-product-compliance
Lightning Source LLC
Chambersburg PA
CBHW020732160426
43192CB00006B/199